C0-AVK-970

CONTEMPORARY BAYESIAN AND FREQUENTIST STATISTICAL RESEARCH METHODS FOR NATURAL RESOURCE SCIENTISTS

THE WILEY BICENTENNIAL–KNOWLEDGE FOR GENERATIONS

Each generation has its unique needs and aspirations. When Charles Wiley first opened his small printing shop in lower Manhattan in 1807, it was a generation of boundless potential searching for an identity. And we were there, helping to define a new American literary tradition. Over half a century later, in the midst of the Second Industrial Revolution, it was a generation focused on building the future. Once again, we were there, supplying the critical scientific, technical, and engineering knowledge that helped frame the world. Throughout the 20th Century, and into the new millennium, nations began to reach out beyond their own borders and a new international community was born. Wiley was there, expanding its operations around the world to enable a global exchange of ideas, opinions, and know-how.

For 200 years, Wiley has been an integral part of each generation's journey, enabling the flow of information and understanding necessary to meet their needs and fulfill their aspirations. Today, bold new technologies are changing the way we live and learn. Wiley will be there, providing you the must-have knowledge you need to imagine new worlds, new possibilities, and new opportunities.

Generations come and go, but you can always count on Wiley to provide you the knowledge you need, when and where you need it!

WILLIAM J. PESCE
PRESIDENT AND CHIEF EXECUTIVE OFFICER

PETER BOOTH WILEY
CHAIRMAN OF THE BOARD

CONTEMPORARY BAYESIAN AND FREQUENTIST STATISTICAL RESEARCH METHODS FOR NATURAL RESOURCE SCIENTISTS

Howard B. Stauffer

Mathematics Department, Humboldt State University, Arcata, California

WILEY-INTERSCIENCE
A JOHN WILEY & SONS, INC., PUBLICATION

Published by John Wiley & Sons, Inc., Hoboken, New Jersey
Published simultaneously in Canada

For general information on our other products and services or for technical support, please contact our Customer Care Department within the United States at (800) 762-2974, outside the United States at (317) 572-3993 or fax (317) 572-4002.

Wiley also publishes its books in variety of electronic formats. Some content that appears in print may not be available in electronic formats. For more information about Wiley products, visit our web site at www.wiley.com.

Wiley Bicentennial Logo: Richard J. Pacifico

Library of Congress Cataloging-in-Publication Data:

Stauffer, Howard B., 1941-
Contemporary Bayesian and frequentist statistical research methods for natural resource scientists/Howard B. Stauffer.
p. cm.
ISBN 978-0-470-16504-1 (cloth)
1. Bayesian statistical decision theory. 2. Mathematical statistics. I. Title.
QA279.5.S76 2008
519.5'42—dc22

2007015575

Printed in the United States of America

10 9 8 7 6 5 4 3 2 1

To my parents,
Howard Hamilton Stauffer and Elizabeth Boyer Stauffer,
and to my family,
wife Rebecca Ann Stauffer,
daughter Sarah Elizabeth Stauffer,
and son Noah Hamilton Stauffer.
Their love and support has sustained me and provided
meaning and joy in my life.

CONTENTS

PREFACE

This book began as a critique against the current misuses of statistics in the natural resource sciences. I had worked for many years as a forestry and wildlife management statistician, in academia, government, and industry. I was frustrated with the frequent misuse of statistical analysis and inference with natural resource data. Hypothesis testing was commonly misused with observational data to compare so-called habitat treatments such as old-growth and young-growth forest habitat for their effects on wildlife species. Such hypotheses were statistical rather than scientific, referring to specific stands of interest. Many null hypotheses were "silly" and clearly not true. Sample datasets were not completely randomized, and "experimental" conditions were not effectively controlled. I was reviewing manuscripts and attending seminars where null hypotheses were being rejected that were clearly false a priori, and effect sizes between treatments, the differences of biological importance that were of interest to wildlife managers, were not even being estimated. More seriously, null hypotheses that were clearly false were being "supported" by hypothesis testing results that failed to reject, in studies where sample sizes were small, effect sizes of importance, and power to detect these effect sizes were not specified, and this power was very likely small.

Natural resource scientists did not clearly understand how to interpret their inferences from frequentist statistical analysis. The indirect logic of frequentist statistical inference, in interpreting the meaning of confidence intervals or the test statistics and p values from hypothesis testing, was proving to be very confusing to natural resource scientists. The challenge of natural resource scientists in explaining such frequentist inferences to managers, attorneys, politicians, and the public was proving to be even more daunting.

I was concerned with the extent of data dredging that was common in the field. I was commonly seeing datasets collected for habitat selection modeling using multiple linear regression or logistic regression analysis with measurements for over 100 covariates and sample sizes under 100. Scientists were not giving enough thought to sampling design and the type of analysis appropriate for their studies, prior to data collection. Stepwise and best-subsets selection methods were being utilized without concern for their potential for overfitting sample datasets with large and unspecified amounts of compounded error.

Then, around 1998/99, several pioneering applied statisticians with many years of experience in the field of wildlife management began to show the way out of this wilderness. Ken Burnham and David Anderson (1998, 2002) published their landmark book advocating the use of a priori model selection and inference using the Akaike information criterion (AIC) as a way of reducing model overfitting and compounding of error with model selection and inference. Doug Johnson's (1999) article critiquing the misuses of hypothesis testing in wildlife management research was published in the Journal of Wildlife Management. These ideas took the wildlife management research community by storm. Ray Hilborn and Mark Mangel (1997) published a seminal book advocating the use of Bayesian statistical analysis and inference in the fields of ecology and fisheries management. Pinhiero and Bates (2000) published an important book describing the applications of mixed-effects modeling in S-Plus. Ramsey and Schafer (2002) warned natural resource scientists about the distinctions between observational and experimental data. Because of some of these influences, a priori model selection and inference has become the accepted dominant paradigm for model selection and inference in the field of wildlife management. The misapplication of hypothesis testing has been reduced. Perhaps even too hastily, the old ways of doing statistics have been discarded in the rush to remain "current." Meanwhile, many other important contemporary methods of applied statistics remain relatively unknown among natural resource scientists, methods such as generalized linear modeling, mixed-effects modeling, and Bayesian statistical analysis and inference.

This book was written to introduce these newer contemporary methods of statistical analysis to natural resource scientists and strike a balance between the old and new ways of doing statistics. Chapter 1 introduces three case studies that illustrate the need for newer contemporary methods of statistical analysis and inference for natural resource science applications. It also reviews some of the most important fundamental methods of traditional frequentist statistical analysis and inference and ends with a brief introduction to the frequentist software S-Plus and R that are used throughout the book. Chapters 2–4 introduce an alternative approach to traditional frequentist statistical analysis and inference, namely, Bayesian statistical analysis and inference. These three chapters provide an introduction to the fundamental concepts of Bayesian statistical analysis, its historical background, conjugate solutions, Bayesian hypothesis testing and decisionmaking, Markov Chain Monte Carlo (MCMC) solutions, and applications in WinBUGS (Windows version of Bayesian statistical inference Using Gibbs Sampling) software. Chapter 5 presents two alternative strategies to model selection and inference, a posteriori model selection and inference, and a priori parsimonious model selection and inference using AIC and the deviance information criterion (DIC). Chapter 6 introduces the ideas of generalized linear modeling (GLM), focusing on the most popular GLM of logistic regression. Chapter 7 presents an introduction to mixed-effects modeling in S-Plus® and R. Chapters 5–7 provide applications with both frequentist and Bayesian statistical analysis and inference approaches, illustrating the strengths and limitations of each approach. Chapter 8 concludes with a summary of the contemporary methods introduced in this book.

This book can be used as a textbook for an intermediate undergraduate or introductory graduate semester course in contemporary research statistics for natural resources sciences. It assumes a minimum prerequisite undergraduate course in introductory statistics that includes the estimation of parameters such as mean and proportion;

hypothesis testing with t tests, F tests for analysis of variance (ANOVA), and chi-square (χ^2) tests; and linear regression analysis. Parts of the book can be read independently along with the introductory Chapter 1, Chapters 2–4 on Bayesian statistical analysis and inference, Chapter 5 on strategies for model selection and inference, Chapter 6 as an introduction to generalized linear modeling, and Chapter 7 on mixed-effects modeling. The book can also be read and used as a refresher manual or a reference book by natural resource scientists. Parts of the book have served as resource materials that I have used for 2-day workshops on topics of statistics such as Bayesian statistical analysis and inference using WinBUGS and capture–recapture analysis using MARK.

I'd like to thank many colleagues who have provided advice, encouragement, and support throughout my career as an applied statistician and influenced a perspective that has led to the writing of this book: David Anderson, Doug Johnson, Barry Noone, Bill Zielinski, C. J. Ralph, Cindy Zabel, Hart Welsh, Cynthia Perrine, Larry Fox, Jan Derksen, Rich Padula, Bryan Gaynor, Ken Mitchell, Sam Otukol, A. Y. Omule, David Gilbert, Les Safranyik, Mark Rizzardi, Yoon Kim, Butch Weckerly, David Hankin, John Sawyer, Mike Messler, Andrea Pickart, Matt Johnson, and Mark Colwell. I'd also like to thank the Wiley staff who were so helpful during the publication process: editor Susanne Steitz-Filler, senior production editor Kris Parrish, and copy editor Cathy Hertz. My career has been a most interesting one, working with natural resource scientists in academia, government, and industry, applying traditional and contemporary ideas in the application of statistical design and analysis to the natural resource sciences. It is up to natural resource scientists to make the most appropriate and effective choices on the applications of statistical analysis to their research problems. It is my hope that this book will help in providing the tools to make that possible.

Howard B. Stauffer

Mathematics Department
Humboldt State University
Arcata, California

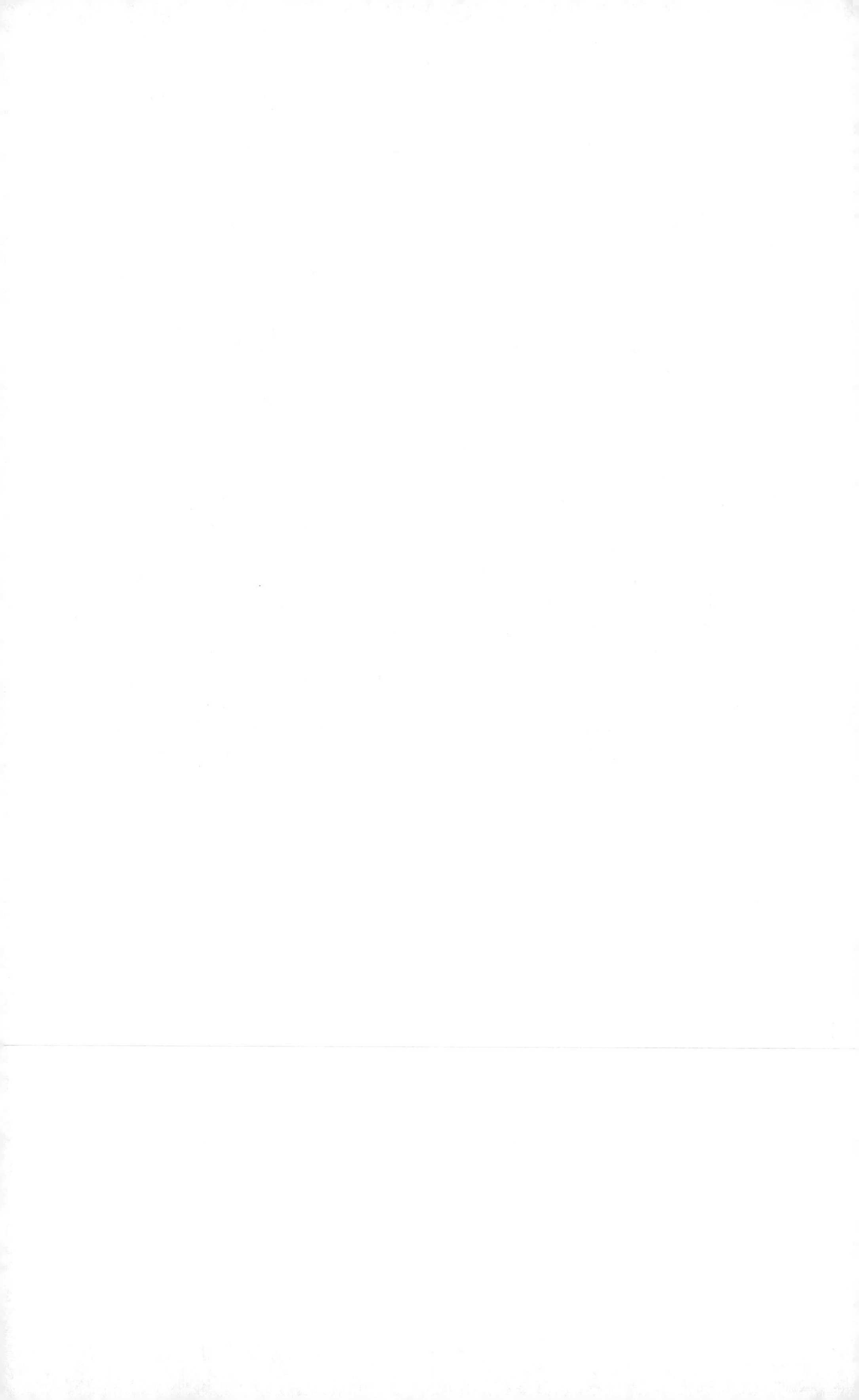

1 Introduction

We will begin this initial chapter by introducing three case studies that illustrate some of the fundamental general statistical problems challenging the contemporary natural resource scientist. We will then present a review and preview of some solution strategies to these general problems. The first solution strategies that we will review are traditional frequentist approaches: parameter estimation from sample surveys, hypothesis testing from experiments, and linear regression modeling. Each of these methods is summarized using a frequentist approach to statistical analysis. We will then preview some more contemporary solution strategies: an alternative Bayesian approach to statistical analysis and other more advanced solutions to the case studies, generalized linear modeling, and mixed-effects modeling using both frequentist and Bayesian approaches to statistical analysis. We will also preview a more contemporary approach to model selection and inference using information-theoretic criteria such as Akaike's information criterion for frequentist statistical analysis and the deviance information criterion for Bayesian statistical analysis. All of these contemporary methods will be discussed in greater detail throughout the remainder of this book and illustrated with examples.

In this initial chapter we include a reminder of the importance of project management in natural resource studies with statistical components. Project management consists of organizing projects into three phases: a planning phase, a data collection phase, and a concluding phase. The planning phase includes an identification of the problem and the objectives of the project, along with a statistical design for the collection of the dataset. The concluding phase includes a statistical analysis of the dataset, along with interpretation and conclusions drawn from the analysis. All of these statistical components—the statistical design, the collection of the dataset, and the statistical analysis—provide essential tools for the solutions to the objectives of the project.

We conclude this initial chapter with an introduction to the frequentist statistical analysis software used throughout the book: the proprietary software S-Plus and its freeware "equivalent" R. The Bayesian statistical analysis software WinBUGS will be introduced in Chapters 2–4 when Bayesian ideas are discussed.

Contemporary Bayesian and Frequentist Statistical Research Methods for Natural Resource Scientists. By Howard B. Stauffer

1.1 INTRODUCTION

In recent years there have been major advances in the methods of statistics used for research in the natural resource sciences. Yet, little of this is known outside selected research circles. Students and scientists in the natural resource sciences have continued to use traditional frequentist methods, such as the estimation of parameters from sample surveys, *t* tests and ANOVA hypothesis testing from experiments, and linear regression modeling. However, extraordinary newer methods are now available that enhance, complement, and extend these basic techniques, methods such as Bayesian statistical inference, information-theoretic approaches to model selection, generalized linear modeling, and mixed-effects modeling. It is the primary objective of this book to introduce these newer contemporary methods to natural resource students and scientists.

This book must begin by emphasizing critical statistical issues that have too often been neglected in natural resource studies in the past. We stress the importance of the planning and concluding phases in a data collection project. We particularly highlight the essential role of statistical design and analysis that help ensure the efficient, powerful, and effective use of data. Our approach throughout the book will be "hands-on," illustrating concepts with examples using the software languages of S-Plus or R for frequentist statistical analysis and WinBUGS for Bayesian statistical analysis.

Let's begin with a description of several case studies that illustrate problems of fundamental interest to contemporary natural resource scientists.

1.2 THREE CASE STUDIES

1.2.1 Case Study 1: Maintenance of a Population Parameter Above a Critical Threshold Level

A fundamental problem of interest to contemporary natural resource scientists is to assess whether a critical population parameter, such as a proportion parameter p, has been maintained above (or below) a specified critical threshold level: $p \geq p_c$ (or $p \leq p_c$)?

Many examples in natural resource science illustrate this problem:

1. A timber company is required to maintain the proportion p of its timberlands occupied by nesting Northern Spotted Owl pairs above a specified threshold level p_c. The threshold p_c is a level determined by biologists to ensure the viability of the local population of owls.
2. Federal managers of a national forest are interested in maintaining the proportion p of forest covered by dense undergrowth below a specified threshold level p_c, to limit the risk of fire.
3. The managers of a national park are interested in maintaining the proportion p of a disease or insect infestation below a specified threshold level p_c to control its spread.

4. Fishery biologists managing a watershed are interested in maintaining the proportional abundance p of a fishery above a specified threshold level p_c of its carrying capacity to ensure its long-term sustainability.
5. A government agency implementing a natural resource conservation policy is interested in ensuring that the proportion p of the public in favor of one of its controversial policies is maintained above a certain threshold level p_c.

Besides the proportion parameter p in the examples presented above, there are many other biological parameters of interest to natural resource managers with similar threshold issues, such as the mean abundance μ, survival rate ϕ_i from year i to year $i+1$, fitness $\lambda_i = N_{i+1}/N_i$ (where N_i and N_{i+1} are the population abundances in years i and $i+1$), ecological diversity index such as the Shannon–Wiener diversity index H, and population total τ.

The failure to maintain the population parameter p above (or below) the threshold level p_c might suggest the need for a "corrective action" decision in the examples listed above, such as

1. Reducing the timber harvesting
2. Applying fire suppression treatment
3. Applying disease or insect treatment
4. Increasing the watershed river flow by releasing more water from a dam
5. Altering the natural resource conservation policy

Alternatively, success at maintaining the population parameter p above (or below) the threshold p_c might suggest a decision of "no action."

In such circumstances, a common approach employed by natural resource scientists is to begin monitoring the population and collecting sample data, say, on an annual basis, in order to assess the status of the population parameter. The intent is to conduct statistical analysis on the sample data and make inferences about the population parameter to determine whether it is above (or below) the threshold, and thus whether corrective action or no action is needed at the management level.

1.2.2 Case Study 2: Estimation of the Abundance of a Discrete Population

Our second case study focuses on the analysis of population count data. Often biological populations, such as birds, amphibians, or mammals, are sampled with discrete measurements such as plot counts, in fixed-area plots called quadrants. The intent is to estimate population size or density in an area using total count estimates of abundance or mean estimates of density.

The analysis consists of estimates of total or mean. Traditional estimates of total or mean are based on the assumption of the normal distribution of the population measurements. For plot counts, however, measurements are discrete and noncontinuous, consisting of nonnegative integers in a skewed distribution. If the biological population is randomly dispersed spatially, a proper model for the analysis should be based on the Poisson distribution rather than the normal distribution for the plot counts. If, however, the population is spatially aggregated or clumped, the analysis should be based on a more general model for the population measurements, such as a negative binomial distribution. Furthermore, the plot counts will likely be sampled without complete certainty of detection. Animals may be within the plot and yet be undetected by the sample surveyor. A rigorous analysis of the population therefore must factor in the Poisson or negative binomial distribution of the plot measurements, sampled with an uncertainty of detection. We will examine such analyses in Chapters 2–4 with Bayesian statistical analysis and Chapters 6–7 with generalized linear models and mixed-effects models.

1.2.3 Case Study 3: Habitat Selection Modeling of a Wildlife Population

In general, it can be quite difficult to estimate the presence or abundance of a wildlife population. Many important biological populations whose presence or abundance needs to be estimated are endangered or locally threatened wildlife species, such as the Northern Spotted Owl and Marbled Murrelet bird populations, Del Norte salamander amphibian population, and grizzly bear mammal population. These endangered species are often of particular importance because they are associated with old-growth ecosystems that are also in danger of extinction. Therefore it is important to monitor these populations, estimating their presence or abundance over time, to assess the status of the old-growth ecosystems. A particularly effective approach to estimating these mobile populations is to model their relationship with habitat.

With **habitat selection modeling**, the presence or abundance of a mobile population species is treated as a dependent response variable. Its relationship with "independent" predictor explanatory habitat variables such as vegetation, geologic, and meteorologic attributes can be assessed with statistical modeling. The intent of the habitat selection modeling is to analyze the relationship between the mobile wildlife population variable and the habitat variables and use it to describe or predict the presence or abundance of the endangered species as a function of the habitat variables. The idea behind the modeling is that many habitat variables can be more easily and less expensively sampled than can the mobile wildlife population.

The relationship in such circumstances is assumed to be associative rather than causal; thus, the modeling is descriptive, based on population monitoring with sample survey data, and not on experimental manipulation to establish evidence for cause and effect. The mobile wildlife population may have access to only a limited amount of habitat attributes and be able to express a restricted preference among what remains. Other habitat attributes that the mobile wildlife population most prefers may no longer be available for selection. Hence the habitat "selection" relationship must be interpreted within this context.

Habitat selection modeling is often based on regression analysis. For continuous-abundance response variables such as biomass, multiple linear regression analysis may indeed be applicable. For discrete-abundance response variables such as population counts, however, Poisson regression or negative binomial regression may be more appropriate. For binary response variables, such as the presence or absence, or occupancy versus nonoccupancy, of a population, logistic regression analysis or some other form of generalized linear modeling may be more appropriate. We will examine these methods of analysis, along with strategies for model selection, in Chapters 5 and 6. Traditional multiple linear regression analysis is discussed in Chapter 5. Logistic regression analysis, Poisson regression analysis, negative binomial regression analysis, and other forms of generalized linear modeling are discussed in Chapter 6.

1.2.4 Case Studies Summary

This book presents various contemporary statistical options available to the natural resource scientist to analyze and interpret sample data for these case studies and other general statistical problems of current interest to natural resource scientists. We will first review more familiar traditional statistical methods of sample survey parameter estimation, experimental hypothesis testing, and multiple linear regression modeling, and then describe the less familiar contemporary methods of Bayesian statistical inference, model selection strategies, generalized linear modeling, and mixed-effects modeling. These methods provide contemporary natural resource scientists with an up-to-date statistical toolbox of methods to tackle many important challenging problems of current interest.

1.3 OVERVIEW OF SOME SOLUTION STRATEGIES

In this section we present both a review of traditional statistical methods and a preview of contemporary statistical methods that provide solutions to the case studies that were presented in the previous section: assessing whether a population parameter has been maintained above (or below) a critical threshold level, the estimation of abundance of a discrete population, and habitat selection modeling. Further details on the contemporary methods will follow in later chapters.

1.3.1 Sample Surveys and Parameter Estimation

A first traditional statistical approach to addressing the fundamental case study problems of Section 1.2 is to conduct a sample survey of the population and collect sample data using a rigorous sampling design. The aim of the survey in case study 1 is to estimate a proportion parameter p or mean parameter μ from the sample data and compare it with a critical threshold level p_c or μ_c. The aim of the survey in case study 2 is to estimate the mean abundance parameter μ of a discrete population. The aim of the survey in case study 3 is to model a mobile wildlife population as a function of habitat attributes and estimate the proportion parameter p or abundance parameter mean μ of the species in the habitat or at a specific site. Ideally, a natural resource scientist would like to use an approximately unbiased **estimator** $\hat{\theta} = \hat{p}$ or $\hat{\mu}$ for the **estimate** of the **parameter** $\theta = p$ or μ, respectively, of minimum **sampling**

error E, with a specified **level of confidence** P (or **level of significance** $\alpha = 1 - P$).

If simple randomly sampled measurements $\{y_i\}$ are continuous and normally distributed with sample size n, the **mean estimator** is given by

$$\hat{\mu} = \sum_{i=1}^{n} \frac{y_i}{n},$$

with **standard deviation**

$$s = \sqrt{\sum_{i=1}^{n} \frac{(y_i - \hat{\mu})^2}{n-1}},$$

standard error

$$\text{se} = \frac{s}{\sqrt{n}},$$

and **sampling error**

$$E = t_{(1-\alpha/2),n-1} \cdot \text{se} = t_{(1-\alpha/2),n-1} \cdot \frac{s}{\sqrt{n}},$$

where $t_{(1-\alpha/2),\,n-1}$ is the t value with $(n-1)$ degrees of freedom at the $(1-\alpha/2)$ percentile with α level of significance.

If simple randomly sampled measurements $\{y_i\}$ are binary and binomially distributed with sample size $n \geq 30$, the **proportion estimator** is given by

$$\hat{p} = \frac{y}{n}, \quad \text{where} \quad y = \sum_{i=1}^{n} y_i,$$

with standard error

$$\text{se} = \sqrt{\frac{\hat{p} \cdot (1-\hat{p})}{n-1}},$$

and sampling error

$$E = t_{(1-\alpha/2),n-1} \cdot \text{se},$$

where $t_{(1-\alpha/2),n-1}$ is the t value with $(n-1)$ degrees of freedom at the $(1-\alpha/2)$ percentile with α level of significance.

Recall that an **unbiased estimator** has the property that the average of all estimates, with repeated sampling, is equal to the parameter value (Fig. 1.1). Confidence levels of $P = 95\%$, 90%, or 80% are commonly used for natural resource survey sampling with levels of significance $\alpha = 1 - P = 5\%$, 10%, and 20%, respectively. A **confidence interval**

$$\text{CI} = [\hat{\theta} - E, \hat{\theta} + E]$$

Figure 1.1. Sample mean density estimates (tickmarks X), based on 100 repeated sample surveys of a normally distributed population with mean density parameter value 50.0. Note that the average of these sample mean density estimates is approximately equal to the mean density parameter value for this unbiased mean estimator.

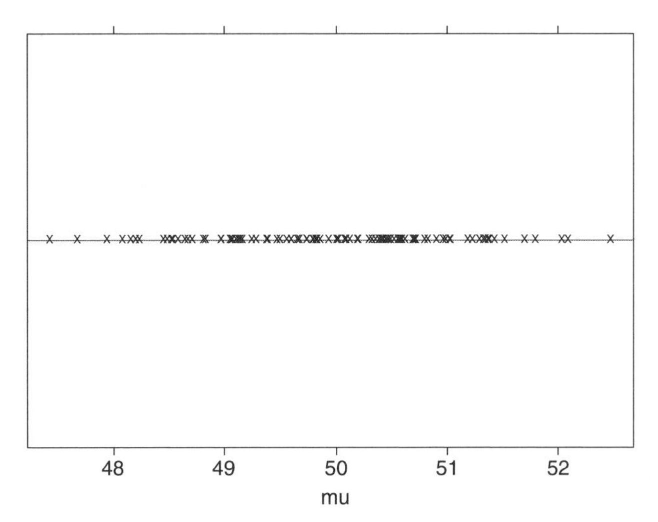

can be calculated and the frequentist inference drawn that there is a probability P that confidence intervals will contain the parameter θ, with repeated sampling (Fig. 1.2).

Note that the logic of frequentist inference is of the form "if (parameter), then probability(data)." It assumes that the parameter is fixed and provides conditional probability properties for statistics from the sample datasets.

Figure 1.2. Twenty 95% confidence intervals estimated from samples obtained from repeated sample surveys of a population with mean parameter value 50.0 ("|"). Note that 19 of these confidence intervals contain the mean parameter, as is expected.

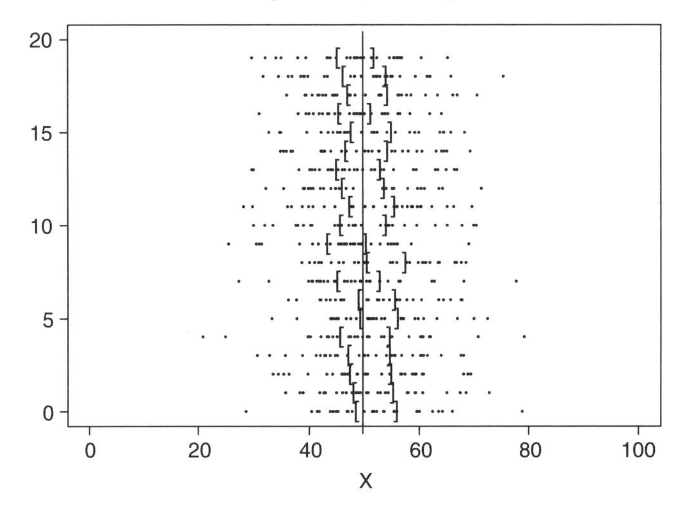

For case study 1, a **decision protocol** should be specified in advance, before the sample data are collected. For example, one such decision protocol would be to compare the estimate $\hat{\theta}$ obtained from the sample data with the critical threshold level θ_c. If the estimate is above (or below) the critical threshold level θ_c, then the recommended management decision would be "corrective action." Otherwise, if the estimate is below (or above) θ_c, the recommended management decision would be "no action."

An alternative decision protocol for case study 1 would be to compare the confidence interval CI with the critical threshold level. If the confidence interval is above (or below) the critical threshold level, then the recommended management decision would be "corrective action." If the confidence interval is below (or above) the critical threshold level, the recommended management decision would be "no action." If the confidence interval overlaps the critical threshold level, the situation would be ambiguous and need to be reassessed, perhaps with an additional survey with larger sample size. The **precision** of the estimate, the size of the sampling error, would obviously affect the results. A larger sample size would reduce the sampling error and hence the size of the confidence interval. Therefore, the population should be sampled with a sample size large enough to reduce the sampling error so that the confidence interval will (hopefully!) fall on one side or the other of the critical threshold level.

Other decision protocols could be chosen for case study 1 using this general approach of sample surveys with parameter estimation and estimation of error. The important point, however, is that a decision protocol for a sample survey should be specified in advance of data collection so that a decision can be made, clearly and unambiguously, at the end of the survey. In Chapters 2–4 we shall see how a Bayesian statistical analysis approach can facilitate the use of a decision protocol.

For case studies 2 and 3, estimates and confidence intervals can be used to make inferences on the abundance of a discrete population, and the population habitat selection probability of presence or mean abundance, respectively. For further review of the basic concepts of sampling design and analysis, see Cochran (1977), Scheaffer et al. (1996), Thompson (1992), Sarndal et al. (1992), Thompson and Seber (1996), Thompson et al. (1998), Stauffer (1982a, 1982b), Hansen and Hurwitz (1943), Horvitz and Thompson (1952), and Gregoire (1998).

1.3.2 Experiments and Hypothesis Testing

A second statistical approach to addressing the problems posed by some of the case studies of Section 1.2 is to conduct an experiment and use frequentist hypothesis testing, developed by Neyman and Pearson (1928a, 1928b, 1933, 1936) and Fisher (1922, 1925a, 1925b, 1934, 1958). With case study 1, for example, the scientist could formulate the null hypothesis

$$H_0 : p = p_c$$

and the one-tailed alternative hypothesis

$$H_A : p < p_c,$$

collect experimental data using a rigorous experimental design, and test the null hypothesis. If the null hypothesis is rejected, the recommended management decision would be "corrective action." If the null hypothesis is not rejected, the recommended management decision would be "no action." Note, that with this approach, the burden of proof would be on "corrective action." If the direction of the alternative hypothesis is reversed, the burden of proof would be on "no action." Regardless, the burden of proof would not be equal for the two hypotheses.

A one-sample z test could be used for the hypothesis testing, with a one-sided alternative hypothesis. If the null hypothesis is true, the test statistic

$$z_{\mathrm{s}} = \frac{\hat{p} - p_{\mathrm{c}}}{\sqrt{\dfrac{\hat{p} \cdot (1 - \hat{p})}{n - 1}}}$$

is standard normally distributed. The **Neyman–Pearson hypothesis testing protocol** requires that a **type I error** α be specified, with confidence $P = 1 - \alpha$ (say, $\alpha = 5\%$ and $P = 95\%$), prior to the data collection and analysis, assuming the null hypothesis to be true. If the test statistic z_{s}, calculated from the experimental dataset, is in the rejection region, the α percentile left tail of the standard normal distribution (equivalent to $p < \alpha$), then the null hypothesis would be rejected. Otherwise, the null hypothesis would not be rejected.

Note again, that the logic of frequentist inference is of the form "if (parameter), then probability (data)." It assumes the null hypothesis that the parameter is fixed and provides conditional probability properties for statistics from the experimental datasets.

To review the basic concepts of experimental design and hypothesis testing, see Hicks (1993), Kuehl (1994), Dowdy and Wearden (1991), Sokal and Rohlf (1995), Zar (1996), Winer et al. (1991), Cohen (1988), Siegel and Castellan (1988), Conover (1980), Daniel (1990), PASS (2002), and nQuery (2002). However, beware of the overuse and misuse of hypothesis testing in natural resource science, particularly with observational, rather than experimental, datasets; see Johnson (1999), Anderson et al. (2001, 2002), Ramsey and Schafer (2002), and Robinson and Wainer (2002).

1.3.3 Multiple Linear Regression, Generalized Linear Modeling, and Model Selection

Habitat selection modeling can often be used effectively to address case study 3 of Section 1.2. Mobile wildlife populations may be difficult to sample directly, but their responses often vary with habitat attributes that are more easily sampled. For instance, the response of endangered wildlife species associated with old-growth habitat may tend to be larger in value in late-seral-stage rather than early-seral-stage habitat. In such circumstances, it may be useful, more cost-effective, and less time-consuming to fit and compare habitat selection models that describe the mobile biological population response as a function of various habitat variables.

The objective is to express the wildlife population response as a function of variables describing the vegetation, geologic, and climatic characteristics of its habitat. If the population response is continuous with normally distributed error, the population may be described by a multiple linear regression model.

If, however, the population response measurements are categorical, such as binary with values of 0 or 1 (e.g., "present" or "absent," "occupied" or "unoccupied," "alive" or "dead," "yes" or "no"), or discrete with integer values such as plot counts with error that may not be normally distributed, then it may be possible to "link" the response measurements to a linear function described by a generalized linear model (GLM) such as logistic regression.

Multiple linear regression models and generalized linear models can be used to describe the response values at specific sites as a function of habitat characteristics. With such a modeling approach to statistical analysis, model selection strategies are required to effectively compare models for goodness of fit to sample datasets and to avoid overfitting and compounding of error. The utility of such an approach depends on the goodness of fit of the best-fitting models to the population and their predictive accuracy. Chapter 5 includes a basic review of multiple linear regression and a description of contemporary strategies that can be used for model selection and inference. To further review the details of multiple linear regression, see Appendix A at the end of this book, or see Seber (1977), Draper and Smith (1981), Hocking (1996), Ryan (1997), Cook (1998), and Cook and Weisberg (1999). Generalized linear modeling is introduced in Chapter 6.

1.3.4 A Preview of Bayesian Statistical Inference

Bayesian statistical analysis and inference provides an important alternative approach to frequentist statistical analysis and inference for natural resource scientists, yet it has become practical and accessible for general use only relatively recently. Traditional frequentist statistical analysis and inference provides probabilities for sample datasets, based on assumptions for parameters, with an interpretation of results in the context of repeated surveys or experiments. Although frequentist statistical analysis methods are well known by natural resource scientists, these methods are often incorrectly applied with inferences that are frequently misunderstood (Johnson 1999, 2002). If properly applied and correctly interpreted, frequentist statistical analysis provides rigorous standards for inferences: the unbiased and minimum error properties of estimators, the accuracy probabilities of confidence intervals, and the type I and type II errors of hypothesis testing.

Alternatively, Bayesian statistical analysis and inference provides probabilities for parameters, based on sample datasets (Iversen 1984, Berger 1985). Inferences from Bayesian statistical analysis are directly applicable to parameters that are of central interest to natural resource scientists. Unfortunately, Bayesian statistical inference has not been of leading interest to a majority of statisticians and practicing natural resource scientists in the past because its use has been impractical and inaccessible until very recently (as of 2007). Bayesian statistical analysis and inference requires an assumption of a prior distribution for the parameters. Using the sample dataset and a likelihood model for the dataset, Bayesian statistical analysis provides a posterior distribution for the parameters, based on the prior distribution, the dataset, and the model. The posterior distribution thus updates a scientist's understanding of the parameters. The Bayesian approach combines previous information about the parameter with an analysis of the sample dataset to obtain an updated assessment of the parameters. The posterior distribution provides probabilities for the parameters that can be useful for natural resource scientists and managers. They can utilize summary statistics of the posterior distribution, such as the mean, median, mode, standard deviation, and percentiles, or use the entire posterior distribution itself to evaluate the parameters. A probability region, the smallest middle interval encompassing 95% of the posterior distribution, provides a Bayesian 95% **credible interval**. This interval can be directly interpreted as the region within which the parameter is likely to be found, with 95% probability. Thus, with Bayesian statistical inference, there is no need to interpret the results indirectly as a frequentist does, in terms of probabilities of datasets with repeated surveys or experiments. The logic of Bayesian statistical analysis provides probabilities for parameters, given the data, in contrast to the logic of frequentist statistical analysis, which provides probabilities for datasets, given the parameter.

Bayesians must, however, bear responsibility for the appropriate selection of priors and the standards of results. As with frequentist results, Bayesian results must be assessed for goodness of fit to assess the reliability of model predictions. Priors influence posteriors, particularly with small-sample datasets, and must be chosen judiciously. Until relatively recently, Bayesian solutions to complex problems were seldom computable. However, owing to a collection of computer simulation algorithms, the Markov Chain Monte Carlo (MCMC) algorithms developed in the mid-twentieth century (Bremaud 1999, Carlin and Louis 2000, Congdon 2001, Gill 2002, Link et al. 2002) and to public-domain software such as WinBUGS that is now downloadable from the Web (Spiegelhalter et al. 2001), natural resource scientists now have the resources needed for the practical use of Bayesian statistical inference.

This book describes and illustrates both frequentist and Bayesian paradigms for statistical analysis and inference, emphasizing the advantages and disadvantages of each in particular contexts. It is the practical perspective of this book that the contemporary natural resource scientist should be familiar with both. Bayesian statistical inference is introduced in Chapters 2–4 and applied comparatively, along with frequentist statistical inference, in Chapters 5–7, with other important contemporary research methods of statistical analysis.

1.3.5 A Preview of Model Selection Strategies and Information-Theoretic Criteria for Model Selection

With either frequentist or Bayesian statistical analysis approaches, a rigorous and theoretically justifiable approach to model fitting, selection, and inference is required. Traditionally, with multiple linear regression modeling, analysts have used statistics such as parameter coefficient estimates and their significance, the coefficient of determination R^2, the residual standard error $s_{y|x}$, the ANOVA F test, the adjusted R^2, and Mallows' C_p to evaluate the relative fit of models (Seber 1977, Draper and Smith 1981, Hocking 1996, Ryan 1997, Cook and Weisberg 1999). These statistics test various assumptions of the model fit, such as whether the model is statistically equivalent to the null model. They do not directly assess the issue of whether the model is the best fitting to the sample dataset. Unfortunately, they also sometimes tend to overfit the model to the sample dataset, with compounding of error. We shall say more about this later on. We recommend a more modern information-theoretic approach to model fitting, using Akaike's information criterion (AIC), the corrected Akaike information criterion ($\mathrm{AIC_c}$), or the Bayesian information criterion (BIC) with frequentist statistical analysis (Burnham and Anderson 1998, 2002), and the deviance information criterion (DIC) with Bayesian statistical analysis (Spiegelhalter et al. 2001, Carlin and Louis 2000). These criteria provide a more rigorous and theoretically justified approach to model fitting that avoids the overfitting of models to the sample dataset and the compounding of error.

Akaike's information criterion was developed relatively recently by the Japanese mathematician Hirotugu Akaike (1973, 1974). It is an information-theoretic measurement of the relative **Kullback–Leibler distance (KL distance)** between a model and the reality. The **Akaike's information criterion (AIC)** is the linear Taylor series approximation of the relative KL distance, whereas the **corrected Akaike information criterion ($\mathrm{AIC_c}$)** is a second-order Taylor series approximation. Since $\mathrm{AIC_c}$ is more precise, we recommend that it be used in preference to AIC, particularly for datasets with small numbers of samples. The best-fitting model in a collection of models has the lowest AIC or $\mathrm{AIC_c}$ value.

For any probabilistic statistical model with a likelihood function $\mathcal{L}$ (more on this in Chapters 2, 5, and 6), AIC and $\mathrm{AIC_c}$ are defined using the deviance = $D = -2 \cdot \log(\mathcal{L}) = -2 \cdot l$

$$\begin{aligned} \mathrm{AIC} &= D + 2 \cdot k \\ &= -2 \cdot \log(\mathcal{L}) + 2 \cdot k \end{aligned}$$

and

$$\begin{aligned} \mathrm{AIC_c} &= D + 2 \cdot k + 2 \cdot \frac{k \cdot (k+1)}{n-k-1} \\ &= -2 \cdot \log(\mathcal{L}) + 2 \cdot k + 2 \cdot \frac{k \cdot (k+1)}{n-k-1}, \end{aligned}$$

where k = the number of parameters in the model and n = the sample size.

The AIC and $\mathrm{AIC_c}$ criteria for multiple linear regression are given by the formulas

$$\begin{aligned} \mathrm{AIC} &= n \cdot \log\left(\hat{\sigma}^2 \cdot \frac{n-p-1}{n}\right) + 2 \cdot k \\ &= n \cdot \log\left(\hat{\sigma}^2 \cdot \frac{n-k+1}{n}\right) + 2 \cdot k \end{aligned}$$

and

$$\begin{aligned}\mathrm{AIC_c} &= n \cdot \log\left(\hat{\sigma}^2 \cdot \frac{n-p-1}{n}\right) + 2 \cdot k + 2 \cdot \frac{k \cdot (k+1)}{n-k-1} \\ &= n \cdot \log\left(\hat{\sigma}^2 \cdot \frac{n-k+1}{n}\right) + 2 \cdot k + 2 \cdot \frac{k \cdot (k+1)}{n-k-1},\end{aligned}$$

where $\hat{\sigma} = s_{y|x}$ is the residual standard error, p is the number of covariates or explanatory variables, and $k = p + 2$ is the number of parameters (including the covariates coefficients, the intercept, and σ). For the linear regression model with the parameters β_0, β_1, and σ, $p = 1$ and $k = 3$.

The $\mathrm{AIC_c}$ criterion penalizes a model with too many covariates from overfitting the sample data. It determines the most parsimonious model, the one with an optimum mix of minimal bias and maximal precision. As the number of parameters increases, models more closely fit sample datasets, reducing the bias. However, as the number of parameters per sample increases, the precision of the parameter estimates tends to decrease. The $\mathrm{AIC_c}$ criterion moderates this process, striking the most optimal compromise between reduced bias and maximal precision.

The corrected Akaike information criterion measures the amount of noise, or **entropy**, in the sample data, separating it from the **signal** or **information**. It is a relative measure of the KL distance between the model and the reality. The absolute measure of the entropy is the calculated $\mathrm{AIC_c}$ plus a constant. The constant remains unknown since the reality is unknown. However, since each model has the same constant, $\mathrm{AIC_c}$s may be compared to determine the relatively best-fitting model. The reader should be warned, however, that this fit is relative. Goodness-of-fit tests must additionally be used in the concluding analysis to assess the absolute fit of the best-fitting models with the lowest $\mathrm{AIC_c}$s.

The $\mathrm{AIC_c}$ is most applicable to models of realities that are complex and infinite- or high-dimensional, as are most natural resource populations. For such complex realities, finite-dimensional models will necessarily be at best only an approximation. For realities that are finite-dimensional, of fairly low dimension, such as $k = 1$–5, with k fixed as the sample size n increases, "dimension-consistent" criteria such as the Bayesian information criterion are more applicable (Burnham and Anderson 1998, 2002).

The **Bayesian information criterion (BIC)**, developed by Schwarz (1978), also uses a formula based on the deviance or log likelihood and "penalizes" models for the overuse of covariates

$$\begin{aligned}\mathrm{BIC} &= D + k \cdot \log(n) \\ &= -2 \cdot \log(\mathcal{L}) + k \cdot \log(n).\end{aligned}$$

For multiple linear regression models, BIC is given by

$$\begin{aligned}\mathrm{BIC} &= n \cdot \log\left(\hat{\sigma}^2 \cdot \frac{n-p-1}{n}\right) + k \cdot \log(n) \\ &= n \cdot \log\left(\hat{\sigma}^2 \cdot \frac{n-k+1}{n}\right) + k \cdot \log(n).\end{aligned}$$

The BIC is derived using Bayesian assumptions of equal priors for each model and vague priors on the parameters (Burnham and Anderson 1998), with the objective of predicting rather than understanding the process of a system. The BIC penalizes more heavily for increases in the number of parameters and hence sometimes tends to select models that are underfit with excessive bias and precision. For natural resource modeling, most realities are complex and infinite-dimensional; hence, AIC_c is the more appropriate criterion for comparing statistical models in the natural resource sciences.

1.3.6 A Preview of Mixed-Effects Modeling

Data collected from monitoring do not always fulfill the assumptions of independence required for the use of many traditional statistical methods: with parameter estimation in sample surveys, with hypothesis testing in experiments, and with model fitting and selection in multiple linear regression or generalized linear modeling. Rather, data are often dependent, clustered or grouped by location or time, or collected from permanent plots or at sites repeatedly over time or from subpopulations of larger populations (e.g., as with meta-population data). Traditionally, scientists have often been discouraged from collecting dependent or pseudoreplicated datasets to avoid these problems with the analysis. But dependencies in biological populations are quite common, and it may be difficult or impossible to avoid collecting data with such dependencies. It would make more sense to collect data with such dependencies and account for them in the analysis. This is now possible with **mixed-effects modeling**, which incorporates both traditional **fixed effects** describing the influences of **treatments** on the population and **random effects** describing dependencies in the data created by groupings. This is achieved in an efficient manner with mixed-effects modeling, incorporating variance components into the models to describe the random effects due to the clusters. We will provide an introduction to mixed-effects modeling using the powerful utilities now available in S-Plus and R for frequentist analysis and WinBUGS for Bayesian analysis in Chapter 7.

1.4 REVIEW: PRINCIPLES OF PROJECT MANAGEMENT

In this section of this introductory chapter, we remind the reader of the critical importance in a natural resource data collection project of practicing the principles of sound project management. A data collection project consists of three phases: a planning phase at the beginning, a data collection phase in the middle, and a concluding phase at the end.

It is very important at the beginning of a project to devote sufficient attention to the planning phase in order to develop a rigorous and effective statistical design for the data collection. To properly determine the appropriate statistical design, the problem, objectives, and methods of analysis for the project must be clearly specified prior to data collection. Far too often in natural resource data collection projects, the problem and objectives are not specified clearly enough in quantitative terms. If a project, for example, has the objective of examining the downward trend of a declining species, the amount of downward trend that is biologically important should be specified in advance of data collection. Then the sample size for the data collection should be calculated to ensure with a high probability that the biologically significant trend will be detected with statistical significance if it exists.

It is unfortunately also far too common in data collection projects to wait until after the data have been collected to decide on the method of analysis. As the method of analysis required by the objectives may impose restrictions on the statistical design of the data collection, it should be determined prior to data collection. The method of analysis may require a certain amount of precision or power to realize the objectives of the project, and this will require a sufficiently large sample or experimental dataset. Hence, the sample size or numbers of replicates must be determined prior to data collection.

It is also vitally important to devote sufficient attention to the concluding phase at the end of a project, to allow time for a comprehensive analysis and thoughtful interpretation and conclusions. Comprehensive analysis will seldom be "turnkey," which can be finished in a few hours or days, but rather a far more lengthy process. The analysis process may consist of examining a range of candidate models to determine the best fitting. This collection of models may be small and finite or wide-ranging in number. We shall describe alternative strategies for the model selection and inference in Chapter 6.

It is tempting in a natural resource data collection project to devote a majority of the time to data collection. This unfortunately may leave an insufficient amount of time at the beginning of the project for planning and at the end of the project for a thorough analysis and thoughtful period of reflection on the conclusions and interpretation of the results. A good principle in general is to spend equal amounts of effort on all phases of the project. This will help ensure that the results are efficient, powerful, and effective. The aim is to extract the biological "information" from the data, separating the "signal" from the "noise" in as optimal a manner as possible. Time and effort devoted to rigorous design and analysis in a data collection project are indeed well spent.

1.5 APPLICATIONS

The practical application of theory is where the learning process really crystallizes. This is particularly true with statistical analysis. We will use a hands-on approach to emphasize the practical use of statistics with the application of theory. We will use both simulated and real-world sample datasets as examples and encourage the reader to do likewise with other datasets.

We encourage readers to analyze their own datasets while progressing through the book. Readers are especially encouraged to conduct a project while reading this book, designing, collecting, and analyzing their own sample or experimental datasets. Address a biologically interesting question, but one sufficiently limited in scope that it can be investigated with a data collection project completed within the timeframe of the reading of this book. For example, examine the abundance level or trend of a local biological population of interest such as a bird or plant population. Write a 2–5-page proposal for the project. Keep the project small and realistically simple; be especially careful to keep the scope of the project within a realistic timeframe.

We will use simulated datasets to illustrate many of the ideas. The reader will thus be able to compare the statistical results with the known "realities" used to generate the simulated datasets. We will also provide real-world datasets for the analyses of important species such as the Northern Spotted Owl, Siskiyou Mountains salamander, and beach layia, some of which are endangered. These datasets will provide practical, realistic experience. The interested reader can also obtain additional sample and experimental datasets on the Web and in many other standard references, such as Sokal and Rohlf (1995), Zar (1996), and Ramsey and Schafer (2002). These references contain excellent examples, problems, and datasets.

1.6 S-Plus® AND R ORIENTATION I: INTRODUCTION

1.6.1 Orientation I

In this section, we provide an introductory orientation to the statistical software used for the frequentist analysis throughout the book, the research-oriented proprietary statistical modeling software S-Plus® (2000) and the de facto equivalent freeware R (R Development Core Team 2005). S-Plus was first developed in the 1970s at Bell Labs by Rick Becker, John Chambers, and Allan Wilks with the goal of defining a language to perform repetitive tasks in data analysis. These authors did not consider the original language to be primarily statistical and most of the statistical functionality was added later. S-Plus provides a flexible, interactive, integrated modern environment for statistical analysis, with particular emphasis on the modeling of linear systems. Much of the syntax of S-Plus is reminiscent of the Unix and C environments.

The current version of R is the result of a collaborative effort with contributions from all over the world. The R language was initially written by Robert Gentleman and Ross Ihaka of the Statistics Department of the University of Auckland. It is freeware, the de facto equivalent of S-Plus.

Details on S-Plus can be found in the S-Plus `Help` menu, the S-Plus Manual (S-Plus 2000), or Krause and Olson (2000). Details on R can be found in the R `Help` menu. S-Plus and R are object-oriented languages with datasets and functions consisting of objects. Objects can be created and manipulated by the user and added to the standard library of defaulted objects that are available in S-Plus and R. To begin this introductory S-Plus–R session, the user should sign onto either S-Plus or R. In S-Plus, the user should enter the `Commands Window` mode from the `Window` menu to begin at the > prompt. In R, the user should begin at the > prompt in the R Console. Although S-Plus and R do have some menu features, the S-Plus command mode and R Console options will be emphasized throughout this book, leaving it to readers to explore the menu options. We will present frequentist software code that is sufficiently general for use in either S-Plus or R and will be sufficiently careful to point out where there are differences.

1.6.2 Simple Manipulations

Let's begin by illustrating simple data manipulation capabilities in S-Plus and R, starting with arithmetic using +, −, *, /, and ∧ for addition, subtraction, multiplication, division, and exponentiation, respectively. Proceed with the following code (Fig. 1.3a), typing the first line, pressing the <ENTER> key, receiving the second line response from S-Plus or R, typing the third line and pressing the <ENTER> key to receive the fourth-line response, and so on. The index [1] at the beginning of each response line refers to the first indexed entry of the arithmetic operation response, a vector of size 1.

For the next illustration, the c concatenation, rep repetition, 1:k integer sequence, and seq general sequence commands aid in constructing data objects in

Figure 1.3. Command code for S-Plus and R Orientation I. (**a**) Simple manipulation: arithmetic; (**b**) Simple manipulations: creation of vectors; (**c**) Simple manipulations: object removal; (**d**) Data structures: vectors, data frames, lists; (**e**) Data structures: testing; (**f**) Data structures: coercion; (**g**) Random numbers: normal distribution; (**h**) Random numbers: sampling; (**i**) Directory structure; (**j**) Functions and control structures: functions code. (**k**) Functions and control structures: functions execution; (**l**) Linear regression analysis.

(**a**)

```
> 9 + 3 <ENTER>
[1] 12
> 9 - 3 <ENTER>
[1] 6
> 9 * 3 <ENTER>
[1] 27
> 9 / 3 <ENTER>
[1] 3
> 9 ^ 3 <ENTER>
[1] 729
```

(**b**)

```
> c(2,5,4,9)
[1] 2 5 4 9
> rep(5,4)
[1] 5 5 5 5
> 2:9
[1] 2 3 4 5 6 7 8 9
> seq(0,1,.1)
[1] 0.0 0.1 0.2 0.3 0.4 0.5 0.6 0.7 0.8 0.9 1.0
```

(**c**)

```
> x <- c(2,4)
> x
[1] 2 4
> rm(x)
> x
Error: Object "x" not found
```

Figure 1.3. *Continued.*

(d)

```
> v1 <- c(2,3,5)
> v1
[1] 2 3 5
> v2 <- c(5,3,7)
> v2[3]
[1] 7
> v3 <- c("low","medium","high")
> v4 <- c(T,F,T,T)
> m1 <- cbind(v1,v2)
> m1
     v1 v2
[1,]  2  5
[2,]  3  3
[3,]  5  7
> d1 <- data.frame(v1,v3)
> d1
  v1     v3
1  2    low
2  3 medium
3  5   high
> list1 <- list(v1,v4)
> list1
[[1]]:
[1] 2 3 5
[[2]]:
[1] T F T T
```

(e)

```
> is.numeric(v1)
[1] T
> is.numeric(v3)
[1] F
```

(f)

```
> v1
[1] 2 3 5
> as.character(v1)
[1] "2" "3" "5"
> is.numeric(as.character(v1))
[1] F
```

Figure 1.3. *Continued.*

(g)

```
> y <- rnorm(10,5,1)
> y
 [1] 6.444356 3.777367 3.974797 5.568687 5.875846
 [6] 3.441874 5.037888 7.006315 4.543586 2.957072
> pnorm(4,5,1)
[1] 0.1586553
> qnorm(.1586553,5,1)
[1] 4
> dnorm(5,5,1)
[1] 0.3989423
```

(h)

```
> x <- 1:10
> x
 [1]  1  2  3  4  5  6  7  8  9 10
> sample(x,5)
[1] 3 8 1 9 7
> sample(x,15)
Error in sample(x,15): Population not large enough
for given sample size
> sample(x,15,replace = T)
 [1]  7  4  7  2  7  7  9  7  6  2  3  4  6 6 10
```

(i)

```
> d1 <- data.frame(v1,v3)
> d1
  v1     v3
1  2    low
2  3 medium
3  5   high
> rm(v1,v3)
> v1 # v1 now exists only internal to d1
Error: Object "v1" not found
> search()
 [1] "C:\\Program Files\\sp2000\\users\\stauffer\\_Data"
 [2] "C:\\Program Files\\sp2000\\splus\\_Functio"
 [3] "C:\\Program Files\\sp2000\\stat\\_Functio"
 [4] "C:\\Program Files\\sp2000\\s\\_Functio"
 ...
> attach(d1)
> search()
 [1] "C:\\Program Files\\sp2000\\users\\stauffer\\_Data"
 [2] "d1"
 [3] "C:\\Program Files\\sp2000\\splus\\_Functio"
 [4] "C:\\Program Files\\sp2000\\stat\\_Functio"
 ...
> v1 # v1 now exists at level 2 in the directory
[1] 2 3 5
> detach(2)
> v1 # again, v1 now exists only internal to d1
Error: Object "v1" not found
```

Figure 1.3. *Continued.*

(j)

```
function(x)
{
# function: add - adds the values in vector x
# author: Jill Analyst
# date:   January 1, 2007
sum <- 0.0
for (i in 1:length(x))
   {sum <- sum + x[i]}
return(sum)
}
```

(k)

```
> v1
[1] 2 3 5
> add(v1)
[1] 10
```

(l)

```
> x <- runif(20,2,8)
> y <- 10+1.5*x+rnorm(20,0,1)
> output <- lm(y~x)
> summary(output)
Coefficients:
                  Value Std. Error t value Pr(>|t|)
(Intercept) 10.1862  0.8926    11.4113  0.0000
          x  1.4600  0.1655     8.8207  0.0000
Residual standard error: 1.179 on 18 degrees of freedom
Multiple R-Squared: 0.8121
F-statistic: 77.8 on 1 and 18 degrees of freedom, the p-value is
     5.937e-008
> plot(x,y) # See Fig. 1.4.
> abline(10.1862,1.4600)
```

S-Plus and R (hereafter we omit reference to the <ENTER> key at the end of each input command where this is clear from the context) (Fig. 1.3b). All of these data object reponses are vectors of values, starting with the first entry indexed by `[1]` at the beginning of the output line. S-Plus and R are case-sensitive. Use the `Help` menu `Search S-Plus Help` in S-Plus or the `Help` menu `R functions (text)` ... option in R to learn more about the commands and their syntax.

The `<-` operation assigns values to objects, and the `rm` command removes or deletes objects in both S-Plus and R (Fig. 1.3c). In S-Plus, the underscore `_` can be substituted for `<-` to indicate assignment. The list of objects currently available at the top-level directory of S-Plus and R can be viewed by using the `objects()` command.

1.6.3 Data Structures

In S-Plus and R, datasets are organized into simple data structures of type **numeric** for quantitative or numerical values, **factor** for qualitative or categorical values, **character** for character strings, and **logical** for objects with logical values true `T` or false `F`. These simple atomic types can be combined into complex data structures such as one-dimensional **vectors** of simple values of the same type, two-dimensional **matrices** of columns of vectors of the same type and length, two-dimensional **data frames** with columns of vectors of possibly different types but the same size, and **lists** of simple and complex types of varying size. Vectors are matrices, matrices are data frames, and data frames are lists, but the converse is rarely true. Let's combine simple values into a vector with the concatenate `c` command, vectors into matrices with the `matrix` command, column bind `cbind` command, or row bind `rbind` command, vectors into data frames with the `data.frame` command, and data structures into lists with the `list` command (Fig. 1.3d). Notice how the row entries in the matrix `m1` are indicated by `[1,]`, `[2,]`, and `[3,]`; the column vector entries in the data frame `d1` are indicated by `v1` and `v3`; and the entries in the list `list1` are indicated by `[[1]]` and `[[2]]`. The type of values in data structures can be tested with the `is.numeric`, `is.character`, `is.logical`, `is.vector`, `is.matrix`, `is.data.frame`, or `is.list` commands (Fig. 1.3e). Data structures can sometimes be coerced into other types of more complex objects and values with `as.matrix` to convert a vector into a matrix, `as.data.frame` to convert a matrix into a data frame, `as.list` to convert a data frame into a list, and `as.character` to convert a numeric vector into a character vector (Fig. 1.3f).

1.6.4 Random Numbers

Probability density, cumulative probability, quantile, and random values can be generated in S-Plus and R by using the `d`, `p`, `q`, and `r` prefixes with common probability distributions such as the normal, uniform, gamma, exponential, *t*, *F*, chi-square (χ^2), lognormal, Poisson, negative binomial, and binomial distributions using `norm`, `unif`, `gamma`, `exp`, `t`, `f`, `chisq`, `lnorm`, `pois`, `nbinom`, and `binom` suffixes and their specified parameters (Fig. 1.3g). For example, in the figure, y is a vector with 10 numeric values randomly sampled from the normal distribution $N(5, 1)$ with mean $\mu = 5$ and standard deviation $\sigma = 1$. The cumulative probability at 4, quantile of 0.1586553, and density value at 5 for $N(5, 1)$ are also calculated.

The `sample` command can be used to generate random data from a vector, without or with replacement (Fig. 1.3h). The default of this command is to sample without replacement.

1.6.5 Graphs

Graphs can be obtained for numeric values using the `dotplot` (`stripchart` in R) and `hist` commands on vectors, `plot` command on pairs of vectors, and `pairs` command on matrices or data frames of columns. You can also create an $n \times m$ trellis of graphs with n rows and m columns using the `par(mfrow=c(n,m))` command.

1.6.6 Importing and Exporting Files

Data files can be imported or exported as objects using the `Import Data` and `Export Data` options in the `File` menu in S-Plus. Data files are imported as data frames unless specified otherwise. In R, data can be imported and exported using `copy and paste` commands in the data file and the `Data editor` option in the `Edit` menu. Alternatively, in R, text files of data with the `txt` extension `data.txt` from the defaulted R-2.2.1 folder can be imported using the `read.table("data.txt")` command.

1.6.7 Saving and Restoring Objects

You can save and restore your directory of objects or individual objects in S-Plus by using `data.dump(objects(), "directory and filename")` or `data.dump(c("object1","object2",...), "directory and filename")` commands and `data.restore("directory and filename")` commands. Alternatively, in S-Plus, you can use the `Workspace Save` and `Workspace Open` options in the `File` menu. In R, you can save and restore your directory of objects by using the `Save Workspace` and `Load Workspace` options in the `File` menu.

1.6.8 Directory Structure

The S-Plus and R directory structures are hierarchical in structure and can be viewed using the `search()` command (Fig. 1.3i). The directory of an object can be determined using the `find (object)` command. An object such as a data frame or list can be opened and closed, with internal objects such as vectors made available at a specified directory level (defaulted to level 2), by using the `attach(object, level)` and `detach(level)` commands. The # symbol in a line of code indicates to S-Plus or R that the remainder of the command is a comment.

1.6.9 Functions and Control Structures

Function subprograms can be created as objects using the `fix(name of object)` command. If a mistake is made in creating and editing the subprogram object, an error message will be issued. Use the `fix()` command to return to editing to correct the mistake in the object before signing off from S-Plus or R. Otherwise the function object will not be saved. The standard programming control structures are available in S-Plus and R: (1) sequential; (2) conditional, with `if (test) then {} else {}` syntax; (3) repetition, with `for (comparison or name in values) expr` or `for (comparison or name in values) expr else`

Figure 1.4. Linear regression model fit to the sample dataset $y = 10 + 1.5 * x +$ error where error $\sim N(0,1)$ with $x \sim$ Unif (2,8) and $n = 20$ samples.

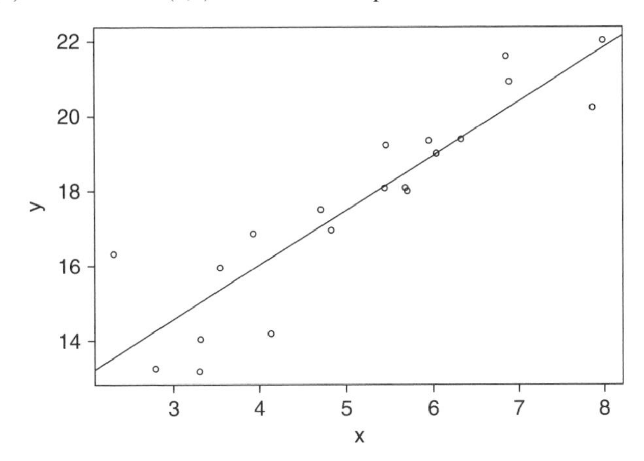

`expr` syntax; and (4) subprograms. Objects enter subprogram as parameters in the initial `function(parameters)` statement and exit in a final `return (objects)` statement. Braces {} are used to delineate blocks of code. We illustrate with a function `add` that adds the values in the vector x (Fig. 1.3j). This function `add` could be created using the `fix(add)` command and executed (Fig. 1.3k).

1.6.10 Linear Regression Analysis in S-Plus and R

We conclude this introductory section on S-Plus and R by illustrating the use of linear regression modeling, with vectors of independent values x and dependent response values y (Fig. 1.3l). The sample data (x, y) were simulated from a "reality" using the linear relationship $y = 10 + 1.5 \cdot x$ with normal errors $e \sim N(0, 1)$ having mean $\mu = 0$ and standard deviation $\sigma = 1$. The output results of the linear regression analysis are statistically compatible with the simulated reality (Fig. 1.4).

1.7 S-Plus AND R ORIENTATION II: DISTRIBUTIONS

We encourage the reader at this point to review the most important distributions that are fundamental to statistical analysis: the uniform and normal distributions for continuous data, the Poisson distribution for count data, and the binomial distribution for binary data. In this section, we review these distributions in S-Plus and R.

1.7.1 Uniform Distribution

Let us start with an example where we generate $n = 30$ randomly located plots in a forest stand illustrating the application of the **uniform distribution**. This can be accomplished by first circumscribing the forest stand with a rectangular area, say, $m_1 \times m_2 = 200 \times 100$ units in size. Then we use S-Plus or R to generate n' plot centers (x_i, y_i), $i = 1, 2, \ldots, n'$, within this 200×100 rectangular area that circumscribes the forest stand. We choose $n' > n$ large enough so that $n = 30$ of the plots will be within the forest stand subarea, say, $n' = 50$. Then we select the first $n = 30$ plot centers from the list of $n' = 50$ that are within the forest stand subarea of the rectangular area. To determine the plot locations, generate $n' = 50$ x_i values from the uniform distribution Unif(x; 0, 200) with limits 0 and 200 and $n' = 50$ y_i values from the uniform distribution Unif(y; 0, 100) with limits 0–100 (Fig. 1.5a). Histograms of the x and y samples and the plot locations are displayed in Figs. 1.6a–c. The histograms of the samples from the uniform distributions are not, of course, entirely flat because of the small sample size. Generate samples from other uniform distributions and examine their distribution with the hist command. The uniform distribution is a constant, flat distribution with an equal probability of sampling any value within its range of limits, so it is a useful distribution for generating simple random sample locations of values from continuous variables. The uniform distribution Unif(y; θ_1, θ_2) of the continuous random variable y has two parameters, θ_1 and θ_2, that define the limits of the range of y (Fig. 1.6d).

1.7.2 Normal Distribution

Next, we illustrate the **normal distribution** $N(y;\ \mu,\ \sigma)$ of the continuous random variable y with the two parameters μ, the mean average, and the standard deviation σ, by generating a simulated sample from a population of tree diameters in a community forest. Let's assume that this population consists of even-aged tree diameters that are normally distributed with mean 30″ and standard deviation 5″. Begin by simulating a large sample of 10,000 measurements and looking at the histogram that approximates closely that of the entire population distribution. Then generate a sample of size $n = 50$ (Fig. 1.5b). Our analysis first examines the histogram of a large sample that approximates the population (Fig. 1.6e). Then it examines the values at the 2.5 and 97.5 percentiles of this large sample dataset and compares these sampled percentiles with the actual theoretical quantiles of the population using the qnorm command. It also calculates the cumulative probabilities 2 and 1 standard deviations away from the mean (i.e., at diameter at breast height $d_{\mathrm{bh}} =$ 20, 40, 25, and 35, respectively) using the `qnorm` command. Finally we plot the d_{bh} normal distribution N(d_{bh}; 30,5) within the range [20, 40] using the `dnorm` command (Fig. 1.6f).

We encourage the reader to generate samples from other normal distributions and look at their sample histograms, quantiles, cumulative probabilities, and density curves. Many continuous natural resource datasets are approximately normal. Recall also that the Central-Limit Theorem states that, if the data are normally distributed, the mean estimates, with repeated sampling, are also normally distributed.

Figure 1.5. Command code for S-Plus and R Orientation II: distributions. (**a**) Uniform distribution. (**b**) Normal distribution. (**c**) Poisson distribution. (**d**) Binomial distribution: Bernoulli trials. (**e**) Binomial distribution: binomial trials. (**f**) Binomial distribution: other examples. (**g**) Binomial distribution: yet other examples. (**h**) Simple random sampling.

(**a**)
```
> x <- runif(50,0,200)
> round(x,2)  # Round to two digits of accuracy.
 [1] 179.21  75.49  56.99  93.85   6.30  25.04
 [7]   5.69 134.85 144.23  ...
 ...
 [49]  64.55  13.48
> hist(x)  # See Fig. 1.6a.
> y <- runif(50,0,100)
> round(y,2)
 [1] 51.07 66.36 51.08 54.26 51.97 45.82 54.34
 [8] 79.85 27.99 32.90  ...
... [50] 63.68
> hist(y)  # See Fig. 1.6b.
> plot(x,y)  # See Fig. 1.6c .
```

(**b**)
```
> dbh <- rnorm(10000,30,5)
> hist(dbh)  # See Fig. 1.6e.
> sort(dbh)[250]  # This is the sample 2.5 percentile.
[1] 20.22898
> sort(dbh)[9750]  # This the sample 97.5 percentile.
[1] 39.64929
> qnorm(.025,30,5)  # This is the population 2.5 percentile.
[1] 20.20018
> qnorm(.975,30,5)  # This is the population 97.5 percentile.
[1] 39.79982
> pnorm(20,30,5)  # This is the population cumulative probability at d_bh = 20.
[1] 0.02275013
> pnorm(40,30,5)  # This is the population cumulative probability at d_bh = 40.
[1] 0.9772499
> pnorm(25,30,5)  # This is the population cumulative probability at d_bh = 25.
[1] 0.1586553
> pnorm(35,30,5)  # This is the population cumulative probability at d_bh = 35.
[1] 0.8413447
> sample.dbh <- rnorm(50,30,5)  # This is a random
 sample of 50 dbh values that are normally distributed with μ = 30 and σ = 5(N(d_bh ; 30,5)).
> sample.dbh
 [1] 39.39297 35.16121 27.17841 25.52627 27.79735
 [6] 28.98155 34.67952 25.39373  ...
 ...
 [46] 30.34656 18.95812 31.54154 22.42361 24.00949
> dbh <- seq(20,40,.05)
> normal <- dnorm(dbh,30,5)  # These are density values for the incremental d_bh values.
> plot(dbh,normal)  # See Fig. 1.6f.
```

(**c**)
```
> count <- 0:30
> poisson <- dpois(count,0.5)
> plot(count,poisson)  # See Fig. 1.6g.
```

Figure 1.5. *Continued.*

(d)
```
> n <- 30
> p <- 0.40
> x <- rbinom(n,1,p)
> x
 [1] 0 1 0 0 0 1 1 0 0 1 0 0 1 0 0 0 0 0 1 0 1 0
 [23] 0 0 1 1 0 0 1 0
> y <- sum(x)
> y
[1] 10
> phat <- y/n
> phat
[1] 0.3333333
```

(e)
```
> y <- rbinom(10000,30,0.40)
> y
 [1]   8 13 13 12 13 11  7 10 13 11 ...
   ... 16 11 10 13 14 11 13   ...
 [9997] 15 13 15 13
> hist(y)    # See Fig. 1.6k.
```

(f)
```
> y <- rbinom(10000,30,0.10)
> hist(y)    # See Fig. 1.6l.
> y <- rbinom(10000,30,0.90)
> hist(y)    # See Fig. 1.6m.
> y <- rbinom(10000,10,0.50)
> hist(y)    # See Fig. 1.6n.
```

(g)
```
> y <- 0:20
> binomial <- dbinom(y,20,0.70)
> plot(y,binomial) # See Fig. 1.6o.
> sample <- rbinom(30,1,0.40)
> sample
 [1] 1 1 0 1 0 0 0 0 0 1 1 1 0 0 0 1 1 1 0 0 0
     1 0 1 0 1 1 1 1
> sum(sample)
[1] 15
> sample <- rbinom(1,30,0.40)
> sample
[1] 12
```

(h)
```
> frame <- 1:1000
> sample(frame,30)
 [1] 983 957 351 453 105 843 71 677 ...
 [27] 813 161 240 862
```

1.7.3 Poisson Distribution

Next, we examine the **Poisson distribution** Pois($y =$ count;λ) of the discrete random variable y with mean average parameter λ for count data of populations that are randomly dispersed. We start by examining Poisson data with a mean average parameter of $\lambda = 0.5$ (Fig. 1.5c). The Poisson distribution is defined for the discrete random variable that assumes nonnegative integer values only, and the analysis examines it for the first 31 $y =$ count values, $0, 1, 2, \ldots, 30$. For small values of the parameter, such as $\lambda = 0.5$, the Poisson distribution has an "exponentially declining" shape, with a mode at the origin, asymptotically approaching the ordinal axis as count $->$ 0. As the mean average parameter λ increases, the Poisson becomes unimodal with mode at the mean or $\lambda - 1$, again asymptotically approaching the ordinal axis as count $->$ 0 (Figs. 1.6g–j).

The Poisson distribution models count data for species that are randomly distributed spatially (Pielou 1969, Hilborn and Mangel 1997, Rice 1995). The Poisson distribution Pois($y =$ count; λ) is characterized by one parameter only, the mean $\lambda > 0$. The variance σ^2 of the Poisson distribution is equal to its mean λ: $\sigma^2 = \lambda$. If the species is aggregated, or clustered spatially, as is often the case for natural resource data, the variance is greater than the mean, $\sigma^2 > \mu$, and the random variable $y =$ count is better modeled with the negative binomial distribution. The negative binomial distribution NB($y =$ count; n, a) for the discrete random variable y has two parameters, n and a, where the mean average $\mu = n/a$ and the variance $\sigma^2 = n/a + n/a^2 = \mu + \mu^2/n$ (Hilborn and Mangel 1997). If the species is regularly or uniformly distributed spatially, the variance is less than the mean, $\sigma^2 < \lambda$. The expression $I = \sigma^2/\mu$ is sometimes used as an **index of nonrandomness** for count data (Pielou 1969). Hence $I < 1$, $I = 1$, or $I > 1$ depending on whether the population is regular, random, or aggregated, respectively.

1.7.4 Binomial Distribution

We conclude this section by examining the binomial distribution. The **binomial distribution** $B(y; n, p)$ is the probability distribution of the discrete random variable $y =$ the number of successes in a binomial experiment consisting of a sequence of n independent **Bernoulli trials**, each of which has two possible outcomes, "success" or 1, and "failure" or 0, with probabilities p and $q = 1 - p$, respectively. Note that the random variable can assume the values $y = 0, 1, 2, \ldots, n$. A Bernoulli trial is a binomial experiment with $n = 1$ trial. First we look at an example in S-Plus and R of a binomial experiment with $n = 30$ and $p = 0.40$ (Fig. 1.5d). This binomial experiment was simulated in S-Plus and R as a sequence of $n = 30$ independent Bernoulli trials. The analysis calculated the random variable value y as the sum of the Bernoulli outcomes, that is, the number of successes. The analysis also calculated the proportion estimate

$$\text{phat} = \hat{p} = y/n = 0.33$$

from the sample. This estimate $\hat{p} = 0.33$ approximates the parameter value $p = 0.40$ for the population.

Next, we look at the distribution of the binomial random variable y itself in the binomial experiment $B(y;\ n = 30,\ p = 0.40)$, simulating 10,000 such experiments (Fig. 1.5e). Note that the distribution looks bell-shaped (Fig. 1.6k). It can be

Figure 1.6. Simulated randomly generated data from distributions, with S-Plus and R code. (**a**) Histogram of sample data x from the uniform distribution Unif(x; 0, 200) with parameter limits 0 and 200 and sample size $n = 50$ from the code

```
> x <- runif(50,0,200)
> hist(x)
```

(**b**) Histogram of sample data y from the uniform distribution Unif(y; 0, 100) with parameter limits 0 and 100 and sample size $n = 50$ from the code

```
> y <- runif(50,0,100)
> hist(y)
```

(**c**) Plot of randomly sampled (x, y) plot centers in a 200×100 area from plots (a) and (b) above from the code

```
> plot(x,y)
```

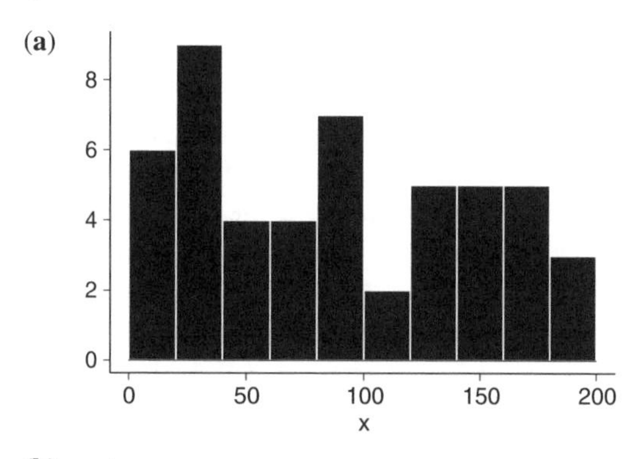

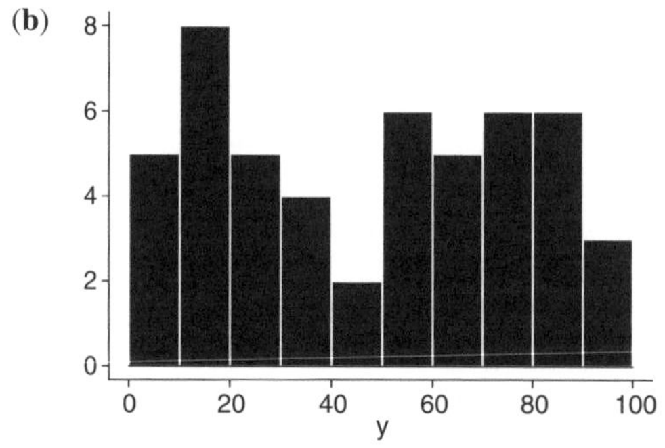

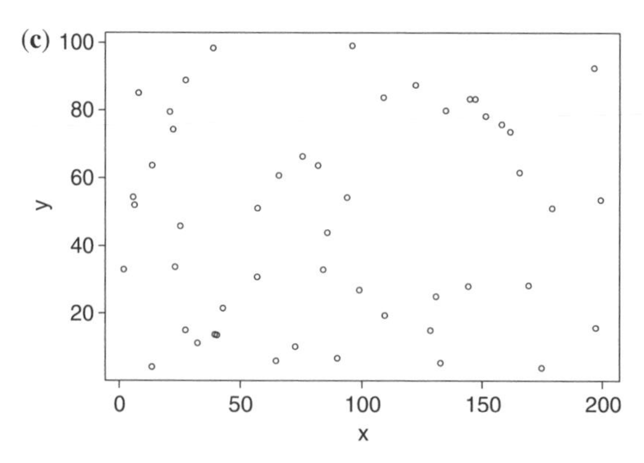

Figure 1.6. *Continued.*
(**d**) Density function of the uniform distribution Unif(y; $\theta_1 = 0$, $\theta_2 = 200$).

(**e**) Histogram of $n = 10{,}000$ tree diameters (d_{bh} values) sampled in S-Plus or R from a normal distribution $N(d_{bh}; \mu, \theta)$ with mean $\mu = 30$ and standard deviation $\sigma = 5$ from the code

```
> dbh <- rnorm(10000,30,5)
> hist(dbh)
```

(**f**) Density curve of the tree diameter (d_{bh}) normal distribution $N(d_{bh}; \mu, \theta)$ with mean $\mu = 30$ and standard deviation $\theta = 5$ from the code

```
> dbh <- seq(20,40,0.05)
> normal <- dnorm(dbh,30,5)
> plot(dbh,normal)
```

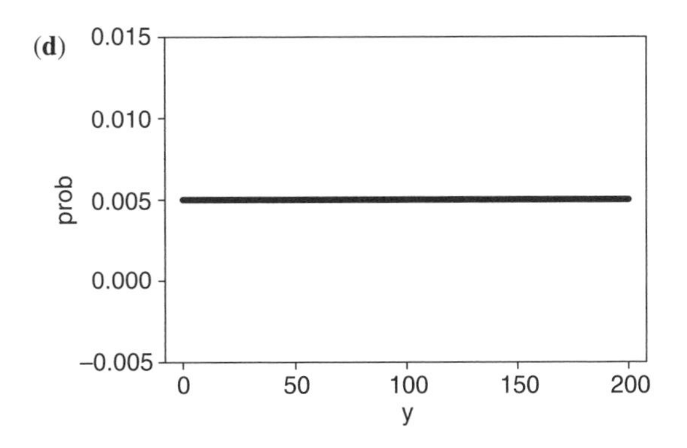

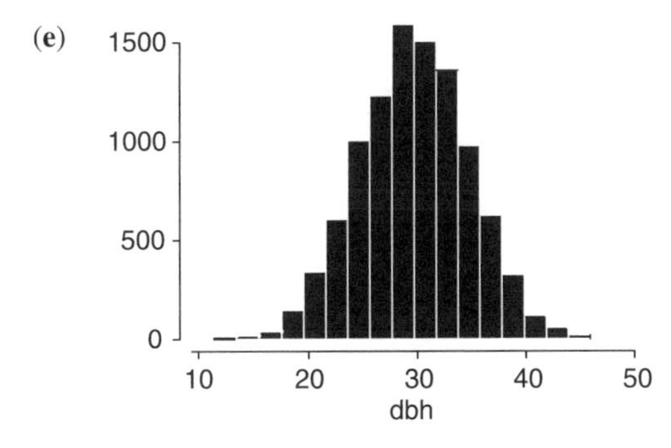

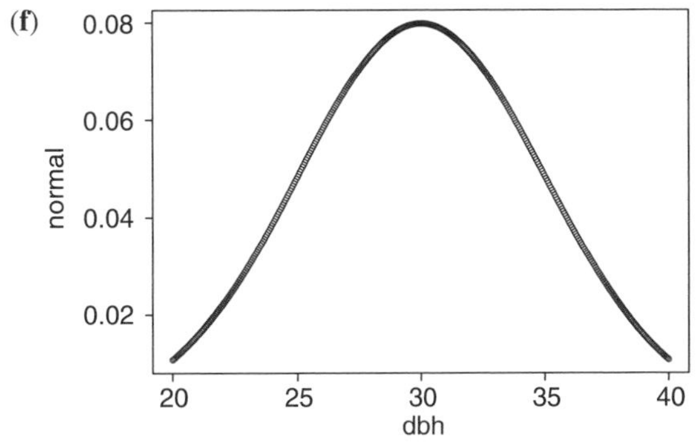

Figure 1.6 *Continued.*
(**g**) Distribution of the count data Poisson distribution *P*(count; λ) with mean λ = 0.5 from the code

```
> count <- 0:30
> poisson1 <- dpois(count,0.5)
> plot(count,poisson1)
```

(**h**) Distribution of the count data Poisson distribution *P*(count; λ) with mean λ = 1.0 from the code

```
> count <- 0:30
> poisson2 <- dpois(count,1.0)
> plot(count,poisson2)
```

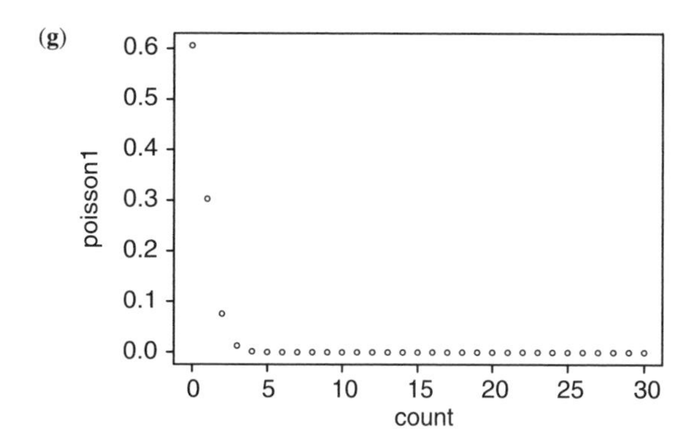

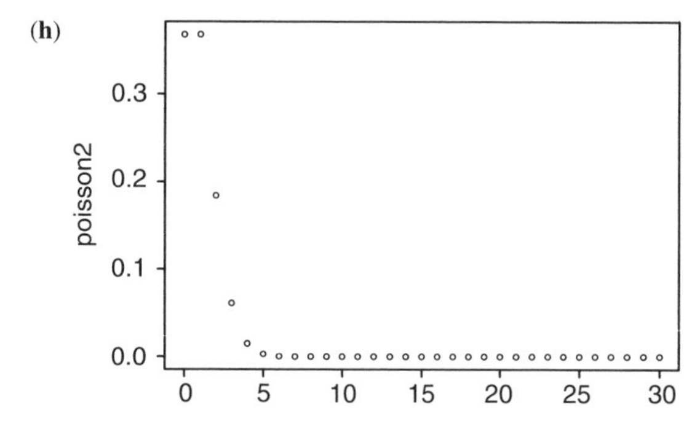

shown that the binomial distribution is approximately normally distributed if the following conditions hold:

1. $n \cdot p \geq 5$
2. $n \cdot q \geq 5$.

The mean of *y* is

$$\mu = n \cdot p = 30 \cdot 0.40 = 12$$

and the variance is

$$\sigma^2 = n \cdot p \cdot q = 30.0 \cdot 40 \cdot 0 \cdot 60 = 7.2.$$

Figure 1.6. *Continued.*
(**i**) Distribution of the count data Poisson distribution $P(\text{count}; \lambda)$ with mean $\lambda = 3.0$ from the code

```
< count c <- 0:30
< poisson3 <- dpois(count,3.0)
< plot(count,poisson3)
```

(**j**) Distribution of the count data Poisson distribution $P(\text{count}; \lambda)$ with mean $\lambda = 5.0$ from the code

```
< count <- 0:30
< poisson4 <- dpois(count,5.0)
< plot(count,poisson4)
```

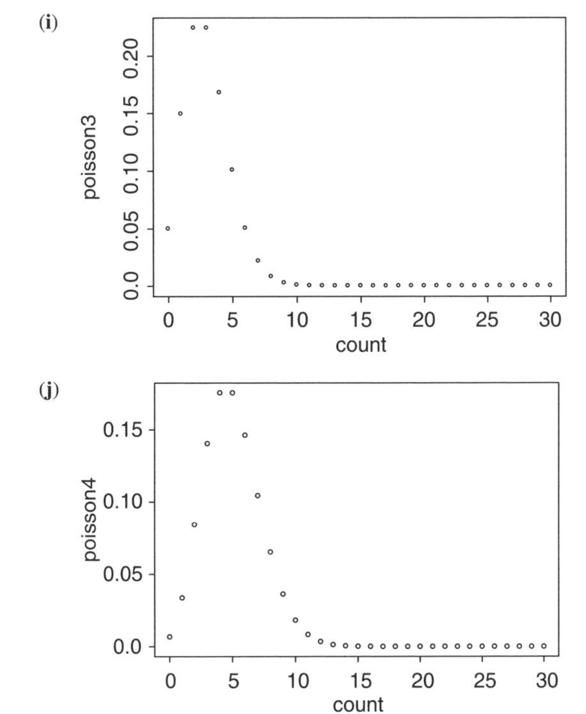

The mode of y occurs close to the mean since the distribution is approximately normal and symmetric. If $p = 0.50$, the binomial distribution is symmetric. Let's look at some other examples of binomial distributions (Figs. 1.5f and 1.6l–n). Note that the distributions are skewed for $B(y;\ 30,\ 0.10)$ with $p = 0.10$ and $B(y; 30, 0.90)$ with $p = 0.90$, and symmetric for $B(y;\ 10,\ 0.50)$ with $p = 0.50$. They are nonnormal for $B(y;\ 30,\ 0.10)$ since $n \cdot p = 3 < 5$ and $B(y;\ 30,\ 0.90)$ since $n \cdot q = 3 < 5$. Note also that the modes are close to the distribution means $\mu = n \cdot p = 3$, 27, and 5, respectively, at the means if the distribution is symmetric and toward the tail if the distribution is skewed.

Figure 1.6. *Continued.*

(**k**) Histogram of 10,000 samples in S-Plus or R from the binomial distribution B(y;n,p) with $n = 30$ and $p = 0.40$ from the code

```
< y <- rbinom(10000,30,0.40)
< hist(y)
```

(**l**) Histogram of 10,000 samples in S-Plus or R from the binomial distribution B(y;n,p) with $n = 30$ and $p = 0.10$ from the code

```
< y <- rbinom(10000,30,0.10)
< hist(y)
```

(**m**) Histogram of 10,000 samples in S-Plus or R from the binomial distribution B(y;n,p) with $n = 30$ and $p = 0.90$ from the code

```
< y <- rbinom(10000,30,0.90)
< hist(y)
```

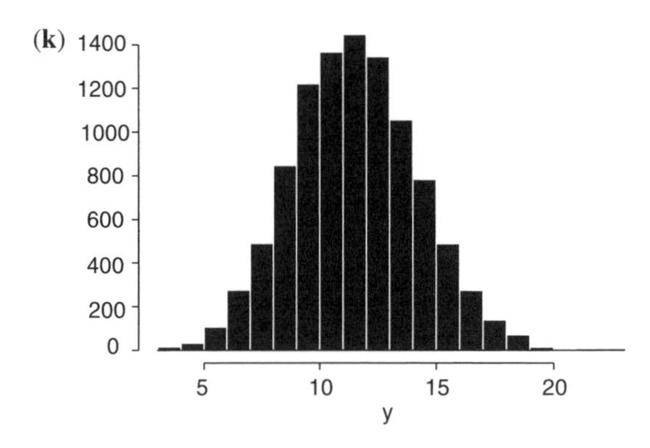

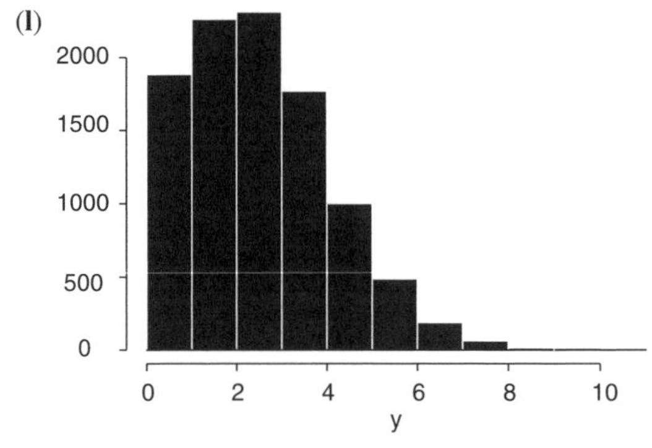

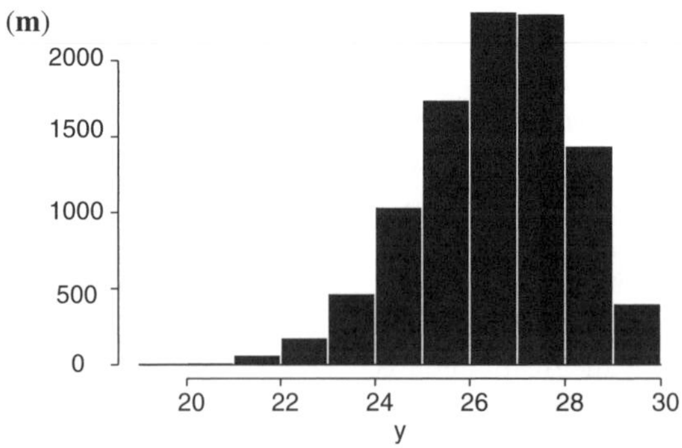

Figure 1.6. *Continued.*
(**n**) Histogram of 10,000 samples in S-Plus or R from the binomial distribution B(y;n,p) with $n = 10$ and $p = 0.50$ from the code

```
< y <- rbinom(10000,10,0.50)
< hist(y)
```

(**o**) Density function of the binomial distribution B (y;n,p) with $n = 20$ and $p = 0.70$ from the code

```
< y <- 0:20
< binomial <- dbinom(y,20,0.70)
< plot(y,binomial)
```

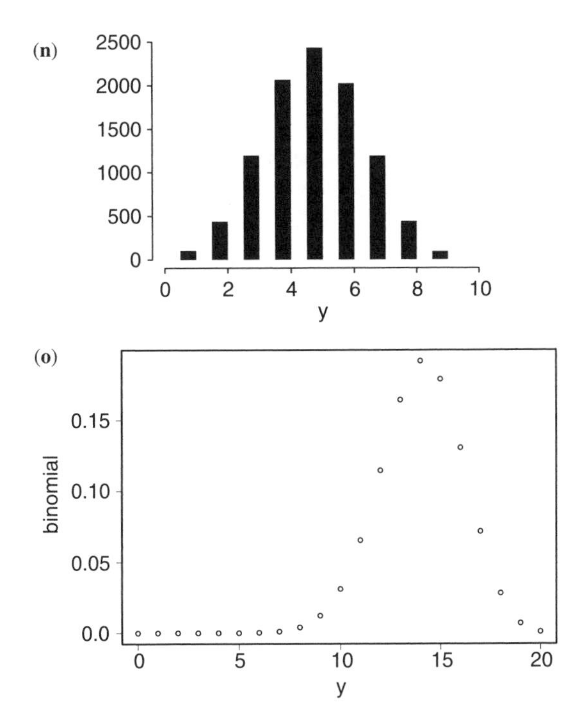

The interested reader can further investigate the binomial distribution by looking at density functions such as $B(y;\, n = 20,\, p = 0.70)$ and examining samples of realistic size, simulating either a sequence of Bernoulli trials or one binomial experiment (Figs. 1.5g and 1.6o).

1.7.5 Simple Random Sampling

Let's look at one final example in S-Plus and R to conclude this section, one that illustrates an additional application of random sampling. Let's randomly select a sample of size $n = 30$ from a finite frame of sampling units for a population labeled by integers from 1 to $N = 1000$ (Fig. 1.5h). The sampling is without replacement. To sample with replacement, use the command `sample (frame, 30, replace = T)`.

1.8 S-Plus AND R ORIENTATION III: ESTIMATION OF MEAN AND PROPORTION, SAMPLING ERROR, AND CONFIDENCE INTERVALS

1.8.1 Estimation of Mean

This chapter continues with further S-Plus and R examples, illustrating its use for the estimation of parameters. Let's first simulate the simple random sampling of data $\{y_i\}$ from a normally distributed population $y \sim N(\mu = 20, \sigma = 5)$ with a sample size of $n = 30$ and calculate a confidence interval for the mean average parameter (Figs. 1.7a and 1.8). Notice the proximity of the estimates $\bar{y} = 20.73598$ and $s = 2.09667$ to the parameters $\mu = 20$ and $\sigma = 2$ that they are estimating. This sample dataset produces one of the 95% CIs that correctly includes the mean parameter $\mu = 20$.

Figure 1.7. Command code for S-Plus and R Orientation III: Estimation of mean and proportion, sampling error, and confidence intervals. (**a**) Estimation of mean. (**b**) Estimation of proportion.

(**a**)
```
> y <- rnorm(30,20,5) # See Fig. 1.8.
> y
 [1]  22.47808 19.44215 23.00900 26.26603 30.81094
 [6]  20.83952 13.63786 34.21628 19.51049 17.87346
 [11] 14.48432 20.60656 11.03379 24.49875 16.68419
 [16] 11.53640 16.87080 27.87817 21.28147 11.47134
 [21] 18.96305 23.70433 25.33535 25.58838 21.86278
 [26] 20.25139 17.78660 27.64230 16.99594 19.51984
> hist(y)
> ybar <- mean(y)
> ybar
[1] 20.73598
> s <- stdev(y) # Use sd(y) in R
> s
[1] 5.614986
> n <- 30
> se <- s/sqrt(n)
> se
[1] 1.025151
> t <- qt(0.975,n-1)
> t
[1] 2.04523
> E <- t*se
> E
[1] 2.09667
> left.limit <- round(ybar-E,2)
> right.limit <- round(ybar+E,2)
> CI <- paste("[",left.limit,",",right.limit,"]",sep="")
> CI
[1] "[18.64,22.83]"
```

Figure 1.7. *Continued.*

(b)
```
> y <- rbinom(30,1,0.2)
> y
 [1] 0 1 0 0 0 0 0 0 1 0 0 0 1 1 0 1 1 0 0 0 0 1
 [23] 0 0 0 1 1 1 0 1
> hist(y)    # See Fig. 1.9
> n <- 30
> y <- sum(y)
> phat <- y/n
> phat
[1] 0.3666667
> se <- sqrt(phat*(1-phat)/(n-1))
> se
[1] 0.08948555
> t <- qt(0.975,n-1)
> E <- t*se
> E
[1] 0.1830185
> left.limit <- round(phat-E,2)
> right.limit <- round(phat+E,2)
> CI <- paste("[",left.limit,",",right.limit,"]",sep="")
> CI
[1] "[0.18,0.55]"
```

The `t.test(y)$conf.int` command in S-Plus and R also can be used to calculate 95% confidence intervals of simple random samples $\{y\}$. To use an alternative confidence level P in the command, such as $P = 0.90$, insert the `conf.level = P` option in the command (e.g., `t.test(y,conf.level=0.90)$conf.int`).

Figure 1.8. Histogram of continuous sample data $y \sim N\ (\mu = 20,\ \sigma = 5)$ with sample size $n = 30$.

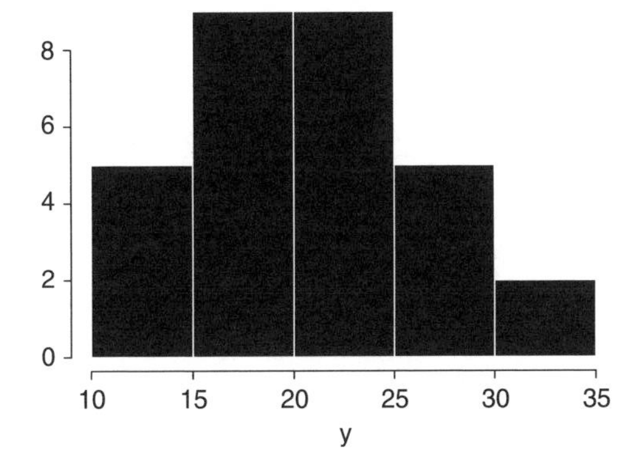

Figure 1.9. Histogram of binary sample data $y \sim B(n = 30,\ p = 0.20)$ with sample size $n = 30$.

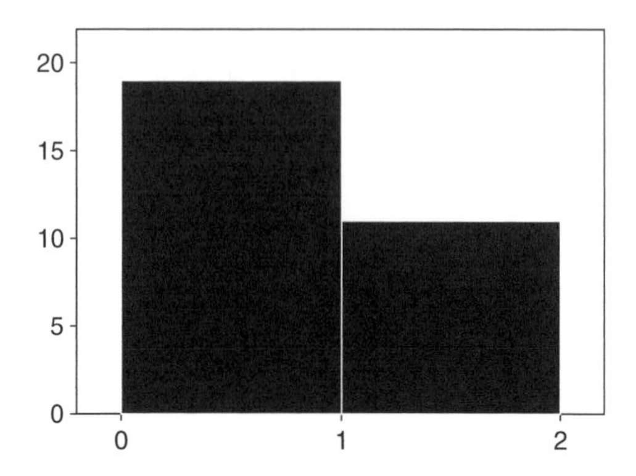

1.8.2 Estimation of Proportion

Next let's simulate the simple random sampling of binary data $\{y_i\}$ from a binomial distribution $B(y;\ n = 30,\ p = 0.20)$, where $y = \Sigma y_i$ with a sample size of $n = 30$ (Fig. 1.7b and 1.9). Note here the relatively large amount of sampling error $E =$ 8.9% for the estimate $\hat{p}$, due to the relatively small sample size $n = 30$ for the proportion estimator. The confidence interval from this sample includes the parameter $p = 0.20$, as is expected of 19 out of 20 confidence intervals, with repeated SRS sampling of the population.

1.9 S-Plus AND R ORIENTATION IV: LINEAR REGRESSION

To conclude this introductory chapter, let's illustrate linear regression modeling in S-Plus and R with an example. Generate a sample of size $n = 80$ from a linear "reality"

$$y = \beta_0 + \beta_1 \cdot x + e = 50 + 1.2 \cdot x + e,$$

where the error $e \sim N(\mu = 0, \sigma = 20)$ is normally distributed, and fit it with a linear regression model (Figs. 1.10a and 1.11).

The estimated statistics from the analysis are as follows:

1. The estimates of the constant and slope coefficients $\beta_0 = 50$ and $\beta_1 = 1.2$ respectively are $\hat{\beta}_0 = 53.8435$ and $\hat{\beta}_1 = 1.1893$. Both are highly significant with $p < 0.00005$. The 95% confidence intervals, based on these estimates and the estimated standard errors, both contain the respective parameters.
2. The coefficient of determination is $R^2 = 0.9288$.
3. The residual standard error is $s_{y|x} = 20.33$, which is close to the parameter $\sigma =$ 20 that it is estimating.
4. The ANOVA F statistic $F_s = 1018$ is highly significant with $p < 0.00005$.
5. AIC = 712.9258.

Figure 1.10. Command code for S-Plus and R Orientation IV: Linear Regression. (**a**) Linear regression model, with a constant; (**b**) Linear regression analysis, without a constant; (**c**) Comparison of models.

(**a**)
```
>x <- runif(80,0,200)
>y <- 50+1.2*x+rnorm(80,0,20)
>plot(x,y)   # See Fig. 1.11
>output1 <- lm(y~x)  # This is the full model.
>summary(output1)
...
Coefficients:
              Value Std. Error  t value Pr(>|t|)
(Intercept) 53.8435  3.9109     13.7676  0.0000
          x  1.1893  0.0373     31.9001  0.0000
Residual standard error: 20.33 on 78 degrees of freedom
Multiple R-Squared: 0.9288
F-statistic: 1018 on 1 and 78 degrees of freedom,
   the p-value is 0
...
>AIC(output1)
[1] 712.9258
```

(**b**)
```
>output2 <- lm(y~1)  # This is the null model.
>summary(output2)
...
Coefficients:
               Value Std. Error  t value Pr(>|t|)
(Intercept) 155.3716   8.4639    18.3571  0.0000
Residual standard error: 75.70 on 79 degrees of
   freedom
Multiple R-Squared: 1.385e-029
F-statistic: -Inf on 0 and 79 degrees of freedom,
   the p-value is NA
...
>AIC(output2)
[1] 922.3147
>output3 <- lm(y~x-1)  # This is the ratio model.
>summary(output3)
...
Coefficients:
    Value Std. Error t value Pr(>|t|)
x  1.6070  0.0399    40.3046  0.0000
Residual standard error: 37.41 on 79 degrees of freedom
Multiple R-Squared: 0.9536
F-statistic: 1624 on 1 and 79 degrees of freedom,
   the p-value is 0
...
>AIC(output3)
[1] 809.5328
```

Figure 1.10. *Continued.*

(c)
```
>anova(output1,output2)
Analysis of Variance Table
Response: y
Terms Resid. Df       RSS  Df Sum of Sq  F Value Pr(F)
1     x       78   32232.2
2     1       79 452744.4  -1 -420512.2 1017.614   0
>anova(output1,output3)
Analysis of Variance Table
Response: y
Terms Resid. Df       RSS  Df Sum of Sq  F Value Pr(F)
1     x       78   32232.2
2   x - 1     79 110559.6  -1 -78327.37 189.5475   0
```

We compare this full linear model with two other more parsimonious models, the null model without a linear term and the ratio model without a constant (Fig. 1.10b). Both of these models have coefficient estimates that are at odds with reality, $\hat{\beta}_0 = 155.3716$ for $\beta_0 = 50$ for the null model and $\hat{\beta}_1 = 1.6070$ for $\beta_1 = 1.2$ for the ratio model. The R^2 statistic $R^2 < 0.00005$ is very low for the null model, the $s_{y|x}$ statistics $s_{y|x} = 75.70$ and 37.41 are high and far from the reality $\sigma = 20$, and the AICs = 922.3147 and 809.5328 are much larger than that of the full model. The conclusion is that the full linear regression model is the best fitting model, with a high R^2, lower $s_{y|x}$, and lower AIC, as expected. The nested first and second models and first and third models are also compared in S-Plus and R with an ANOVA F test (Seber 1977, Draper and Smith 1981, Hocking 1996, Ryan 1997, Cook and Weisberg 1999) (Fig. 1.10c). The second and third models cannot be compared with the F

Figure 1.11. Scatterplot of the (x, y) sample dataset generated from the reality $y = 50 + 1.2 \cdot x + e$, where $e \sim N\ (\mu = 0, \sigma = 20)$ and sample size $n = 80$.

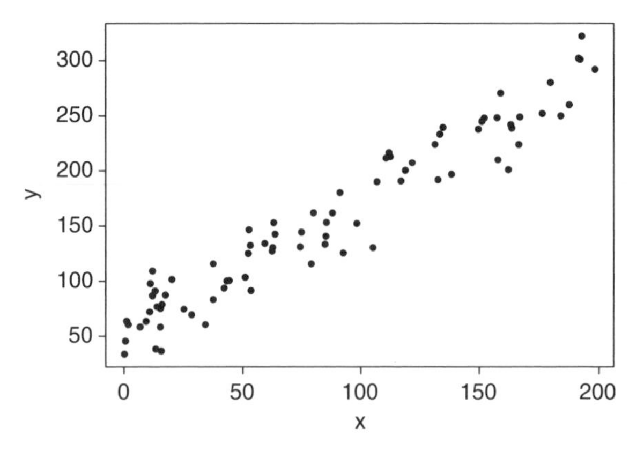

test since they are not nested. The ANOVA F-test results indicate that the full linear regression model is a statistically significantly improvement compared to the other two models. The second and third nonnested models can, however, be compared using AIC, which indicates that the ratio model is better, fitting than the constant null model ($AIC_{model3} = 809.53$ vs. $AIC_{model2} = 922.31$).

1.10 SUMMARY

We began this chapter by introducing three case studies of fundamental general importance to natural resource scientists. These case studies provide a framework for the methods of solution presented throughout this book. The first case study posed the problem of maintaining a population parameter above a critical threshold level. The parameter could be a proportion parameter such as the proportion of a timber ownership that is occupied by nesting pairs of Northern Spotted Owls. Or, it could be other parameters of interest such as mean abundance μ, survival rate ϕ_i from year i to year $i+1$; fitness $\lambda_i = N_{i+1}/N_i$, where $N_i\,(N_{i+1})$ is the population abundance in year $i\,(i+1)$; ecological diversity index H, such as the Shannon–Wiener diversity index; or population total τ. The second case study posed the problem of estimating discrete population abundance of a biological population such as a bird, amphibian, or mammal population. The third case study posed the problem of habitat selection modeling of a mobile biological population such as an endangered bird, amphibian, or mammal population. The habitat can be described by vegetation, geologic, and climatic attributes.

We then presented a review and preview of solutions to the problems introduced by the case studies: a review of traditional methods such as the estimation of parameters and error, hypothesis testing, and multiple linear regression modeling; and a preview of more contemporary methods such as Bayesian statistical inference, model selection and inference, generalized linear modeling, and mixed-effects modeling. These methods provide contemporary natural resource scientists with a statistical toolbox of methods of solution to tackle many of their most important and challenging current problems. The remainder of this book focuses on descriptions and illustrations of these contemporary methods.

We emphasize the importance of the planning and concluding phases of natural resource data collection projects, and particularly the critical role of statistical design and analysis. The datasets used throughout the book consist of both simulated datasets that allow comparison of statistical results with known realities and real-world datasets that provide realistic practical experience. The book encourages a hands-on approach to the use of statistics, including examples and problems in each chapter as topics are introduced. This first chapter concluded by providing an introduction to S-Plus and R, the statistical software tool for the frequentist statistical analysis applications throughout the book. WinBUGS, the statistical software tool for Bayesian analysis, will be introduced in later chapters.

PROBLEMS

1.1 Estimation and confidence interval solutions for the mean parameter: case study 1. Conduct frequentist statistical analysis in S-Plus or R of the tree diameter data (i.e., d_{bh} = diameter at breast height) $d = \{d_i\}$ of sample size $n = 30$ in Fig. 1.12 collected using simple random sampling. Calculate an estimate $\bar{d}$ of the mean tree diameter δ of the forest from this simple random sample along with the standard deviation s, standard error SE, sampling error E, and $P = 1 - \alpha = 0.95 = 95\%$ confidence interval CI $= [\bar{d} - E, \bar{d} + E]$. Use the S-Plus or R commands `mean, stdev` (in S-Plus) or `sd` (in R), and `t. test` on the data vector object d to calculate the statistics. Provide frequentist statistical

Figure 1.12. Tree diameter sample data d (diameter at breast height $= d_{bh}$): simulated with normal distribution $N(d;\ \mu = 25.0, \sigma = 3.0)$.

```
dbh
29.6
22.7
28.6
23.1
27.3
28.3
23.7
28.4
25.0
24.5
27.7
22.7
27.6
22.7
27.0
21.2
26.8
28.4
21.2
23.6
24.6
24.0
22.0
28.2
27.0
22.8
28.4
27.7
31.9
18.9
```

inference statements about the unbiasedness and minimum error properties of the mean d_{bh} estimate $\bar{d}$ and the confidence property of the 95% CI. Examine the histogram of the d_{bh} data d and discuss whether these frequentist inference properties, based upon the basis normal distribution theory, of the estimate and the CI are reasonable. Using the threshold mean d_{bh} parameter value $\delta_c = 26.0$, provide a decision for the problem described in case study 1 in terms of (**a**) the mean d_{bh} estimate $\bar{d}$ and (**b**) the confidence interval CI as to whether the mean d_{bh} δ of the forest has exceeded the threshold. Are these decision results (a) and (b) in conflict with each other? If the confidence level of the confidence interval is reduced to 80%, 67%, or 50%, is decision (b) above altered? Why are these confidence levels particularly of interest? How do these reductions of confidence levels affect the error rates of the results? How do you explain the differing results of the decisionmaking? In Chapter 5 we will demonstrate how one might use habitat variables x to more effectively model normal data y with multiple linear regression modeling.

1.2 Estimation and confidence interval solutions for the proportion parameter: case study 1. Examine the presence–absence data of sample size $n = 60$ with $y = 34$ ones (1s) and $(n - y) = 26$ zeros(0s) in sample plots collected using simple random sampling. Conduct an analysis of this dataset using the binomial model $y \sim B(y;\ n,\ p)$. Calculate the estimate $\hat{p} = y/n$ of the proportion p of the habitat occupied by the endangered species. Calculate the standard error

$$\mathrm{SE} = \sqrt{\frac{\hat{p} \cdot (1 - \hat{p})}{n - 1}},$$

sampling error $E = t_{1-\alpha/2,\, n-1} \cdot \mathrm{SE}$, and $P = 1 - \alpha = 95\%$ confidence interval $\mathrm{CI} = [\hat{p} - E,\ \hat{p} + E]$. Provide frequentist statistical inference statements about the properties of the proportion estimate $\hat{p}$ and the 95% CI. Using the threshold proportion parameter value $p_c = 65\%$, provide a decision for the threshold problem described in case study 1 based on (**a**) the proportion estimate $\hat{p}$ and (**b**) CI for the proportion parameter p. Are these decision results (a) and (b) in conflict with each other? If the confidence level of the confidence interval is reduced to 80%, is decision (b) above altered? If the confidence level is reduced to 50%, is decision (b) altered? How do these reductions of confidence levels affect the error rates of the results? How do you explain the differing results of the decisionmaking? In Chapter 6 we will demonstrate how one might use habitat variables x to more effectively model binomial data y with generalized linear modeling.

1.3 Hypothesis testing solutions: case study 1. Conduct Neyman–Pearson hypothesis testing on the sample tree diameter dataset $d = \{d_i\}$ in Fig. 1.12. Use the null hypothesis

$$H_0 : \delta = 26.0$$

and one-sided alternative hypothesis

$$H_A : \delta < 26.0$$

with a confidence level $P = 0.95 = 95\%$. Use the command `t.test(d, mu=δc, alternative="less")` in S-Plus or R for the hypothesis testing. Provide a frequentist statistical inference statement for your results. In terms of the histogram of the data d, discuss whether the frequentist inference property of the hypothesis testing, based on normal theory, is reasonable. Using the threshold mean d_{bh} parameter $\delta_c = 26.0$, provide a decision based on hypothesis testing for the problem described in case study 1. Do the results change for confidence levels of 80%, 67%, or 50%? Why or why not? How do these decisionmaking results, based on one-tailed hypothesis testing, agree or disagree with those based on the parameter estimation and CIs of Problem 1.1? Explain the differences.

1.4 Bayesian statistical analysis solutions: case study 1. Conduct a Bayesian statistical analysis on the sample tree diameter dataset $d = \{d_i\}$ in Fig. 1.12 as follows. Suppose that a Bayesian statistical analysis of sample dataset d with a normal model, based on a uniform prior $\text{Unif}(\delta; 0, 60)$, provides a posteriori distribution for δ that is normal $N(\delta; \mu_{post}, \sigma_{post})$ with mean $\mu_{post} = 25.5$ and standard deviation $\sigma_{post} = 3.1$. Using the threshold mean d_{bh} parameter $\delta_c = 26.0$, provide a decision based on Bayesian statistical inference for the problem described in case study 1. What is the assessment of risk of an incorrect decision of whether the threshold has been exceeded with this approach to statistical analysis? How do the results of this approach differ from the frequentist approaches of Problems 1.1 and 1.3?

1.5 Estimation of population abundance with count data: case study 2. Conduct a frequentist statistical analysis using S-Plus or R of the bird count data $c = \{c_i\}$ in Fig. 1.13. Estimate the mean $\bar{c}$, standard deviation s, standard error $\text{SE} = s/\sqrt{n}$, sampling error $E = t_{1-\alpha/2,\, n-1} \cdot \text{SE}$, and 95% confidence interval $\text{CI} = [\bar{c} - E,\ \bar{c} + E]$. Examine the histogram of the count data and discuss whether the frequentist inference properties, based on normal distribution theory, of the estimate and CI are reasonable.

1.6 Habitat selection modeling with linear regression: case study 3. Conduct habitat selection modeling analysis of the dataset (x, y) in Fig. 1.14 with independent predictor habitat variable x and dependent wildlife response variable y using linear regression modeling. The habitat variable x measures the amount of old-growth habitat on the sample plot. The response variable y measures the amount of biomass of an endangered species on the sample plot. Assume that the data were collected using simple random sampling. Use the linear modeling command `output <- lm(y~x)` and `summary(output)` in S-Plus or R to conduct the linear regression analysis and examine the results. Examine the

Figure 1.13. Count data: Poisson simulated count data with mean $\lambda = 1.25$ and sample size $n = 50$.

```
counts
2
1
0
0
2
0
2
1
2
1
1
0
5
1
0
1
4
1
0
4
0
1
2
2
0
4
2
2
1
0
2
0
3
0
1
0
1
2
1
1
2
3
0
0
2
2
1
1
1
2
```

Figure 1.14. Habitat selection modeling data (x, y): simulated with habitat variable $x \sim$ Unif(0,100) and wildlife variable $y = 30.0 + 1.2^{*}x + e$ with $e \sim N(0, 10.0)$.

x	y
14.9	51.9
35.4	72.3
17.4	68.4
30.0	62.3
83.6	141.8
38.2	82.7
65.2	89.4
25.4	69.6
5.9	28.3
51.5	94.2
31.6	54.9
13.1	45.9
31.4	72.7
98.9	156.0
30.4	52.5
38.2	77.4
94.6	151.6
21.8	67.8
17.9	48.8
45.4	58.3
76.5	121.0
80.2	129.2
10.6	37.9
98.5	147.3
66.1	131.1
1.8	40.6
29.0	59.1
98.4	131.6
99.4	168.1
59.9	108.6
46.1	64.9
48.3	99.2
58.8	93.6
10.9	48.5
1.1	34.7
90.7	146.4
28.0	56.0
89.1	121.7
32.1	92.9
77.1	111.8
99.2	149.1
14.4	60.8
67.1	90.2
19.2	52.7
99.1	140.7
38.1	78.9
4.3	25.8
31.1	75.5
94.3	146.1
8.7	47.9

estimates $\hat{\beta}_0$ and $\hat{\beta}_1$ and t tests of the coefficients of the linear regression model

$$y = \beta_0 + \beta_1 \cdot x + e,$$

where β_0 is the y-intercept constant coefficient, β_1 is the coefficient of x, and $e \sim N(e;\ \mu = 0,\ \sigma)$ is the normally distributed error with mean $\mu = 0$ and standard deviation σ. Are these estimates statistically significant at the 95% confidence level? Examine the residual standard error estimate $s_{y|x}$ of σ, the R-squared statistic, and the F-test statistic. Also examine the AIC of the modeling results using `AIC(output)` in S-Plus or R. Do these statistics, coupled with the plot of the data (use `plot(x,y)` followed by `abline` `(`$\hat{\beta}_0, \hat{\beta}_1$`)` in S-Plus or R), suggest that the modeling results provide a good fit of the model to the sample dataset?

2 Bayesian Statistical Analysis I: Introduction

The best way to have a good idea is to have lots of ideas.

—Linus Pauling

In this chapter we will present an introduction to Bayesian statistical inference. We begin with an introduction containing an historical overview and a discussion of some of the limitations of frequentist statistical inference for natural resource applications. We then discuss several different approaches to model fitting, concluding with the Bayesian approach. We discuss the fundamental concepts of Bayesian statistical inference, including the Bayes' Theorem for conditional probability. Bayesian statistical inference uses prior and posterior distributions, along with likelihood functions derived from sample datasets and models, to assess parameters. We conclude this chapter by discussing the range of options available for choices of priors, including noninformative priors, Jeffreys' priors, and conjugate priors. Conjugate priors provide closed-form mathematical solutions for Bayesian analyses of many important types of natural resource datasets, such as binary data with the binomial model, count data with the Poisson model, and continuous data with the normal model.

2.1 INTRODUCTION

2.1.1 Historical Background

Frequentist statistical inference has been the dominant paradigm for statistical analysis for many years. Parameter estimation, hypothesis testing, and statistical modeling applications in natural resource sciences have been based primarily on frequentist statistical analysis, developed by Fisher (1922, 1925a, 1925b, 1934, 1958), Neyman and Pearson (1928a, 1928b, 1933, 1936), and other leading statisticians

Contemporary Bayesian and Frequentist Statistical Research Methods for Natural Resource Scientists. By Howard B. Stauffer

of the early and midtwentieth century. Bayes' Theorem, about conditional probability, provides the foundation for Bayesian statistical inference. Although Bayes formulated his famous theorem way back in the late eighteenth century, Bayesian statistical analysis solutions in general have remained elusive, if not intractable, for mathematicians until relatively recently. Hence, the practical use of Bayesian statistical inference in natural resource science has been limited.

Bayesian statistical analysis is derived from the Bayes Theorem, originally published posthumously by the Reverend Thomas Bayes in 1763. Gill (2002) comments that, contrary to the current requirement of "publish or perish," Bayes "perished before publishing." It is doubtful that Bayes realized at that time that mathematical statisticians would one day develop an alternative approach to statistical analysis and inference based on his idea. A small group of Bayesian statisticians pursued this alternative approach in the twentieth century despite a notable lack of enthusiasm from other statisticians in the frequentist mainstream. Bayesians have, however, historically been relegated to a minority status at most universities, although a few hotspots for Bayesians have provided sanctuary, at academic research centers such as Duke University and University of Minnesota in the United States and Imperial College in London, England. Until relatively recently, Bayesian statistical analysis has not been regarded by the vast majority of statisticians as the most reasonable approach to statistical analysis because of its intractability to solution, its potential susceptibility to the use of "subjective" prior information, and other less legitimate reasons. Consequently, Bayesian statistical inference has played a minor role in the statistical curriculum of most universities and has remained inaccessible to most natural resource scientists.

However, the advent of computers in the midtwentieth century provided the technology for a revolution in the development and use of Bayesian statistical analysis. The mathematician Nicholas Metropolis (Metropolis et al. 1953), working in collaboration with the nuclear physicist Edward Teller in the 1950s, developed a Monte Carlo simulation technique that generates samples in a stationary Markov chain, providing general Bayesian solutions for statistical problems. Gibbs sampling (Gelfand and Smith 1990) and an extension due to Hastings (1970) provided further contributions to the general solution. These developments, along with other contributions, have provided a collection of Bayesian iterative solutions to statistical problems known as the *Markov Chain Monte Carlo* (MCMC) algorithms (Gamerman 1997, Bremaud 1999, Draper 2000, Link et al. 2002). These methods have been programmed and made available on the Web, as of this printing, as WinBUGS freeware (Spiegelhalter et al. 2001). WinBUGS is readily accessible and user-friendly for practicing natural resource scientists. We will introduce these MCMC algorithms and the use of the WinBUGS software in Chapter 4.

The practical availability of these new tools for Bayesian statistical analysis, along with a growing realization of the limitations to the use of frequentist methods, provide natural resource scientists the unique opportunity to use alternative methods for analyzing challenging datasets. But first, let's review some of the key problematic issues of concern with the use of frequentist inference for the analysis of natural resource data. The interested reader is encouraged to explore the Bayesian ideas that are introduced in the next three chapters in more depth in the publications by Gill (2002), Hilborn and Mangel (1997), Iversen (1984), Congdon (2001), Berry and Strangl (1996), Carlin and Louis (2000), and Link et al. (2002).

2.1.2 Limitations to the Use of Frequentist Statistical Inference for Natural Resource Applications: An Example

We begin by presenting an example illustrating some of the problems with the use of frequentist statistical inference for many natural resource applications. The example will suggest some of the advantages of the alternative use of Bayesian statistical inference. Suppose that we wish to determine whether there is a difference in the abundance of a particular critical wildlife species between some old-growth and clear-cut redwood forest stands in northern California, and have randomly collected observational data from each type of forest. Recall that observational data, in contrast to experimental data, are not collected with a completely randomized design. Practically speaking, the so-called old growth and clear-cut treatments were not randomly applied to randomly chosen experimental units from all forest stand habitat located throughout northern California, but instead conveniently sampled from stands that were available. We shall follow Johnson's (1999, 2002) advice and avoid the misuse of hypothesis testing with such observational data, instead modeling it with descriptive analysis and limiting inferences to the specifically sampled old-growth and clear-cut redwood forest stands subpopulations, rather than treating results as predictive and making inferences to a larger population. We note that the hypotheses for such observational data sampled from specific stands are statistical rather than scientific, and the null hypotheses are "silly" and clearly untrue.

There are some serious disadvantages to the use of frequentist statistical analysis in this context that the Bayesian approach can remedy. Perhaps the most important disadvantage to the use of frequentist statistical analysis for the natural resource scientist with studies of this type is in the understanding of its inference. It is difficult to visualize the frequentist inference concept of repeatedly sampling such dynamic populations. The stands in this example are unique, in time and place, and such an "experiment" is unlikely to be replicable. The nature of many existing biological populations of interest to natural resource scientists is that they are in flux, endangered, and declining. Conditions are dynamic and will likely never be static. The dominant current ecological paradigm is one of change and lack of stability. Frequentist statistical inference, however, provides probabilities for datasets based on the concept of repeated surveys or experiments in a static population. In natural resource science, however, such conditions are often unrealistic, even conceptually, to conceive. Frequentist statistical inference is nonintuitive and confusing to understand in this context. Natural resource scientists and managers view such a study as a single event in a sequential cumulative process of scientific exploration and would prefer to make an inference that is a direct Bayesian probability assessment for a parameter, based on the sample dataset, rather than make a frequentist probability assessment for the dataset with repeated sample surveys or experiments, based upon the parameter.

It is difficult in such a natural resource context to understand the frequentist meaning of the probability of data. Recall the frequentist interpretation of a 95% confidence interval (CI) in a sample survey. The frequentist inference is that, for a fixed parameter value, repeated simple random sample surveys of the population will produce CIs containing the population parameter with a probability of 95%. Recall the frequentist interpretion of the test statistic that is calculated from the sample dataset in an experiment. The frequentist inference is that, if the null hypothesis H_0 is true with a confidence level $P = 95\%$ and a complementary type I error $\alpha = 1 - P = 5\%$, repeated experiments of the population will produce test statistics in the nonrejection region with a probability of 95% and in the rejection region with a probability of 5%. It is difficult to grasp this frequentist interpretation for the probabilities of datasets in the context of repeated surveys or experiments when only one survey or experiment will actually occur in a natural resource study. The frequentist inference is an indirect assessment for the parameter, a conditional probability statement for the data with repeated surveys or experiments, given the parameter. What the natural resource scientist requires is a direct assessment for the parameter, a conditional probability statement for the parameter, given the sample dataset.

Another major objection to the use of frequentist statistical analysis with natural resource datasets is the independence of its analysis. Frequentist statistical analysis usually formally ignores prior information known about a population. This is so counter to the Western scientific method, where the assessment of populations is based on previously accumulated knowledge gained from experience, study, and experimentation. Natural resource scientists would like to utilize previous studies and incorporate those results more formally into their analysis. This is particularly true with the adaptive management and monitoring of endangered species. With population monitoring, the objective is to proceed sequentially, assessing a population and its progress repeatedly, with sample datasets and analysis. The objective is to utilize an analysis strategy for reassessing the condition of a population in a sequential, cumulative manner.

There are other drawbacks to the frequentist approach that we need not highlight here. We refer the interested reader to Johnson (1999, 2002), Anderson et al. (2002), and Robinson and Wainer (2002) for further details. To summarize, however, there is a need in natural resource science for an alternative approach to statistical analysis and inference, one that addresses these disadvantages of the frequentist approach, the indirectness of its inferences, and the independence of its analysis. In these next three chapters, we will provide an introduction to an alternative approach that will address these needs of natural resource scientists: Bayesian statistical analysis and inference.

2.2 THREE METHODS FOR FITTING MODELS TO DATASETS

In this section, we present an overview of three important statistical methods useful for fitting models to independent data: least-squares (LS), maximum-likelihood (ML), and Bayesian parameter estimation (Hilborn and Mangel 1997). These methods use optimality criteria to estimate parameters in fitting models to sample datasets. It is important to understand each method and appreciate their interrelationships. The first two methods, LS and ML estimation, are frequentist approaches used for many popular classical models, such as the mean and proportion estimators, ANOVA, and multiple linear regression. The third Bayesian method can be viewed as a generalization of the ML method. We assert that, in many instances, it is particularly suited for the statistical analysis of natural resource datasets, for populations that are in flux under conditions of adaptive management.

2.2.1 Least-Squares (LS) Fit—Minimizing a Goodness-of-Fit Profile

The first method of parameter estimation, **least-squares (LS) fit**, minimizes a goodness-of-fit profile (Rice 1995, Hocking 1996, Hilborn and Mangel 1997, Ryan 1997)

$$\mathrm{GOF}(\phi \mid y) = \sum_{i=1}^{n} (y_i - f(\phi, y))^2,$$

the sum of the squared residual errors between the independent predictor data y and the predicted model response $f(\phi, y)$, expressed as a function of the parameter vector ϕ and the data y. The goodness-of-fit profile $\mathrm{GOF}(\phi \mid y)$ is a function of the parameter vector ϕ, conditional to the sample dataset $y = \{y_1, y_2, \ldots, y_n\}$. The LS fit method selects the value of the parameter ϕ that minimizes the goodness-of-fit profile GOF, the sum of the squared error terms.

For example, consider the mean model $f(\mu, y) = \mu$ for the mean parameter μ and the dataset $y = \{y_1, y_2, \ldots, y_n\}$ of continuous real measurements y_i:

$$\mathrm{GOF}(\mu \mid y) = \sum_{i=1}^{n} (y_i - \mu)^2.$$

It can be easily demonstrated using calculus, setting the derivative of the GOF profile with respect to μ equal to 0, that the minimum value occurs at the arithmetic mean for the data y, the classical estimator for the mean parameter

$$\hat{\mu} = \frac{\sum_{i=1}^{n} y_i}{n}.$$

Consider a second example, the linear regression model $f(\beta_0, \beta_1, (x, y)) = \beta_0 + \beta_1 \cdot x$ for parameters β_0 and β_1 and the independent dataset $(x, y) = \{(x_1, y_1), (x_2, y_2), \ldots, (x_n, y_n)\}$ of pairs of continuous real measurements (x_i, y_i):

$$\mathrm{GOF}(\beta_1, \beta_0 | (x, y)) = \sum_{i=1}^{n} (y_i - (\beta_0 + \beta_1 x_i))^2.$$

It can be shown using calculus, setting partial derivatives of the GOF profile with respect to β_0 and β_1 equal to 0, that the minimum value occurs at the classical parametric solution of normal equations for the linear regression model:

$$\hat{\beta}_1 = \frac{\sum_{i=1}^{n} x_i \cdot y_i}{\sum_{i=1}^{n} x_i^2},$$

$$\hat{\beta}_0 = \bar{y} - \hat{\beta}_1 \cdot \bar{x}.$$

Least-squares (LS) fit is the method commonly used for linear regression and multiple linear regression estimation and for estimators used in ANOVA. Least-squares estimation is distribution-free; there are no distributional assumptions required for the error terms, or probability models required for the LS estimation. However, with some mild probability distributional assumptions on the residuals, the Gauss–Markov Theorem holds: LS estimators for the linear regression and multiple linear regression models are *BLUE*, providing *b*est *l*inear *u*nbiased *e*stimators. In other words, of all linear estimators for linear regression and multiple linear regression models having independent and identically distributed residuals with mean 0 and constant error variance (homoscedasticity), LS estimators provide unbiased estimators for the coefficients $\beta_0, \beta_1, \ldots, \beta_k$ with the minimum variance (Seber 1977, and Smith Draper 1981, Cook and Weisberg 1999, Gill 2002). The LS estimators are unbiased; the averages of the estimates, with repeated sampling, equal the coefficient parameters. Furthermore, of all such unbiased linear estimators, these estimators have the minimum variance. In other words, the LS estimators for linear and multiple regression are the most efficient unbiased estimators.

Least-squares estimation can be generalized to generalized least-squares estimation (GLS), for LS estimation of generalized linear regression and generalized multiple linear regression models with nonconstant error variance (heteroscedasticity), using weighted LS estimation. This more general approach uses weights for the squared residuals that are inverses of the error variances, to compensate for varying error variances. Under these more general assumptions, it can also be shown that GLS is also BLUE. The S-Plus and R commands for LS and GLS estimation are `lm` and `gls`, respectively.

2.2.2 Maximum-Likelihood (ML) Fit—Maximizing the Likelihood Profile

The second method of parameter estimation, ML fit, maximizes the likelihood profile (Dobson 1990, Rice 1995, Hilborn and Mangel 1997, Burnham and Anderson 2002)

$$\mathcal{L}(\phi \mid y) = \prod_{i=1}^{n} \mathcal{L}(\phi \mid y_i),$$

where the individual likelihood values $\mathcal{L}(\phi|y_i) = P(y_i|\phi)$ are the model probabilities for the independent predictor data points y_i conditional to the parameter vector values ϕ. The likelihood profile function itself is a function of the parameter vector ϕ conditional to the data y. Statisticians commonly maximum this function with respect to ϕ by maximizing its log transformation, the log-likelihood function. Since the log transformation is monotonically increasing, it provides the same maximal solution and the log-transformed function is more tractable to differentiation since it is a sum rather than a product of terms. The log-transformed function can be differentiated and set equal to 0 to find the maximum with respect to ϕ, or it can be solved using numerical techniques. Furthermore, the log-transformed function has some nice statistical properties.

For example, if we assume the normal model

$$N(y_i; \mu, \sigma) = \frac{1}{\sigma \cdot \sqrt{2\pi}} \cdot \exp\left[-\frac{(y_i - \mu)^2}{2\sigma^2}\right]$$

for the continuous real measurements y_i with mean parameter μ and standard deviation parameter $\sigma > 0$, the likelihood profile for the mean model becomes

$$\begin{aligned} \mathcal{L}(\mu|y) &= \prod_{i=1}^{n} \frac{1}{\sigma \cdot \sqrt{2\pi}} \cdot \exp\left[-\frac{(y_i - \mu)^2}{2\sigma^2}\right] \\ &= \frac{1}{\sigma^n \cdot (2\pi)^{n/2}} \cdot \exp\left[-\sum_{i=1}^{n} \frac{(y_i - \mu)^2}{2\sigma^2}\right]. \end{aligned}$$

Assuming a constant σ, the ML solution is identical to the LS solution for the estimator of the mean parameter μ, since the maximum of the likelihood function is the minimum of the positive value of its exponent.

For the linear regression model, assuming normally distributed errors for the data points (x_i, y_i), the likelihood profile is given by

$$\begin{aligned} \mathcal{L}(\beta_0, \beta_1|(x, y)) &= \prod_{i=1}^{n} \frac{1}{\sigma \cdot \sqrt{2\pi}} \cdot \exp\left[-\frac{(y_i - (\beta_0 + \beta_1 \cdot x_i))^2}{2\sigma^2}\right] \\ &= \frac{1}{\sigma^n \cdot (2\pi)^{n/2}} \cdot \exp\left[-\sum_{i=1}^{n} \frac{(y_i - (\beta_0 + \beta_1 \cdot x_i))^2}{2\sigma^2}\right]. \end{aligned}$$

Again, assuming a constant parameter σ, the ML solution for the estimators of the parameters β_0 and β_1 is identical to that of the LS fit. The LS and ML estimators used for σ are slightly different, however:

$$\hat{\sigma}_{\text{LS}} = \frac{\sum_{i=1}^{n} (y_i - (\hat{\beta}_0 + \hat{\beta}_1 \cdot x_i))^2}{n - 2}$$

and

$$\hat{\sigma}_{\mathrm{ML}} = \frac{\sum_{i=1}^{n}(y_i - (\hat{\beta}_0 + \hat{\beta}_1 \cdot x_i))^2}{n}.$$

The ML method provides estimators that are asymptotically (i.e., as the sample size becomes large, increasing to infinity) unbiased of minimum variance. As demonstrated, the LS mean estimator $\bar{y} = \sum_{i=1}^{n} y_i)/n$ is also the ML mean estimator for normal data. The LS β coefficient estimators for linear regression are also the ML estimators for data with normally distributed error. The ML estimator for variance $\hat{\sigma}^2 = (1/n) \cdot \sum_{i=1}^{n}(y_i - \bar{y})^2$ with the mean model for normal data, with simple random sampling, is slightly biased with correction factor $n/(n-1)$. The bias becomes insignificant with large sample size. Multiplying the ML estimator for variance by the correction factor converts it to the unbiased LS estimator. The generalized linear models (GLMs) presented in Chapter 6 are based on ML estimators.

Maximum-likelihood estimation examines the likelihood, based on a probability model such as the normal, Poisson, or binomial probability distribution for sample data, and uses the maximum for the estimated value. Note that this maximum would be the mode if the likelihood function were scaled to become a probability distribution for the parameter. It can be shown that estimates of the error for the ML parameter estimator are related to the curvature of the likelihood function at the mode, described by the negative reciprocal of the second derivative, as will be described in Chapter 6. Hence it can be stated that frequentist ML estimation provides the mode and curvature of the likelihood function at the mode as a "likelihoodist" estimation of the parameter for the best-fitting model.

We shall next demonstrate how this compares with the third method for fitting models to datasets, the Bayesian approach to parameter estimation. This third method of parameter estimation, Bayesian statistical inference, uses the entire likelihood function, along with a prior distribution, to calculate a posterior distribution for the parameter.

2.2.3 Bayesian Fit—Bayesian Statistical Analysis and Inference

Bayesian statistical analysis is based on an alternative approach to model fit and estimation of parameters (Iversen 1984, Berger 1985, Carlin and Lewis 2000, Draper 2000, Congdon 2001, Gill 2002). The **Bayesian method** is based on an assumption of a **prior distribution** $\pi_{\mathrm{prior}}(\beta)$ representing initial understanding, or belief, about the parameter β, prior to data analysis. If little is known about the value of the parameter, this prior distribution could be "noninformative," such as a flat, uniform distribution, assuming equal probability for the parameter value within a realistic range. Alternatively, the prior could be more informative such as a normal distribution or other plausible distribution that represents an initial assessment of what is known about the parameter, prior to data collection and analysis. The initial assessment could be based on the results of previous studies. The prior affects the analysis results, so it must be chosen carefully to be realistically and conservatively plausible, based on prior evidence regarding the probabilities for the parameter value. If there is controversy about the choice of priors, a collection of plausible priors representing a realistic range of possibilities could be examined in the analysis for their effects on the posterior results, with a **sensitivity analysis**.

The Bayesian method also requires the use of the **likelihood function $\mathcal{L}(\beta)$** for the parameter β, the likelihood profile based on the sample dataset and an assumed model. Commonly assumed models for natural resource datasets include the normal model for continuous data, the Poisson and negative binomial models for count data, the binomial model for binary data, and the multinomial model for mark–recapture datasets. Recall that the likelihood, representing a function of the parameter given the data, is based on the probability distribution of the data at each point, given the parameter. We have previously seen that this same likelihood function is analyzed with frequentist ML fit.

For the Bayesian method, however, the likelihood function $\mathcal{L}(\beta)$ is multiplied by the prior $\pi_{\text{prior}}(\beta)$ to obtain a new function, the product $p(\beta) = \mathcal{L}(\beta) \cdot \pi_{\text{prior}}(\beta)$ of the likelihood and prior (see Bayes Theorem in Section 2.3). This new function is then scaled to obtain a probability distribution function, with a total sum in the case of a discrete parameter, or integral in the case of a continuous parameter, over the range of the parameter that is equal to 1. This scaled probability distribution function is called the posterior distribution for the parameter. The posterior distribution represents a revised, updated understanding, or assessment, of the parameter. It is based, as we have seen, on the prior distribution of the parameter, the data, and the model assumed for the data.

Hence the frequentist uses the data and a model for the data to obtain a likelihood profile and calculates the maximum of the likelihood profile for the ML estimate of the parameter, along with an estimate of error based on the curvature of the likelihood profile at the maximum value. The Bayesian, on the other hand, combines the data and the model with a prior distribution for the parameter to obtain a scaled posterior distribution for the parameter, using the entire posterior distribution for an updated assessment of the parameter.

The Bayesian can then assess the parameter by examining the posterior distribution and its descriptive statistics, such as the mean, median, mode, standard deviation, and percentiles. The Bayesian can calculate a **credible interval** for the parameter, using boundaries of a middle percentile range for the parameter. A 95% (or P%) credible interval, for example, can be based on the 2.5% and 97.5% [or $(100 - P)/2$ and $100 - (100 - P)/2$] percentile points of the posterior distribution. More rigorously, Bayesian credible intervals are based on the highest posterior density (HPD), the smallest interval encompassing at least 95% (or P) of the probability distribution (Carlin and Louis 2000, pp. 36–37). The credible interval is the Bayesian analog to the frequentist confidence interval. There is, however, an entirely different interpretation for the meaning of the Bayesian credible interval; there is a 95% (or P%) probability of the parameter occurring within the specified 95% (or P%) credible interval, assuming the prior distribution, the sample dataset, and the model. The Bayesian interpretation is a direct probabilistic statement about the parameter, rather than the frequentist interpretation of a confidence interval, a probabilistic statement about the data, based on the concept of repeated sampling or experimentation of the population.

2.2.4 Examples

Let's look at some examples to illustrate these ideas. We will examine three prototype examples of natural resource datasets, continuous, count, and binary datasets, with the normal, Poisson, and binomial models, respectively. For these examples, we will use very small sample datasets with just a few points, to illustrate clearly the concepts, leaving it to the reader to extrapolate these results to larger, more realistic datasets.

The first example is based on the normal model for continuous data, with the mean parameter. Let's consider sampling the community forest of a small town in northern California, measuring tree diameters at breast height (i.e., d_{bh}, at 4.5 ft above the ground). Assume that this forest is a 100-year-old even-aged second-growth forest of predominantly redwood and Douglas fir trees that have diameters generally ranging between 20 and 40 in., with preliminary estimates of a mean around 30 in. The city forester who knows the forest well insists that no tree in the forest exceeds 60 in. d_{bh}. None of the existing d_{bh} data collected by the city in previous years exceeded 60 in., and no large old-growth trees or snags still exist in the forest. Let's collect $n = 100$ simple random sample measurements with the objective of assessing the mean tree diameter for the forest. We do not have any prior information about the mean of the tree diameters in this community forest, so we will use a noninformative prior distribution that is uniform, with a range between 0 and 60. The range of this uniform distribution encompasses the practical range of d_{bh} measurement values for the forest, and hence of the mean d_{bh} parameter.

We will also assume that the d_{bh} measurements for trees in this forest are approximately normally distributed, a reasonable assumption since it is an even-aged mature stand. So let's calculate the likelihood function from our data and multiply the likelihood by the prior, scaling the result to produce the posterior distribution. Since the prior is constant and the range of the most significant part of the likelihood function falls well within the range of the prior, the posterior will resemble the likelihood function, truncated at 0 and 60, and scaled to a probability distribution function. The frequentist ML estimate is the mode of this function. The Bayesian solution is provided by the entire posterior distribution. This posterior distribution can be used to make probability inferences about the mean d_{bh} parameter for trees in this community forest population, with statistics of the posterior such as the mean, median, mode, median, and standard deviation, and percentiles such as the 95% credible interval given by the 2.5% and 97.5% percentile limits.

To illustrate specifically, let's examine a simple example, a sample d_{bh} dataset consisting of just two measurements, 28 and 34 in. Let's assume that $\sigma = 5$ to illustrate the method. This is a reasonable estimate for σ since the range of the data varies generally between 20 and 40 in. and approximately 95% of, normal data falls within $4 \cdot \sigma$

Figure 2.1. Goodness-of-fit profile of the mean μ for the dataset $y = \{28, 34\}$ using S-Plus or R code

```
> mu < - seq(0,60,0.1)
> gof <- (28-mu)^2 + (34-mu)^2
> plot(mu,gof)
```

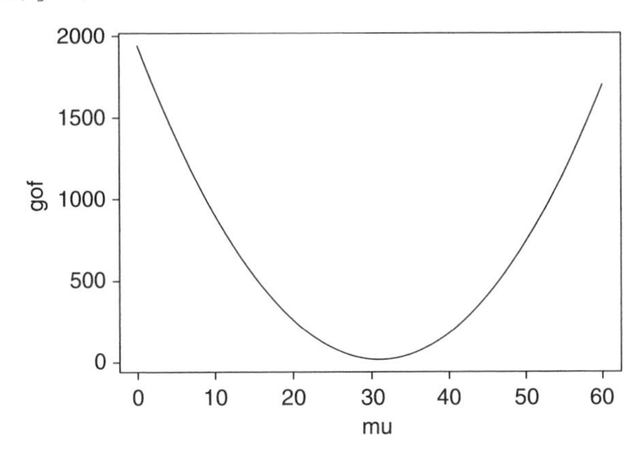

of this range. The goodness-of-fit profile for μ is (Fig. 2.1)

$$\mathrm{GOF}(\mu \mid y) = (28 - \mu)^2 + (34 - \mu)^2.$$

The likelihood profile, or likelihood function, for μ is (Fig. 2.2)

$$\begin{aligned}\mathcal{L}(\mu \mid y) &= \left\{\frac{1}{\sigma \cdot \sqrt{2\pi}} \cdot e^{[(28-\mu)^2/2\sigma^2]}\right\} \cdot \left\{\frac{1}{\sigma \cdot \sqrt{2\pi}} \cdot e^{[(34-\mu)^2/2\sigma^2]}\right\} \\ &= \frac{1}{\sigma^2 \cdot (2\pi)} \cdot \exp\left\{-\left[\frac{(28-\mu)^2}{2\sigma^2} + \frac{(34-\mu)^2}{2\sigma^2}\right]\right\}.\end{aligned}$$

Note in the figures that the LS and ML estimates for the mean parameter are identical at $\hat{\mu} = 31$. The posterior is the scaled truncated likelihood, since the prior is flat. It represents a probability distribution describing an updated assessment of the mean parameter, based on the sample dataset. If the prior were "informative" and described by a normal distribution, we could well imagine that the posterior of normal data and a normal prior would be normal. This is, in fact, the case; the posterior can also be described using conjugacy theory with normal priors or approximated using Markov chain Monte Carlo (MCMC) methods with the Bayesian statistical software WinBUGS. These methods will be examined later in Section 2.4 and in Chapter 4. The parameter $\sigma = 5$ was assumed to be known in this example for the sake of simplicity. However, the parameter σ could also have been analyzed with this Bayesian approach, using priors, a likelihood function, and posteriors for the two-dimensional parameter space (μ, σ).

Figure 2.2. Likelihood profile of the mean μ for the dataset $y = \{28, 34\}$ with the normal model $N(y; \mu, \sigma = 5)$ using S-Plus or R code

```
> mu < - seq(0,60,0.1)
> likelihood <-
(1/(5^2*(2*pi)))*exp(-((28-mu)^2+(34-mu)^2)/(2*5^2))
> plot(mu,likelihood)
```

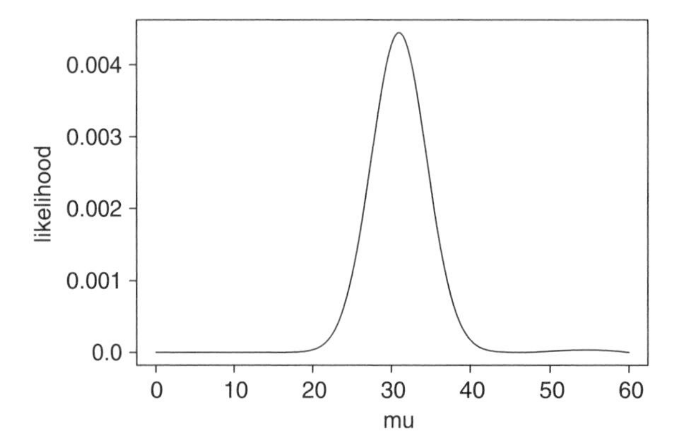

We look next at a Poisson model for a count dataset. Let's imagine conducting a study of curlew abundance in a bay north of San Francisco, estimating mean density using count data collected on $n = 50$ fixed area plots with simple random sampling. We will use the Poisson model as a first approximation for the count data, noting that a necessary assumption for the Poisson model is the random dispersal of the birds on the bay, an assumption that may not in fact be accurate. Recall that the Poisson model has just one parameter, the mean λ, with variance $\sigma^2 = \lambda$. Note that, if the birds are aggregated in clusters with $\sigma^2 > \lambda$, it would be better to use the negative binomial model to incorporate overdispersion into the model. We will discuss that possibility in Chapter 4. Another complication is that the probability of detection may not be equal to 1, nor constant from plot to plot. However, let's ignore these complications for the time being and base the analysis on the Poisson model, assuming random dispersal and complete certainty of detection, for simple illustration. Let's calculate the likelihood function of the mean λ, assuming the Poisson model $\text{Pois}\,(y; \lambda) = e^{-\lambda} \cdot (\lambda^y / y!)$ with mean parameter λ for the sample data $y = \{y_1, y_2, \ldots, y_n\}$ of nonnegative independent integer counts:

$$
\begin{aligned}
\mathcal{L}(\lambda) &= \prod_{i=1}^{n} e^{-\lambda} \cdot \frac{\lambda^{y_i}}{y_i!} \\
&= e^{-n \cdot \lambda} \cdot \frac{\lambda^{\sum_{i=1}^{n} y_i}}{\prod_{i=1}^{n} y_i!}.
\end{aligned}
$$

We multiply this by a prior and scale the product to obtain the posterior. We can again use a noninformative flat prior, say, a uniform distribution ranging between 0 and 20, assuming that the nonnegative count measurements do not exceed 20. If another prior were more plausible, such as a posterior distribution obtained from a Bayesian statistical analysis of a dataset from a previous study of the population, that distribution could be used for the prior. Using the uniform prior between 0 and 20, the posterior will be the scaled likelihood function, truncated at 0 and 20.

Illustrating with a simple example, let's analyze a dataset of three sample data points, plot counts $y_1 = 2$, $y_2 = 0$, and $y_3 = 4$. Then the GOF profile will be (Fig. 2.3)

$$\mathrm{GOF}(\lambda) = (2 - \lambda)^2 + (0 - \lambda)^2 + (4 - \lambda)^2.$$

The likelihood function for λ is (Fig. 2.4)

$$\begin{aligned}\mathcal{L}(\lambda) &= \left[e^{-\lambda} \cdot \frac{\lambda^2}{2!}\right] \cdot \left[e^{-\lambda} \cdot \frac{\lambda^0}{0!}\right] \cdot \left[e^{-\lambda} \cdot \frac{\lambda^4}{4!}\right] \\ &= e^{-3 \cdot \lambda} \cdot \frac{\lambda^{(2+0+4)}}{2! \cdot 0! \cdot 4!} \\ &= e^{-3 \cdot \lambda} \cdot \frac{\lambda^6}{48}.\end{aligned}$$

Note in the figures that the LS and ML estimates for the mean parameter are identical at $\hat{\lambda} = 2$. The posterior is the scaled truncated likelihood since the prior is flat throughout the most significant part of the range of the likelihood function. It represents a probability distribution describing the updated assessment of λ, the mean density parameter, based on the sample dataset. This posterior distribution can be determined exactly using

Figure 2.3. Goodness-of-fit profile of the mean λ for the dataset $y = \{2, 0, 4\}$ using the S-Plus or R code

```
> lambda <- seq(0,20,0.1)
> gof <- (2-lambda)^2 + (0-lambda)^2 +(4-lambda)^2
> plot(lambda,gof)
```

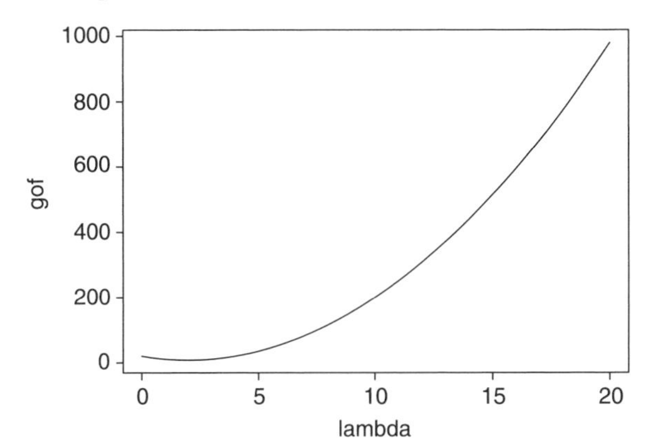

Figure 2.4. Likelihood profile of the mean λ for the dataset $y = \{2, 0, 4\}$ with the Poisson model Pois(y; λ) using the S-Plus or R code

```
> lambda <- seq(0,20,0.1)
> likelihood <- exp(-3*lambda)*lambda^6/48
> plot(lambda,likelihood)
```

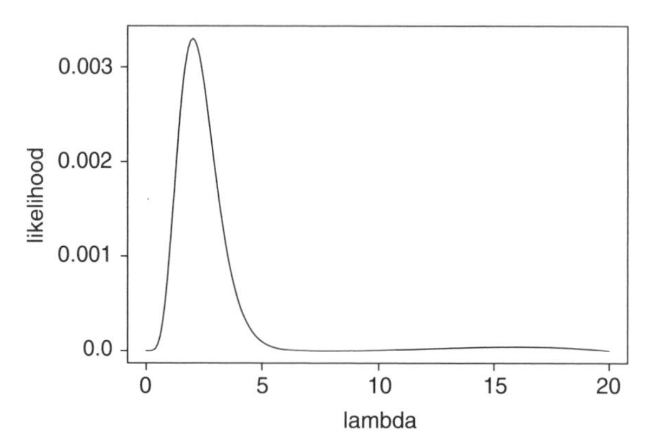

conjugacy theory with gamma priors or approximated using MCMC methods with the Bayesian statistical software WinBUGS (see Section 2.4.2 and Chapter 4).

Let's conclude this section with a third example, examining the ML and Bayesian methods of parameter estimation for binary data using the binomial model. The parameter estimated in this example is the proportion parameter. Let's imagine randomly sampling a binary dataset of presence–absence measurements for Northern Spotted Owls nesting pairs at $n = 80$ 200-hectare (ha) sites in a national forest along the Pacific north coast. Suppose that recent research studies have estimated the home range of Northern Spotted Owl nesting pairs to be approximately 200 ha in size, and use that for the plot size. We will use the binomial model

$$B(y;\ n, p) = \binom{n}{y} \cdot p^y \cdot (1-p)^{n-y}$$

$$= \frac{n!}{y! \cdot (n-y)!} \cdot p^y \cdot (1-p)^{n-y}$$

for the dataset, where y is the sum of the 1s, or occupied sites; n is the sample size; and p is the proportion parameter. The parameter p is the proportion of the habitat that is occupied by nesting pairs. Equivalently, the parameter p is the probability of a random sample containing a nesting pair of Northern Spotted Owls. We again assume complete certainty of detection at the sampled nesting sites. This can be approximately achieved by repeatedly visiting to each site. The likelihood function for this binary dataset is

$$\mathcal{L}(p) = \binom{n}{y} \cdot p^y \cdot (1-p)^{n-y}.$$

Let's again use a noninformative flat prior, a uniform distribution for p ranging between 0 and 1. The posterior distribution is then the scaled product of this likelihood and the noninformative prior, the scaled likelihood function on the interval [0,1].

To illustrate, suppose that we collect data at $n = 80$ sites and measure $y = 20$ to be occupied. Then the likelihood function is (Fig. 2.5)

$$\mathcal{L}(p) = \binom{80}{20} \cdot p^{20} \cdot (1-p)^{80-20} \propto p^{20} \cdot (1-p)^{60}.$$

Note that the ML estimate of the parameter is $\hat{p} = 20/80 = 25.0\%$. The posterior is the scaled likelihood function since the prior is flat. It represents a probability distribution describing an updated assessment of the parameter p. It can be determined exactly with beta priors using conjugacy theory, or approximated using MCMC methods with the Bayesian statistical software WinBUGS (Section 2.4.3 and Chapter 4).

2.3 THE BAYESIAN PARADIGM FOR STATISTICAL INFERENCE: BAYES THEOREM

Bayesian statistical analysis provides a probability distribution representing a natural resource scientist's current assessment of a parameter. The analysis process begins with an assumption of a prior distribution representing the scientist's understanding of the parameter preceding the study. The analysis then combines the prior distribution for the parameter with the likelihood function derived from a sample dataset and a model for the dataset to produce a posterior distribution for the parameter. The posterior distribution is obtained by multiplying the prior distribution function by the likelihood function and scaling this product to provide a probability distribution function, a posterior distribution representing a revised understanding of the parameter. The posterior distribution therefore is a function of the prior

Figure 2.5. Likelihood profile of the proportion parameter for the binary dataset $y = 20$ with the binomial model $B(y;\, n = 80,\, p)$ using the S-Plus or R code

```
> p <- seq(0,1,0.01)
> likelihood <- p^20* (1-p)^60
> plot(p,likelihood)
```

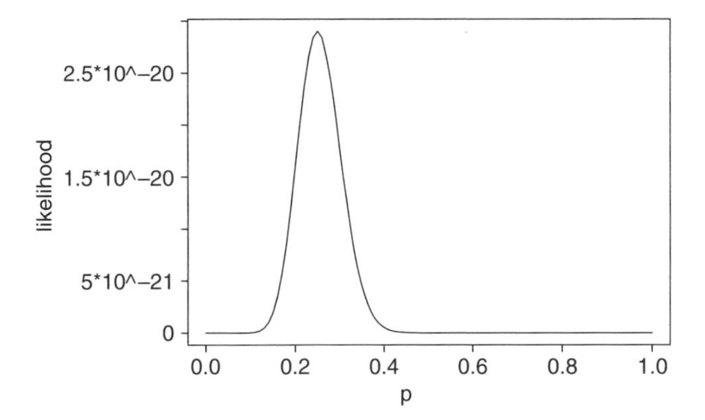

distribution, the sample dataset, and the model that is assumed for the dataset. The entire process is based on a property of conditional probability first recognized by Reverend Thomas Bayes in the eighteenth century (Bayes 1763). This property is now formally stated with the following theorem.

Theorem 2.1: Bayes Theorem. If β is a parameter and D is the dataset, then

1. The discrete parameter conditional probability distribution P, subject to the data, is given by

$$P(\beta|D) = \frac{P(D|\beta) \cdot P(\beta)}{\Sigma P(D|\beta) \cdot P(\beta)}$$

or

$$\text{posterior } (\beta|D) = \frac{\text{likelihood } (\beta|D) \cdot \text{prior } (\beta)}{\Sigma \text{ likelihood } (\beta|D) \cdot \text{prior } (\beta)}$$

2. The continuous parameter conditional probability distribution P, subject to the data, is given by

$$P(\beta|D) = \frac{P(D|\beta) \cdot P(\beta)}{\int P(D|\beta) \cdot P(\beta) \cdot d\beta}$$

or

$$\text{posterior } (\beta|D) = \frac{\text{likelihood } (\beta|D) \cdot \text{prior } (\beta)}{\int \text{likelihood } (\beta|D) \cdot \text{prior } (\beta) \cdot d\beta}.$$

Note that the denominators are the scaling constants that ensure the probability distributions sum (discrete case) or integrate (continuous case) to one. The posteriors represent an updated understanding of the parameter β, conditional to the data D.

The Bayes definition of conditional probability is as follows: the probability P of A conditional to $B = P(A|B) = P(A \cap B)/P(B)$ for subsets of events A and B in a sample space of outcomes of an experiment. An outline of the proof to the Bayes Theorem for the discrete case is as follows (Fig. 2.6):

$$\begin{aligned} P(\beta_i|D) &= P(\beta_i \cap D)/P(D) \\ &= P(D|\beta_i) \cdot P(\beta_i)/P(D) \\ &= P(D|\beta_i) \cdot P(\beta_i)/\Sigma_j P(\beta_j \cap D) \\ &= P(D|\beta_i) \cdot P(\beta_i)/\Sigma_j P(D|\beta_j) \cdot P(\beta_j) \end{aligned}$$

or

$$\text{Posterior } (\beta_i|D) = \frac{\text{likelihood } (\beta_i|D) \cdot \text{prior } (\beta_i)}{\Sigma_j \text{ likelihood } (\beta_j|D) \cdot \text{prior } (\beta_j)}.$$

Figure 2.6. Sample space S of outcomes of an experiment, with discrete parameter values $\beta = \{\beta_1, \beta_2, \beta_3, \beta_4\}$ and with dataset D.

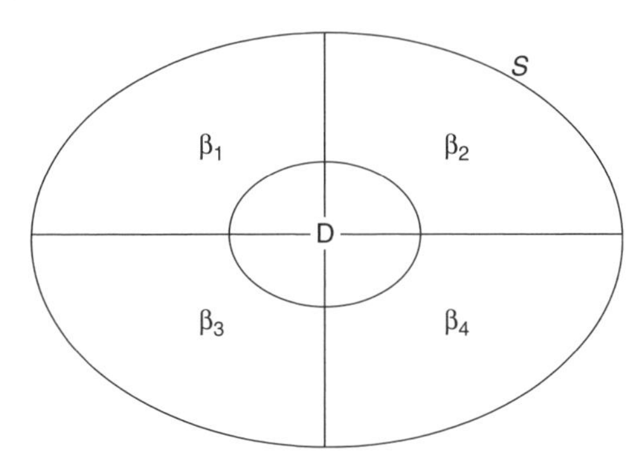

Here S is the sample space of all possible outcomes in an experiment, the β_i constitute all the possible distinct values of the parameter β, and D is a dataset. Reverse β_i and D in the first line and solve for $P(\beta_i \cap D) = P(D \cap \beta_i) = P(D|\beta_i) \cdot P(\beta_i)$ in the numerator to obtain the second line. Substitute $\Sigma_j P(\beta_j \cap D)$ for $P(D)$ the second-line denominator to obtain the third line, since D is the disjoint union of its subsets $\beta_i \cap D$. Finally, use the second-line substitution of $P(\beta_i \cap D) = P(D|\beta_i) \cdot P(\beta_i)$ in the denominator of the third line to complete the derivation in the fourth and fifth lines.

2.4 CONJUGATE PRIORS

Unfortunately, Bayes' Theorem provides a formula that is not always mathematically solvable in closed form, especially for complex formulations. The integrals in the denominator for the continuous case can be extremely challenging if not impossible to solve mathematically in closed form, particularly for higher-dimensional parameter spaces. However, there are many common cases where Bayesian closed-form solutions are possible. For some important types of natural resource datasets and models, it is possible to obtain closed-form solutions by judiciously choosing particular family forms for priors that also provide posteriors. For example, for a continuous dataset modeled by a normal model with known standard deviation σ, a normal prior for the mean parameter μ provides a normal posterior solution. For a count dataset modeled by the Poisson distribution, a choice of gamma prior for the mean parameter λ provides a gamma posterior solution. For a binary dataset modeled by the binomial distribution, a beta prior for the proportion parameter p provides a beta posterior solution. A gamma prior for a dataset such as arrival times modeled by the exponential distribution provides a gamma posterior solution. A Dirichlet prior for a dataset modeled by the multinomial distribution, such as mark–recapture data, provides a Dirichlet posterior solution. These judicious choices of family distributional forms for priors are called conjugate priors.

The closed-form solutions for the posteriors of sample datasets and models with conjugate priors are expressible as transformations of the parameter values of the prior distributions. The transformations are expressable in terms of sample statistics calculated from the data. We illustrate these ideas by presenting several important conjugate priors.

2.4.1 Continuous Data with the Normal Model

Let's consider a continuous dataset from a normally distributed population $N(\mu,\sigma)$. We assume for the sake of simplicity that the parameter σ is known. Let's assume that the prior for μ is normally distributed

$$\mu \sim N(\mu_{pr}, \sigma_{pr}),$$

with mean μ_{pr}, standard deviation σ_{pr}, and precision $\tau_{pr} = 1/\sigma_{pr}^2$.

Bayesians prefer to use precision, the inverse of the variance, rather than variance or standard deviation for a parameter for reasons that will shortly become apparent. We estimate the mean $\bar{y}$ from a sample dataset, with standard deviation σ, standard error se $= \sigma/\sqrt{n}$, and precision $\tau = n/\sigma^2$. Then it can be shown that the posterior distribution for μ is also normally distributed (Iversen 1984, Gill 2002)

$$\mu \sim N(\mu_{post},\sigma_{post})$$

with mean

$$\mu_{post} = \left(\frac{\tau_{pr}}{\tau_{pr} + \tau}\right) \cdot \mu_{pr} + \left(\frac{\tau}{\tau_{pr} + \tau}\right) \cdot \bar{y}$$

and precision

$$\tau_{post} = \tau_{pr} + \tau \text{ (hence, } \sigma_{post} = 1/\sqrt{\tau_{post}}\text{)}.$$

These properties can be derived mathematically since the prior function and likelihood function for the mean both have the same normal form. The interested reader is referred to Hilborn and Mangel (1997) for details.

Thus, the normal posterior distribution for the mean parameter μ has a mean that is the weighted average of the mean of the prior and the estimated mean obtained from the sample dataset. The weights are given by the precisions of the prior and the dataset. The precision of the posterior is the sum of the precisions of the prior and the dataset. Note how the sample size of the dataset directly affects the weights used to calculate the posterior mean and precision. For example, doubling the sample size of the dataset doubles its precision and hence increases the relative role of the sample estimate in the posterior weighted mean and precision accordingly.

For example, let's analyze the tree diameter dataset described in Section 2.2.4. Assume a prior that is normally distributed with mean $\mu_{pr} = 30$ and standard deviation $\sigma_{pr} = 5$ (hence $\tau_{pr} = 1/5^2 = 0.04$). Suppose that the estimate from the sample dataset for the mean is $\bar{y} = 26$. Assume a known $\sigma = 5$ for the dataset population, with standard error $se = \sigma/\sqrt{n} = 5/\sqrt{100} = 0.5$, and precision $\tau = 1/se^2 = 4.0$. Then the posterior distribution for the mean parameter μ has mean

$$\begin{aligned}\mu_{post} &= \left(\frac{\tau_{pr}}{\tau_{pr} + \tau}\right) \cdot \mu_{pr} + \left(\frac{\tau}{\tau_{pr} + \tau}\right) \cdot \bar{y} \\ &= \left(\frac{0.04}{0.04 + 4.00}\right) \cdot 30 + \left(\frac{4.00}{0.04 + 4.00}\right) \cdot 26 \\ &= 0.30 + 25.74 = 26.04\end{aligned}$$

and precision

$$\begin{aligned}\tau_{post} &= \tau_{pr} + \tau \\ &= 0.04 + 4.00 = 4.04.\end{aligned}$$

Note how the sample size $n = 100$ for the dataset affected the precision of the mean estimate and the weights used to calculate the posterior mean. Note also the "shrinkage" of the mean estimate toward the prior value, slight in this example because of the relatively large sample size. Our conclusion is that the original prior assessment of the mean parameter was biased upward, and that the sample dataset provided evidence to revise that assessment downward.

In summary, for continuous data with the normal model, Bayesian statistical analysis using the normal conjugate prior provides the following transformation for the posterior:

$$\begin{aligned}&\text{prior } N(\mu; \mu_{pr}, \sigma_{pr}) \rightarrow (\text{data } y + \text{normal model } N(y; \mu, \sigma)) \\ &\qquad \rightarrow \text{posterior } N(\mu; \mu_{post}, \sigma_{post}),\end{aligned}$$

where

$$\mu_{post} = \left(\frac{\tau_{pr}}{\tau_{pr} + \tau}\right) \cdot \mu_{pr} + \left(\frac{\tau}{\tau_{pr} + \tau}\right) \cdot \bar{y}$$

and

$$\tau_{post} = \tau_{pr} + \tau,$$

with $\tau_{pr} = 1/\sigma_{pr}^2$, $\tau = 1/se^2 = n/\sigma^2$, and $\tau_{post} = 1/\sigma_{post}^2$.

Although this result depends on the assumption of the normality of the prior and the normality and known variance of the data, it is approximately true in general for posteriors; as a first approximation, the precision of a posterior is the sum of the precisions of the prior and the data and the posterior mean is a weighted average of the means of the prior and the data, weighted by precision. This is an intuitively pleasing condition that one would expect as additional datasets are collected.

2.4.2 Count Data with the Poisson Model

Next, let's conduct Bayesian statistical analysis on count data with the Poisson model, using a conjugate prior. The appropriate conjugate distribution to use for these data and model is the gamma distribution $G(x; \alpha, \beta)$: $(0, \infty) \rightarrow (0, \infty)$, with domain and range consisting of positive real numbers, given by

$$G(x; \alpha, \beta) = (\beta^{\alpha}/\Gamma(\alpha)) \cdot e^{-\beta \cdot x} \cdot x^{\alpha-1},$$

where

$$\Gamma(\alpha) = \int_0^{\infty} e^{-t} \cdot t^{\alpha-1} \cdot dt$$

is the gamma function and $\alpha > 0$ and $\beta > 0$ are the parameters (Rice 1995, Wackerly et al. 2002). For positive integers m, $\Gamma(m) = (m - 1)!$, the factorial function. Recall that $0! = 1$, $1! = 1$, $2! = 1 \cdot 2 = 2$, $3! = 1 \cdot 2 \cdot 3 = 6, \ldots$. The gamma distribution is a very flexible and useful distribution, particularly to physical scientists and to Bayesians. Let's make practical use of this distribution by better understanding the function of its two parameters, the shape parameter α and the scale parameter β. If the shape parameter $\alpha < 1$, the gamma distribution G declines exponentially with increasing x. If $\alpha \geq 1$, G is unimodal and skewed to the right with

$$\text{mode} = (\alpha - 1)/\beta.$$

The mean of G is

$$\mu = \alpha/\beta,$$

and the variance is

$$\sigma^2 = \alpha/\beta^2.$$

Conversely, if the mean μ and variance σ^2 are specified, the parameters α and β are given by $\alpha = \mu^2/\sigma^2$ and $\beta = \mu/\sigma^2$. See Fig. 2.7 for examples of the gamma distribution.

A popular choice for a noninformative gamma prior is given by $G(x; \alpha = 10^{-3}, \beta = 10^{-3})$ (Fig. 2.8). This distribution is relatively flat except close to the origin, where parameter values are unlikely to occur, and it has some nice properties as well. (It is an approximation to a Jeffreys prior; see Section 2.5.2.)

Figure 2.7. Gamma distributions $G(x; \alpha, \beta)$ for a selection of parameters α and β using S-Plus or R code (**a**) $\alpha = 0.5$, $\beta = 1$. (**b**) $\alpha = 1$, $\beta = 1$. (**c**) $\alpha = 5$, $\beta = 1$. (**d**) $\alpha = 10$, $\beta = 1$. (**e**) $\alpha = 5$, $\beta = 5$. (**f**) $\alpha = 10$, $\beta = 5$

```
> x<- seq(0, range, 0.01)
> gamma <- dgamma (x, α,β)
> plot(x, gamma, type="l")
```

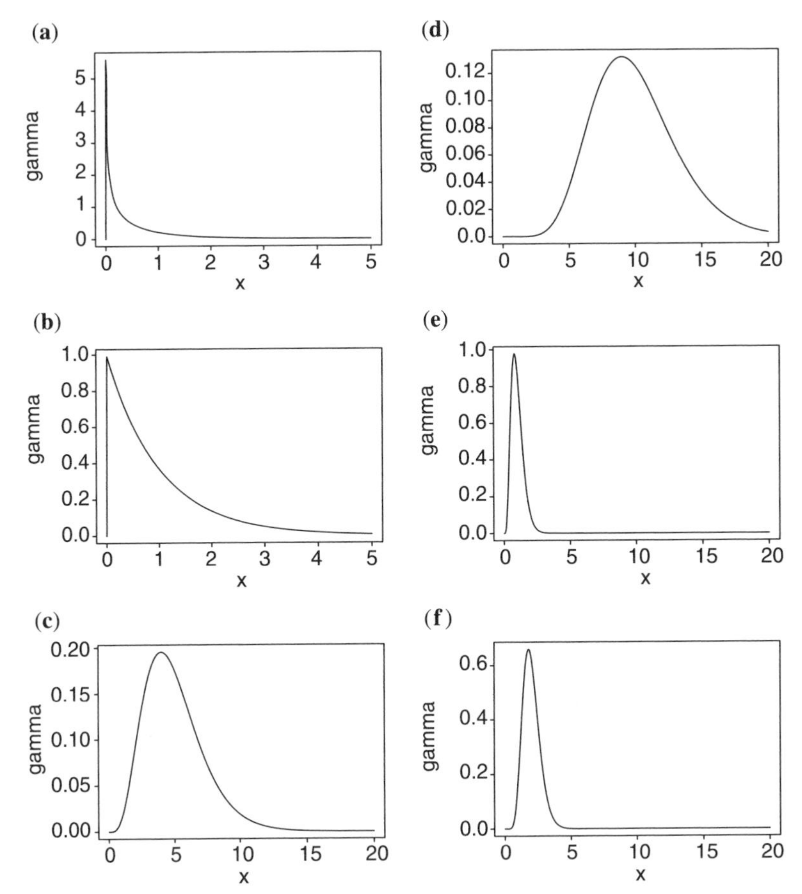

Assuming the Poisson model $\mathrm{Pois}(y; \lambda) = e^{-\lambda} \cdot (\lambda^y/y!)$ for the dataset $y = \{y_1, y_2, \ldots, y_n\}$, it can be shown (Hilborn and Mangel 1997) that the posterior distribution with the gamma prior $G(\lambda; \alpha_{pr}, \beta_{pr})$ for the parameter λ is given by $G(\lambda; \alpha_{post}, \beta_{post})$, where

$$\alpha_{post} = \alpha_{pr} + \Sigma_i y_i,$$
$$\beta_{post} = \beta_{pr} + n.$$

Figure 2.8. The noninformative gamma distribution $G(x; \alpha = 0.001, \beta = 0.001)$.

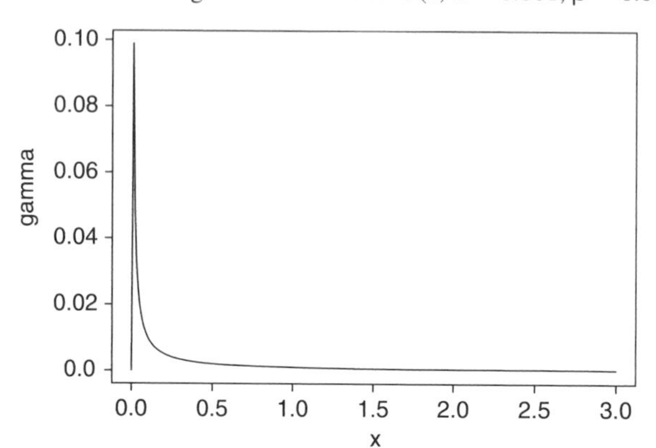

In other words, to obtain the conjugate gamma posterior for count data, simply add the sum of the counts and the sample size to the prior gamma distribution parameters α and β, respectively.

So, suppose that the 50 sites sampled in the second example of Section 2.2.4 for the count data above yielded a total sum of counts $\Sigma_i y_i = 150$ curlews, with an estimated mean of $\bar{y} = 150/50 = 3$ birds per site. If we assume a noninformative gamma prior $G(\lambda; 10^{-3}, 10^{-3})$ for the Poisson mean parameter λ, the posterior gamma distribution will be given by $G(\lambda; 10^{-3}+150, 10^{-3}+50) = G(\lambda; 150.001, 50.001)$ (Fig. 2.9). Note that the posterior distribution for the dataset with this noninformative prior resembles the scaled likelihood function with a mode at the ML estimator value $\bar{y} = 3$.

Figure 2.9. Posterior distribution $G(\lambda; \alpha_{\text{post}} = 150.001, \beta_{\text{post}} = 50.001)$ for the count dataset y with $\Sigma_i y_i = 150$, sample size $n = 50$, and noninformative gamma prior $G(\lambda; \alpha_{\text{pr}} = 0.001, \beta_{\text{pr}} = 0.001)$.

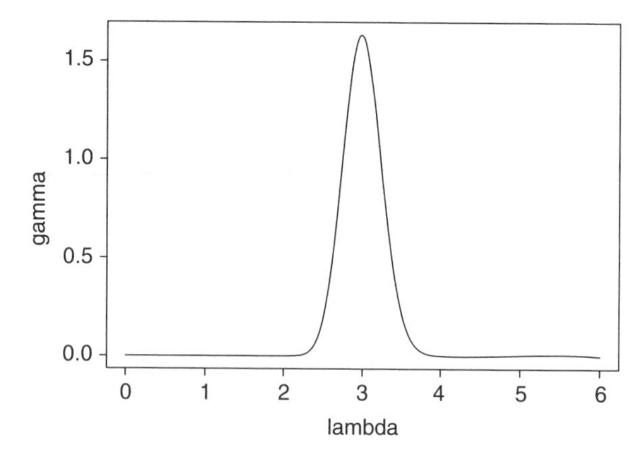

In summary, for count data with the Poisson model, Bayesian statistical analysis using the gamma conjugate prior yields the following transformation for the posterior:

$$\begin{aligned}\text{prior } G(\lambda; \alpha_{\text{pr}}, \beta_{\text{pr}}) &\rightarrow [\text{data } y + \text{Poisson model Pois } (y; \lambda)] \\ &\rightarrow \text{posterior } G(\lambda; \alpha_{\text{post}}, \beta_{\text{post}}),\end{aligned}$$

where

$$\begin{aligned}\alpha_{\text{post}} &= \alpha_{\text{pr}} + \Sigma_i y_i, \\ \beta_{\text{post}} &= \beta_{\text{pr}} + n.\end{aligned}$$

2.4.3 Binary Data with the Binomial Model

Let's conclude this discussion with a third example of conjugacy, analyzing binary data with the binomial model

$$B(y; n, p) = \binom{n}{y} \cdot p^y \cdot (1-p)^{n-y}.$$

The conjugate prior for the parameter p of the binomial model is the beta distribution $\text{BE}(p; \alpha, \beta)$: $[0, 1] \rightarrow [0, \infty)$ given by

$$\text{BE}(p; \alpha, \beta) = (\Gamma(\alpha + \beta)/(\Gamma(\alpha) \cdot \Gamma(\beta))) \cdot y^{\alpha-1} \cdot (1-y)^{\beta-1},$$

where $\Gamma(\alpha) = \int_0^\infty e^{-t} \cdot t^{\alpha-1} \cdot dt$ is the gamma function. The beta distribution is a flexible distribution with two parameters α and β for a random variable p ranging between 0 and 1. As such, it is particularly appropriate to describe a probability parameter such as p. The mean and variance of the beta distribution $\text{BE}(p; \alpha, \beta)$ are given by

$$\begin{aligned}\mu &= \alpha/(\alpha + \beta), \\ \sigma^2 &= \alpha \cdot \beta/[(\alpha + \beta)^2 \cdot (\alpha + \beta + 1)].\end{aligned}$$

Conversely, if μ and σ^2 are known, the parameters α and β are given by

$$\begin{aligned}\alpha &= \frac{\mu^2 - \mu^3 - \mu \cdot \sigma^2}{\sigma^2} \\ \beta &= \frac{1}{\sigma^2} \cdot \mu \cdot (1-\mu)^2 + (\mu - 1).\end{aligned}$$

The mode of the beta distribution is as follows:

$$
\begin{aligned}
\text{mode} = (\alpha - 1)/(\alpha + \beta - 2) \quad &\text{if} \quad \alpha > 1 \quad \text{and} \quad \beta > 1,\\
0 \quad \text{and} \quad 1 \quad &\text{if} \quad \alpha < 1 \quad \text{and} \quad \beta < 1,\\
0 \quad &\text{if} \quad \alpha < 1 \quad \text{and} \quad \beta \geq 1 \quad \text{or}\\
&\text{if} \quad \alpha = 1 \quad \text{and} \quad \beta > 1,\\
1 \quad &\text{if} \quad \alpha \geq 1 \quad \text{and} \quad \beta < 1 \quad \text{or}\\
&\text{if} \quad \alpha > 1 \quad \text{and} \quad \beta = 1, \quad \text{and}\\
\text{does not exist} \quad &\text{if} \quad \alpha = \beta = 1.
\end{aligned}
$$

The flat uniform beta distribution is $BE(p; 1, 1) = 1$. If $\alpha = \beta$, $BE(p; \alpha, \beta)$ is symmetric. If $\alpha = \beta < 1$, BE is concave upward. If $\alpha = \beta > 1$, BE is convex downward. See Fig. 2.10 for examples.

The effect of binary data with the binomial model $B(y; n, p)$ on the conjugate prior $BE(p; \alpha_{pr}, \beta_{pr})$ is simple to describe: add y, the number of 1s, to α and add $n - y$, the number of 0s, to β (Hilborn and Mangel 1997). So the posterior distribution for p is given by $BE(p, \alpha_{post}, \beta_{post})$ where

$$
\begin{aligned}
\alpha_{post} &= \alpha_{pr} + y,\\
\beta_{post} &= \beta_{pr} + (n - y).
\end{aligned}
$$

For example, let's return to our binary dataset of Section 2.2.4. Suppose that we detected the presence of Northern Spotted Owl pairs on $y = 20$ of the $n = 80$ sampled sites. Using a noninformative beta prior $BE(p; 1, 1)$ for p, the posterior prior would be $BE(p; 1 + 20, 1 + 60) = BE(p; 21, 61)$ (Fig. 2.11).

Note that the mode of the posterior occurs at the ML estimate $\hat{p} = 20/80 = 0.25$ as expected with a flat prior. The mean of the posterior occurs at

$$
\hat{\mu} = \frac{\alpha_{post}}{\alpha_{post} + \beta_{post}} = \frac{21}{21 + 61} = 0.256.
$$

In summary, for binary data with the binomial model, Bayesian statistical analysis using the BE conjugate prior yields the following transformation for the posterior:

$$
\begin{aligned}
\text{prior } BE(p; \alpha_{pr}, \beta_{pr}) &\rightarrow (\text{data } y + \text{ binomial model } B(y; n, p))\\
&\rightarrow \text{posterior } BE(p; \alpha_{post}, \beta_{post})
\end{aligned}
$$

where

$$
\begin{aligned}
\alpha_{post} &= \alpha_{pr} + y,\\
\beta_{post} &= \beta_{pr} + (n - y).
\end{aligned}
$$

Figure 2.10. Beta distributions BE(p; α, β) for a range of parameters α and β using the S-Plus or R code (**a**) $\alpha=0.5$, $\beta=0.5$. (**b**) $\alpha = 1$, $\beta = 1$. (**c**) $\alpha = 2$, $\beta = 2$. (**d**) $\alpha = 1$, $\beta = 5$. (**e**) $\alpha = 1$, $\beta = 10$. (**f**) $\alpha = 2$, $\beta = 10$

```
> p <- seq(0, 1,0.01)
> beta <- dbeta (p, α,β)
> plot(p, beta, type="1")
```

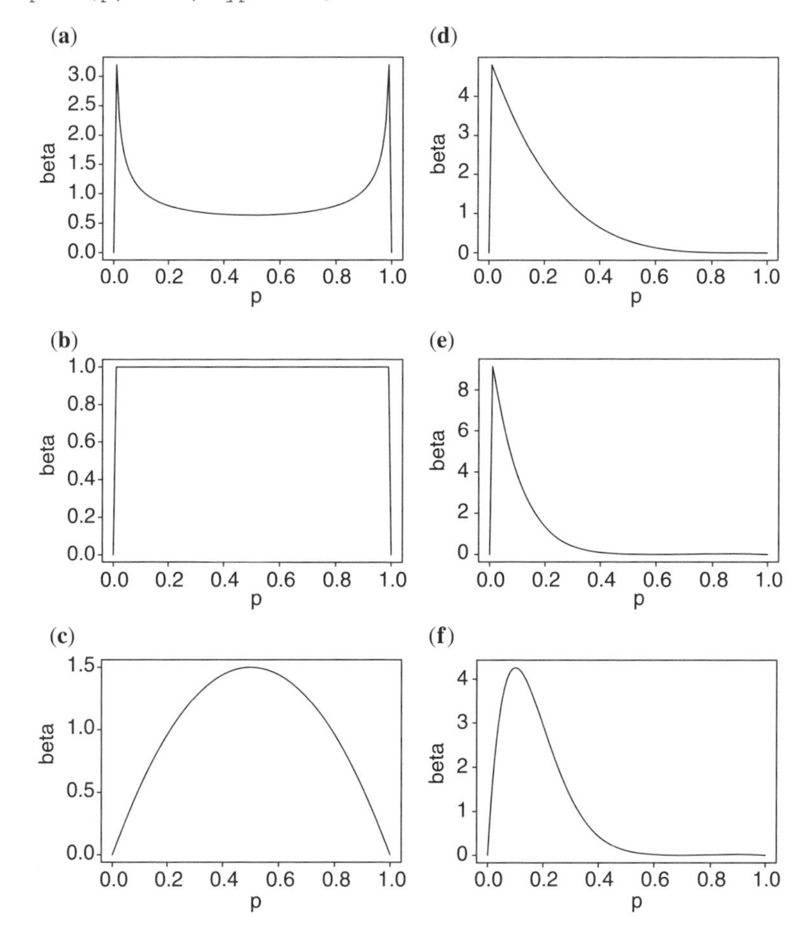

2.4.4 Conjugate Priors for Other Datasets

There are many other datasets and models with conjugate priors. Most notably, multinomial models, useful for capture–recapture datasets, can be analyzed with closed-form Bayesian solutions using a Dirichlet conjugate prior distribution. The multinomial distribution generalizes the binomial distribution to data that are categorical with $k \geq 2$ classes, in contrast to the binomial case where $k = 2$. Analogously, the Dirichlet distribution generalized the beta distribution to the case with k probabilities $p_1, p_2, \ldots, p_k$ for each of k classes of the Dirichlet distribution

Figure 2.11. Posterior distribution BE($p; \alpha = 21, \beta = 61$) for the binary dataset y with $y = 20$ and sample size $n = 80$, and a noninformative beta prior BE($p; \alpha = 1, \beta = 1$).

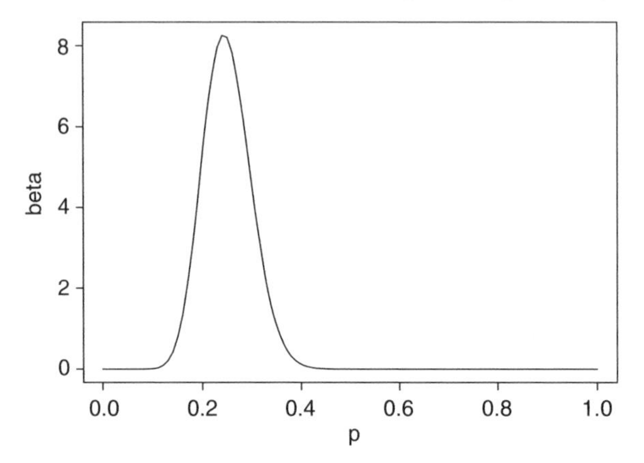

instead of two probabilities $p_1 = p$ and $p_2 = q = (1 - p)$ for the two classes of the binomial distribution.

Another classical case of conjugacy applies to datasets such as arrival times that satisfy the assumptions of the exponential model. In this case, the gamma distribution again provides a conjugate prior.

Aggregated count datasets satisfying the assumptions of the negative binomial model can be analyzed using a beta conjugate prior. Recall that count data for populations that are randomly dispersed should be analyzed with the Poisson model, which has a variance equal to the mean $\sigma^2 = \mu$. If the population is aggregated or clumped with a variance greater than the mean $\sigma^2 > \mu$, the dataset will be overdispersed and should be analyzed with the negative binomial model. We will examine this model in greater detail in Chapter 4.

All of these models are special cases of an important class of models, the exponential family of distributions, all of which have conjugate priors (Gill 2002). The importance of this general class of distributions will become clearer when we examine generalized linear models (GLMs) in Chapter 6. Many common distributions are special cases of the exponential family of distributions, such as the normal, Poisson, negative binomial, binomial, multinomial, exponential, and gamma distributions. Sufficient statistics necessary to calculate ML estimates are guaranteed for the exponential family of distributions. The reader is referred to the texts by Dobson (1990), Hilborn and Mangel (1997), and Gill (2002) for further details.

2.5 OTHER PRIORS

In the previous section, we have demonstrated that many important natural resource datasets and models have associated conjugate priors with closed-form Bayesian solutions for their posteriors. However, for many other more complex datasets and models without conjugate priors, the natural resource scientist can choose from a range of other priors. Priors can have a significant influence on posteriors, particularly for datasets of small sample size, so they must be chosen wisely. Let's examine some of the other types of priors that are of common use: noninformative priors, uniform priors, Jeffreys priors, reference priors, proper and improper priors, vague priors, elicited priors, and empirical Bayes methods of choosing priors.

2.5.1 Noninformative, Uniform, and Proper or Improper Priors

A conservative approach to choosing priors is to assume that little information is known a priori about a parameter and to use a **noninformative prior**. The most common choice of noninformative prior is a **uniform prior**, a flat distribution that assigns the same probability to every value of the parameter. However, problems arise if the parameter is infinite in range and cannot be described by a flat distribution that is a **proper prior** with bounded sum or integral, but rather must be described by an **improper prior** with unbounded sum or integral. This problem need not necessarily be of serious concern if the likelihood of the dataset has a bounded range of support where the function is nonzero. In that case, the posterior may still be a well-defined proper distribution with bounded sum or integral equal to 1 even if the prior is not proper. In many practical situations with natural resource datasets and models, this is the case since the realistic range of most parameters usually is bounded. For instance, mean tree diameters or animal abundance levels for many populations of interest are bounded in range. The mean tree diameter parameter of a forest is bounded by 0 in. and the $d_{bh,max}$ of the largest conceivable tree, so a uniform distribution Unif $(\mu; 0, d_{bh,max})$ with limits 0 and $d_{bh,max}$ can serve as a noninformative prior for the parameter μ. Similarly, the abundance of many animal populations is of limited size N_{max}, so it is bounded by 0 and N_{max}.

2.5.2 Jeffreys Priors

There is an apparent paradox inherent in Bayesian statistical inference in that priors are seldom invariant to transformations of parameters. For example, if a parameter θ is uniformly distributed, it is not generally true that θ^2 is uniformly distributed.

The Bayesian statistician Jeffreys (1961) resolved this apparent difficulty by defining a set of conditions required for a class of priors, now called **Jeffreys priors**, to have posteriors that are invariant to transformations of parameters. The trick is to utilize the shape of the likelihood function of the model describing the dataset and choose a prior on the basis of its curvature. The prior uses the square root of the determinant of the negative expected Fisher information matrix obtained from the Hessian of second derivatives of the likelihood function. The details are technical and beyond the scope of this book. A simple derivation for binary data with the binomial model is given in Gill (2002, pp. 123–125). The more general problem is discussed in Berger (1985). The Jeffreys prior for the proportion parameter p for binary data with the binomial model is the beta distribution $\text{BE}(p; \alpha = 0.5, \beta = 0.5)$ with parameters α

and β equal to 0.5. The Jeffreys prior for the mean parameter λ for count data with the Poisson model is approximated by the gamma distribution $G(\lambda; \alpha = 0.001, \beta = 0.001)$ with parameters α and β equal to 0.001 (Fig. 2.8) (Gill 2002). Jeffreys priors can often be chosen that are noninformative, or approximately uniform, around the significant areas of support of the dataset model likelihood function, where it has nonzero probability, although that is not always the case.

2.5.3 Reference Priors, Vague Priors, and Elicited Priors

Several other types of priors are commonly used by Bayesians, which we will mention here. **Reference priors** are priors, such as conjugate priors, that are convenient to use for certain types of datasets. Other examples of reference priors are **diffuse priors** or **vague priors** such as imprecise normal distributions centered at the expected modes of posteriors with very large standard deviations. These priors are reasonably uniform around the region of interest, yet can be easy to manipulate mathematically.

Elicited priors are priors that are derived from previous knowledge about a parameter. Elicited priors can be constructed from expert opinion, using a specified particular parametric form. Conjugate priors with specified parameter values based on previous experience are examples of elicited priors. It is often a challenge to translate expert opinion into a prior so that a range of priors can be specified, representing the variety of expert opinion. A sensitivity analysis can then be conducted to determine the amount of variation among the posteriors for the range of priors.

2.5.4 Empirical Bayes Methods

Empirical results may be used to describe characteristics of the priors. For example, frequentist statistical estimates from previous studies may be used to specify prior characteristics. Bayesians have shown particular interest more recently in hierarchical models with a hierarchy of parameters at several levels where parameters at each level are expressed in terms of parameters at a lower level. For hierarchical models, parameters at the lowest level, called **hyperparameters**, need to be specified. Frequentist estimates can fulfill this purpose. See Section 4.4.2 for an example of hierarchical modeling and Chapter 5 for an example of a parameter space of models, estimated with frequentist statistics, having AIC weights that can be interpreted as Bayesian posterior probabilities. See Carlin and Louis (2000) for a more detailed account of **empirical Bayes methods**.

2.5.5 Sensitivity Analysis: An Example

Since prior distributions influence posterior distributions, priors must be chosen with care. If the choice of priors is controversial, it is wise to include a sensitivity analysis in the methodology, examining the effects of a realistic range of plausible priors on the posterior distributions and statistics. If the posteriors and their statistics are markedly different for varying priors, the results are sensitive to the choice of priors, and that choice must be made with extreme caution. Alternatively, if the posteriors are similar regardless of priors, the posteriors are robust to the choice of priors, and the choice of priors is not such a critical issue. As the size of the sample dataset used in the analysis increases, the effect of the prior on the posterior is diminished, as demonstrated in Section 2.4.1.

Let's illustrate a simple sensitivity analysis with the binary dataset example of Sections 2.2.4 and 2.4.3 where $y = 20$ Northern Spotted Owl occupied sites were

Figure 2.12. Sensitivity analysis results for the binary dataset with the binomial model $B(y; n, p) = B(20; 80, p)$. (**a**) Noninformative beta priors for p: Jeffreys prior "beta1" = $\text{BE}(p; \alpha = 0.5, \beta = 0.5)$ ("J"), uniform prior "beta2" = $\text{BE}(p; \alpha = 1, \beta = 1)$ ("U"), and skeptical prior "beta3" = $\text{BE}(p; \alpha = 2, \beta = 2)$ ("S"). (**b**) Beta posteriors for beta priors for p: beta posterior "beta4" = $\text{BE}(p; \alpha = 20.5, \beta = 60.5)$ ("J") for Jeffreys prior "beta1" = $\text{BE}(p; \alpha = 0.5, \beta = 0.5)$, beta posterior "beta5" = $\text{BE}(p; \alpha = 21, \beta = 61)$ ("U") for uniform prior "beta2" = $\text{BE}(p; \alpha = 1, \beta = 1)$, and beta posterior "beta6" = $\text{BE}(p; \alpha = 22, \beta = 62)$ ("S") for skeptical prior "beta3" = $\text{BE}(p; \alpha = 2, \beta = 2)$. (**c**) Means, variances, medians, and 95% credible intervals of the posteriors in (b) using the S-Plus or R code (c) Means, variances, medians, and 95% credible intervals of the posteriors in (b) using the S-Plus or R code.

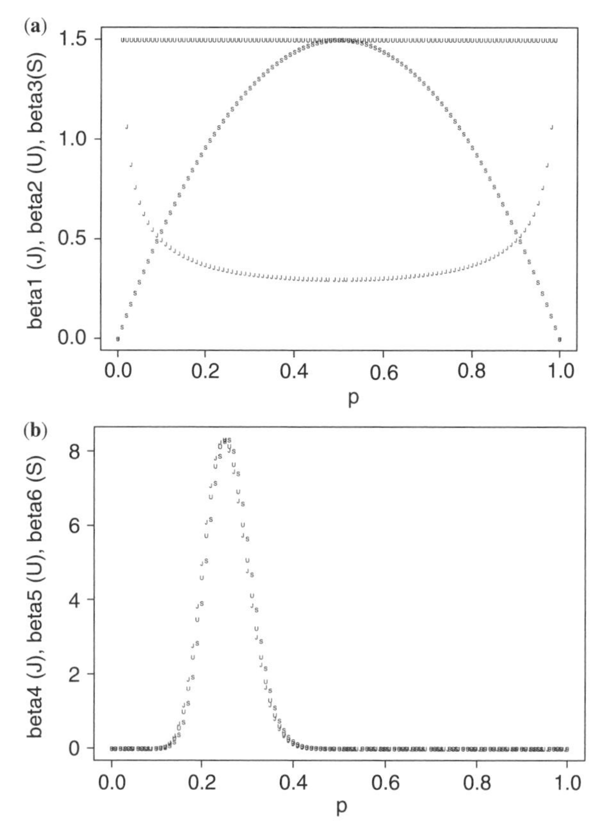

Figure 2.12. *Continued.*

(c)
```
> 20.5/(20.5+60.5) # mean of BE(p; 20.5, 60.5)
[1] 0.2530864
> 21/(21+61) # mean of BE(p; 21, 61)
[1] 0.2560976
> 22/(22+62) # mean of BE(p; 22, 62)
[1] 0.2619048
> (20.5*60.5)/((20.5+60.5)^2*(20.5+60.5+1)) # variance
  of BE(p; 20.5, 60.5)
[1] 0.002302481
> (21*61)/((21+61)^2*(21+61+1)) # variance of BE(p; 21, 61)
[1] 0.00229532
> (22*62)/((22+62)^2*(22+62+1)) # variance of BE(p; 22, 62)
[1] 0.002274243
> qbeta(0.50,20.5,60.5) # median of BE(p; 20.5, 60.5)
[1] 0.2510462
> qbeta(0.50,21,61) # median of BE(p; 21, 61)
[1] 0.2541069
> qbeta(0.50,22,62) # median of BE(p; 22, 62)
[1] 0.2600078
> c(qbeta(0.025,20.5,60.5),qbeta(0.975,20.5,60.5))
  # 95% credible interval limits of BE(p; 20.5, 60.5)
[1] 0.1651318 0.3525829
> c(qbeta(0.025,21,61),qbeta(0.975,21,61))
  # 95% credible interval limits of BE(p; 21, 61)
[1] 0.1681975 0.3552602
> c(qbeta(0.025,22,62),qbeta(0.975,22,62))
  # 95% credible interval limits of BE(p; 22, 62)
[1] 0.1741546 0.3603885
```

detected in $n = 80$ samples. Here the model is binomial $B(y; n, p) = B(y = 20; n = 80, p)$ with the proportion parameter p. Let's choose a plausible range of "noninformative" priors for the sensitivity analysis, using beta conjugate priors for the binomial model. We assume that there is no other information available about p, with no previous studies or plausible alternatives to noninformative priors. Let's examine three noninformative priors: the transformation-invariant Jeffreys prior BE(p; 0.5, 0.5) (Section 2.5.2), the uniform prior BE(p; 1, 1), and a "skeptical" prior BE(p; 2, 2) (Fig. 2.12a). The skeptical prior places a higher weight of credibility, or prior probability, on the parameter p being closer to 50% than nearer to 0% or 100%. By conjugacy theory, the posterior distributions of these priors are BE(p; 20.5, 60.5), BE(p; 21, 61), and BE(p; 22, 62), respectively (Fig. 2.12b). As the figure reveals, their posterior distributions are very similar. Furthermore, their statistics, the posterior means, variances, medians, and 95% credible interval limits are very similar (Fig. 2.12c). The relative differences in the posterior statistics for the different priors are small, bounded by 2%. The dataset was large enough to dominate the differences between the priors in the posteriors. In conclusion, where there is controversy regarding the selection of priors, natural resource scientists can examine such differences with a sensitivity analysis to assess whether they are biologically and statistically important.

2.6 SUMMARY

This chapter provided an introduction to Bayesian statistical inference. We began with some historical background. We discussed some inherent limitations of frequentist statistical inference for natural resource datasets. We presented three methods commonly used by statisticians for fitting models to datasets, least-squares (LS) fit, maximum-likelihood (ML) fit, and Bayesian fit, and illustrated with examples of continuous data with the normal model, count data with the Poisson model, and binary data with the binomial model. We introduced the Bayesian paradigm for statistical inference, with a heuristic proof to the Bayes Theorem. We discussed conjugate priors for datasets and models with closed-form solution, presenting important examples for continuous, count, and binary datasets with normal, Poisson, and binomial models, respectively. We described other alternatives for priors: noninformative priors, uniform priors, Jeffreys priors, reference priors, proper versus improper priors, vague priors, elicited prior, and empirical Bayes methods. We concluded this chapter with a discussion of sensitivity analysis for priors, along with an example.

PROBLEMS

2.1 An occupied Marbled Murrelet nesting site is repeatedly and independently visited m times. There is a probability of at least 50% of detecting the murrelet nesting activity with each site visit.

- **(a)** How many visits m are required to ensure that there is at least a 95% probability of at least one detection of nesting activity, with the m repeated visits?
- **(b)** How is the probability of at least one detection increased, if the site is visited $(m + 1)$ times; that is, what is the probability of at least one detection, with $(m + 1)$ visits?
- **(c)** What is the probability of at least one detection with $(m + 1)$ visits, if the first m visits result in no detections? Solve this conditional probability problem using Bayes' Theorem. How does this conditional probability result compare with the unconditional results? What property of the visit events is illustrated with this result ?

2.2 Bald Eagles are nesting at just one of three possible sites in the Lewiston Lake watershed. The local wildlife officer knows which one of the sites is occupied.

You do not know which site is occupied. You are given one of three possible strategies for deciding which site to visit:

(a) You can choose one of the three sites.

(b) You tentatively select one of the sites. The wildlife officer then identifies one of the other two sites that is unoccupied. You choose the remaining site.

(c) You tentatively select one of the sites. The wildlife officer then identifies one of the other two sites that is unoccupied. You keep your original choice.

Which of the three strategies will maximize your probability of correctly choosing the occupied site?

2.3 Consider the population monitoring scenario as described in Chapter 1. A coastal northern California forest timberlands is being monitored for occupancy or nonoccupancy of Marbled Murrelets, with annual surveys conducted each summer during the nesting season. The sites are chosen independently each year with simple random sampling. Sample sites are visited several times each summer season to ensure a high probability of detection if the murrelets are occupying the sites, so we will assume this conditional probability of detection to be equal to 1. The binary data for the first four summer surveys are as follows (with $y =$ number of occupied sites, $n =$ sample size):

(a) $y_1 = 31, n_1 = 100$

(b) $y_2 = 13, n_2 = 50$

(c) $y_3 = 22, n_3 = 100$

(d) $y_4 = 21, n_4 = 100$

The sampling was reduced the second year because of budgetary constraints. A critical threshold level of 25% occupancy of the timberlands has been mandated by regulatory agencies for the timberlands to be sufficiently occupied by murrelets to ensure the local viability of the species.

(a) Conduct three Bayesian statistical analyses of the first summer's dataset, using conjugate beta priors with $(\alpha, \beta) = (1, 1)$, $(0.5, 0.5)$, and $(2, 2)$, respectively. Compare the means, medians, and 95% credible interval limits of the posteriors for these three priors. Are the results sufficiently robust regardless of priors; that is, are the differences between the posterior statistics for the means, medians, and credible interval limits within an acceptable 3% relative "error" rate? Compute the risk, the probability, of not being above the critical 25% threshold for each prior. Is the risk of not being above the critical 25% threshold below 50% for each prior?

(b) Assume the Jeffreys beta prior with $\alpha = \beta = 0.5$ for the first year's data and conduct a sequential, cumulative Bayesian statistical analysis on the 4 years of monitoring data. Use the posterior from the previous year's analysis for the prior of the following year. What do you conclude is the risk of not being above the critical 25% threshold of occupancy of the forest timberlands of marbled murrelets for each year? Is the risk, of not being above the critical 25% threshold, below 50% for each of the cumulative years of monitoring? Additionally, is there any disturbing trend for the risk over the four years of monitoring and, if so, how might the data be analyzed to take into account this trend? [*Hint*: look at the change in the odds ratio of the risk OR(risk) = probability(risk)/probability(non-risk) for each year's analysis.] We will examine this change in the odds ratio in further detail in Chapter 3.

(c) In a rebuttal by the logging company, their attorney cites a previous sample survey of the company timberlands taken 5 years earlier that determined that $y = 10$ of $n = 30$ sample sites were occupied by murrelets. If these earlier data are used to estimate a beta prior for p with estimated mean $\mu = \hat{p} = (y/n) = \frac{10}{30} = 0.333$ and estimated standard deviation

$$\sigma = \sqrt{\frac{\hat{p} \cdot (1 - \hat{p})}{n - 1}} = \sqrt{\frac{1/3 \cdot 2/3}{29}} = 0.0875$$

for the first year's data analysis, will the annual results of new sequential analysis yield similar or different results for the risk of not being above the critical 25% threshold with a probability below 50%, compared to the previous analysis that did not include the logging company's dataset? If different, demonstrate how the change in the odds ratio of the risk might be used to indicate that there is still a disturbing trend in the risk.

2.4 A Marbled Godwit monitoring survey is conducted in a local marsh conservation habitat. Godwit count data $\{y_1, y_2, \ldots, y_n\}$ are independently collected for each of 3 years on 0.1-ha hectare fixed-area plots using a simple random sampling design. The total number of counts $y = \sum_i y_i$ and sample size n for each year are as follows:

(a) $y_1 = 400$, $n_1 = 100$

(b) $y_2 = 260$, $n_2 = 75$

(c) $y_3 = 140$, $n_3 = 50$

Use an approximate Jeffreys gamma prior with $\alpha = \beta = 0.001$ to conduct a sequential, cumulative Bayesian statistical analysis on the datasets. The monitoring plan for the marsh conservation habitat specifies that a critical mean threshold density level of three godwits per plot be maintained to ensure the local viability of the population. Does the analysis for the monitoring data indicate that the threshold has been exceeded with a probability of 50% or better, for each of the 3 years? Are there any indications of possible trouble ahead? (*Hint*: Look at the odds ratio of the risks.)

2.5 Discuss in general the relative effects of the sample on the mean and precision of the posterior, using the mean parameter of continuous normally distributed data as an approximate guideline. If the prior distribution is noninformative, then

(a) What will be the approximate mean and precision of the posterior?

(b) How will the precision and standard deviation of the posterior be affected by a doubling of the sample dataset?

(c) What must be done to the sample size to half the standard deviation of the posterior?

3 Bayesian Statistical Analysis II: Bayesian Hypothesis Testing and Decision Theory

In this chapter we will present an introduction to Bayesian hypothesis testing and decisionmaking. We will demonstrate how the Bayesian approach can consider more than one or two hypotheses with its method of hypothesis testing. Bayesian statistical inference can also test hypotheses directly, examining the probability of hypotheses conditional to the data rather than the probability of data conditional to the hypotheses. Bayesians can assess either the status of the probability of the hypothesis or the ratio of the odds of the hypothesis, indicating the rate of change of the odds conditional to the data. Bayesians can also assess decisionmaking with a statistical decision theory fully integrated into Bayesian statistical analysis.

3.1 BAYESIAN HYPOTHESIS TESTING: BAYES FACTORS

In Chapter 1 we described various objections to the use of classical frequentist analysis and inference in the context of contemporary natural resource science. The frequentist logic for hypothesis testing is indirect, providing inferences based on the probability of data, conditional to null hypothesis assumptions about the parameter. A direct inference would reverse this logic, providing inferences for the probability of hypotheses, conditional to the data. The frequentist paradigm for hypothesis testing subjects one hypothesis, the null hypothesis H_0, to a test, in contrast to its alternative hypothesis H_A. It would be more realistic in natural resource science to be able to test a multitude of hypotheses, to assess their relative probabilities, given sample data. It would also be more useful to incorporate prior information into hypothesis testing rather than relying solely on independent assessments. In summary, it would be useful for the natural resource scientist to have a method of hypothesis testing that is direct, allows a multitude of hypotheses, and takes advantage of prior

Contemporary Bayesian and Frequentist Statistical Research Methods for Natural Resource Scientists. By Howard B. Stauffer

information. All of these properties can be satisfied with the Bayesian approach to hypothesis testing.

A Bayesian tests a hypothesis H consisting of a subset of parameter values θ in H contained in a parameter space Φ (i.e., $H \subseteq \Phi$) by examining the posterior distribution $\pi_{\text{posterior}} = \pi(\theta|y)$ of the parameter space based upon the experimental dataset y and calculating the posterior probability of the hypothesis

$$\pi_{\text{posterior}}(H|y)$$

and its odds ratio relative to the prior distribution

$$\frac{\pi_{\text{posterior}}(H|y)/(1-\pi_{\text{posterior}}(H|y))}{\pi_{\text{prior}}(H|y)/(1-\pi_{\text{prior}}(H|y))}.$$

Such an odds ratio approach is analogous to the frequentist use of descriptive statistics describing the hypothesis with an odds ratio test. The Bayesian, however, calculates probabilities for the hypothesis conditional to the data, whereas the frequentist calculates probabilities for the data conditional to the hypothesis.

A Bayesian can also test competing hypotheses H_1 and H_2, in a manner analogous to the frequentist likelihood ratio testing protocol, by examining the **Bayes factor** (Jeffreys 1961)

$$\text{BF}(H_1/H_2|y) = \frac{\pi_{\text{posterior}}(H_1|y)/\pi_{\text{posterior}}(H_2|y)}{\pi_{\text{prior}}(H_1)/\pi_{\text{prior}}(H_2)}.$$

The Bayes factor is the odds ratio of the posterior odds to the prior odds of the hypothesis H_1 relative to H_2. Note that H_1 and H_2 need not be nested, as is required with frequentist hypothesis testing using likelihood ratio tests. Jeffreys recommends the following guidelines for degrees of evidence for H_1 and H_2, based on ranges of Bayes factor values:

1. $\text{BF}(H_1/H_2|y) < 1/100$: decisive evidence for H_2
2. $1/100 \leq \text{BF}(H_1/H_2|y) < 1/10$: strong evidence for H_2
3. $1/10 \leq \text{BF}(H_1/H_2|y) < 1/\sqrt{10}$: substantial evidence for H_2
4. $1/\sqrt{10} \leq \text{BF}(H_1/H_2|y) < 1$: minimal evidence for H_2
5. $1 \leq \text{BF}(H_1/H_2|y) < \sqrt{10}$: minimal evidence for H_1
6. $\sqrt{10} \leq \text{BF}(H_1/H_2|y) < 10$: substantial evidence for H_1
7. $10 \leq \text{BF}(H_1/H_2|y) < 100$: strong evidence for H_1
8. $\text{BF}(H_1/H_2|y) > 100$: decisive evidence for H_1

Note that H_2 may be set equal to the set complement of H_1, $H_2 = \Phi - H_1$, the remainder of the parameter space, and we are back to the Bayesian method of testing one hypothesis H_1. Let's look at some examples.

3.1.1 Proportion Estimation of Nesting Northern Spotted Owl Pairs

A habitat conservation plan for a timber company requires that H_1: $p \geq 60\%$ of its ownership be occupied by nesting Northern Spotted Owl pairs to ensure a locally viable population. The prior odds for $H_1/H_2 = \mathrm{H}_1/(\Phi - H_1)$, with H_2: $p < 60\%$ the complement of H_1, based on a flat noninformative prior BE(p;1,1), is $\pi_{\text{prior}}(H_1)/\pi_{\text{prior}}(H_2) = 0.40/0.60 = 0.67$. A random sample survey of $n = 100$ sites of binary data with the binomial model provide $y = 70$ occupied sites and a proportion estimate of $\hat{p} = \frac{70}{100} = 70\%$. The posterior distribution based on the conjugate prior is BE(p;71,31) with posterior odds $\pi_{\text{posterior}}(H_1|y)/\pi_{\text{posterior}}(H_2|y) = 0.979/0.021 = 46.619$ and Bayes factor

$$\mathrm{BF}(H_1/H_2|y) = \frac{\pi_{\text{posterior}}(H_1|y)/\pi_{\text{posterior}}(H_2|y)}{\pi_{\text{prior}}(H_1)/\pi_{\text{prior}}(H_2)}$$

$$= \frac{0.979/0.021}{0.40/0.60} = \frac{46.619}{0.667} = 69.894$$

providing strong evidence for H_1: $p \geq 0.60$. The probabilities for the posterior beta distribution are calculated using the S-Plus or R cumulative distribution command [e.g., `pbeta (0.60,71,31)` for H_2]. The posterior probability of H_1: $p \geq 0.60$, $\pi_{\text{posterior}}(H_1|y) = 97.9\%$, is also compelling evidence for H_1, based on the dataset (Fig. 3.1a).

Alternatively, if the estimate for p from the sampling had been $\hat{p} = \frac{62}{100} = 62\%$, based on $y = 62$ with $n = 100$, then the posterior probability for H_1 would have been $\pi_{\text{posterior}}(H_1|y) = 0.648$ and the Bayes factor would have been $\mathrm{BF}(H_1/H_2|y) = 2.761$, based on the posterior distribution BE(p;63,39) (Fig. 3.1b). This would have been much less compelling evidence in favor of H_1, just minimal evidence, although the posterior probability of H_1: $p \geq 0.60$ would be 65.0%, well over 50%. Note, therefore, that the Bayes factor "measures" the change of the odds from the prior to the posterior distributions, derived from the analysis of the data that reassesses the distribution of the parameter. The Bayes factor is a relative measure, based not solely on the current odds of the hypothesis in the posterior distribution, but on its change relative to the odds in the prior distribution. In conclusion, natural resource scientists should take both Bayes factors and posterior distribution probabilities and odds into account before drawing conclusions about their hypotheses.

3.1.2 Medical Diagnostics

Medical diagnostics provides an example that is perhaps outside the professional realm of the natural resource scientist. It is a context, however, that is familiar to most of us and provides an illustrative example of Bayes factors and Bayesian decision theory that will generalize to a natural resource setting. Consider the

Figure 3.1. Posterior distributions BE(p;α,β) for the binary y with binomial models $B(y;n,p)$ in Section 3.1.1 with posterior probability regions for hypothesis $H_1: p \geq 60\%$. (**a**) BE (p;71,31) with $y = 70$ and sample size $n = 100$, and noninformative prior BE (p;1,1). (**b**) BE (p;63,39) with $y = 62$ and sample size $n = 100$, and noninformative prior BE (p;1,1).

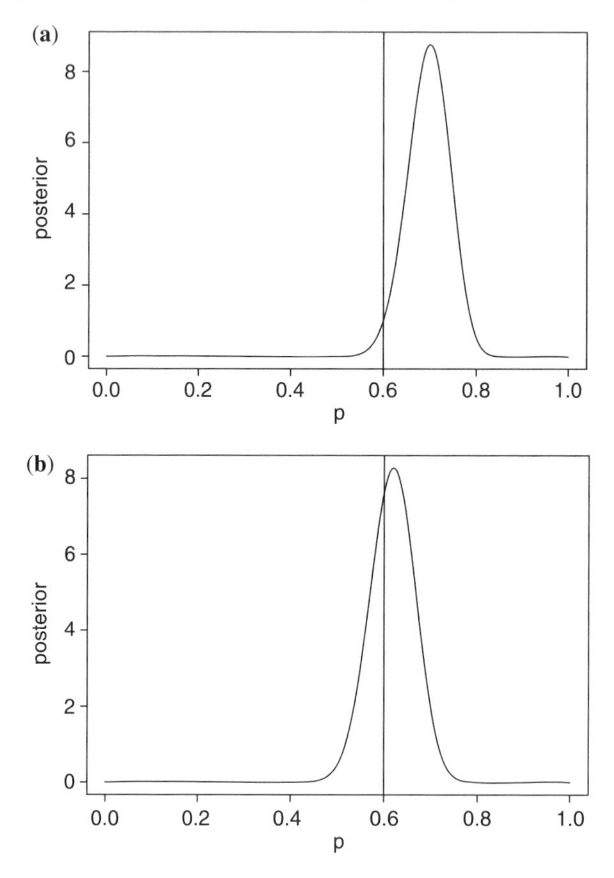

dilemma presented when someone visits a physician's office to report symptoms suggesting the possibility of a certain disease such as lung cancer. Let's say that the patient is a white male who has lived in Los Angeles most of his life, and is a smoker who doesn't exercise. His symptoms are a severe cough, a pain in the chest, and several other abnormal indicators. The physician is able to reference statistics that estimate that a patient with these characteristics and identifiable symptoms has a 10% probability of having the disease. The physician and patient are considering conducting a test, such as a computerized tomography (CT) scan, to obtain additional information on the condition of the patient, to help decide whether to operate. Let's say that the physician also has information, on the basis of previous scientific studies, indicating that the sensitivity and specificity of the test, the CT scan, are 95% and 80%, respectively. The **sensitivity** of the test is the probability of a positive result, if the patient has the disease. The **specificity** of the test is the probability of a negative result, if the patient doesn't have the disease. So there is a 5% chance of error of a negative test result if the patient has the disease and a 20% chance of error of a positive test result if the patient does not have the disease.

The initial 10% statistic for the probability of the patient having the disease provides a Bayesian prior probability distribution for a parameter space. The parameter space D is categorical, finite, and discrete, with two values, say, D_0 for the non-diseased state and D_1 for the diseased state. The dataset T consists of two values, $T = T^-$ for a negative test result, and $T = T^+$ for a positive test result. If the patient does not have the disease, then $D = D_0$ and the probability distribution for the data consists of two values:

$$p(T^-|D_0) = \text{specificity},$$
$$p(T^+|D_0) = (1 - \text{specificity}).$$

If the patient does have the disease, then $D = D_1$ and the probability distribution for the data is given by

$$p(T^-|D_1) = (1 - \text{sensitivity}),$$
$$p(T^+|D_1) = \text{sensitivity}.$$

The prior distribution for D, based on the researched statistics, is given by

$$\pi_{\text{prior}}(D_0) = 0.90,$$
$$\pi_{\text{prior}}(D_1) = 0.10.$$

From Bayes' Theorem, the posterior distribution for D, with negative test results T^-, is given by

$$\begin{aligned}
\pi_{\text{posterior}}(D_0|T^-) &= \frac{p(T^-|D_0) \cdot \pi_{\text{prior}}(D_0)}{p(T^-|D_0) \cdot \pi_{\text{prior}}(D_0) + p(T^-|D_1) \cdot \pi_{\text{prior}}(D_1)} \\
&= \frac{\text{specificity} \cdot \pi_{\text{prior}}(D_0)}{\text{specificity} \cdot \pi_{\text{prior}}(D_0) + (1 - \text{sensitivity}) \cdot \pi_{\text{prior}}(D_1)} \\
&= \frac{0.80 \cdot 0.90}{0.80 \cdot 0.90 + 0.05 \cdot 0.10} = 0.993
\end{aligned}$$

and

$$\begin{aligned}
\pi_{\text{posterior}}(D_1|T^-) &= \frac{p(T^-|D_1)\cdot\pi_{\text{prior}}(D_1)}{p(T^-|D_0)\cdot\pi_{\text{prior}}(D_0)+p(T^-|D_1)\cdot\pi_{\text{prior}}(D_1)} \\
&= \frac{(1-\text{sensitivity})\cdot\pi_{\text{prior}}(D_1)}{\text{specificity}\cdot\pi_{\text{prior}}(D_0)+(1-\text{sensitivity})\cdot\pi_{\text{prior}}(D_1)} \\
&= \frac{0.05\cdot 0.10}{0.80\cdot 0.90+0.05\cdot 0.10} = 0.007.
\end{aligned}$$

For positive test results T^+, the posterior distribution for D is given by

$$\begin{aligned}
\pi_{\text{posterior}}(D_0|T^+) &= \frac{p(T^+|D_0)\cdot\pi_{\text{prior}}(D_0)}{p(T^+|D_0)\cdot\pi_{\text{prior}}(D_0)+p(T^+|D_1)\cdot\pi_{\text{prior}}(D_1)} \\
&= \frac{(1-\text{specificity})\cdot\pi_{\text{prior}}(D_0)}{(1-\text{specificity})\cdot\pi_{\text{prior}}(D_0)+\text{sensitivity}\cdot\pi_{\text{prior}}(D_1)} \\
&= \frac{0.20\cdot 0.90}{0.20\cdot 0.90+0.95\cdot 0.10} = 0.655, \\
\pi_{\text{posterior}}(D_1|T^+) &= \frac{p(T^+|D_1)\cdot\pi_{\text{prior}}(D_1)}{p(T^+|D_0)\cdot\pi_{\text{prior}}(D_0)+p(T^+|D_1)\cdot\pi_{\text{prior}}(D_1)} \\
&= \frac{\text{sensitivity}\cdot\pi_{\text{prior}}(D_1)}{(1-\text{specificity})\cdot\pi_{\text{prior}}(D_0)+\text{sensitivity}\cdot\pi_{\text{prior}}(D_1)} \\
&= \frac{0.95\cdot 0.10}{0.20\cdot 0.90+0.95\cdot 0.10} = 0.345.
\end{aligned}$$

So, if the test is positive, the posterior probability of having the disease is 34.5%, whereas if the test is negative, the posterior probability of having the disease is just 0.7%.

The Bayes factor is obtained by examining the ratio of the prior and posterior odds

$$\begin{aligned}
\text{BF}(D_1/D_0|T^+) &= \frac{\pi_{\text{posterior}}(D_1|T^+)/\pi_{\text{posterior}}(D_0|T^+)}{\pi_{\text{prior}}(D_1)/\pi_{\text{prior}}(D_0)} \\
&= \frac{0.345/0.655}{0.10/0.90} = 4.74, \\
\text{BF}(D_1/D_0|T^-) &= \frac{\pi_{\text{posterior}}(D_1|T^-)/\pi_{\text{posterior}}(D_0|T^-)}{\pi_{\text{prior}}(D_1)/\pi_{\text{prior}}(D_0)} \\
&= \frac{0.007/0.993}{0.10/0.90} = 0.063.
\end{aligned}$$

So the odds of having the disease increase by approximately 5-fold if the test results are positive, and decrease by approximately 20-fold if the test results are negative. Is it worthwhile to take the test? This is a decision for the patient, in consultation with the physician. However, this decision may also depend on the financial costs and the health risks involved with this process, and it would make sense to consider these factors as well. Bayesian statistical analysis and inference has been nicely integrated with decision theory to provide a rigorous answer to this question. We will introduce this idea in the next section.

Before leaving this topic, however, let's point out that, for this simple example of medical diagnostics with a two-valued parameter space and a two-valued dataset, there is a simple general formula for the Bayes factor. Returning to the formulas above for the posterior distribution probabilities, the Bayes factors reduce algebraically to simple functions of sensitivity and specificity

$$\begin{aligned}
\mathrm{BF}(D_1/D_0|T^+) &= \frac{\pi_{\mathrm{posterior}}(D_1|T^+)/\pi_{\mathrm{posterior}}(D_0|T^+)}{\pi_{\mathrm{prior}}(D_1)/\pi_{\mathrm{prior}}(D_0)} \\
&= \frac{\mathrm{sensitivity}\cdot\pi_{\mathrm{prior}}(D_1)/(1-\mathrm{specificity})\cdot\pi_{\mathrm{prior}}(D_0)}{\pi_{\mathrm{prior}}(D_1)/\pi_{\mathrm{prior}}(D_0)} \\
&= \frac{\mathrm{sensitivity}}{(1-\mathrm{specificity})}, \\
\mathrm{BF}(D_0/D_1|T^-) &= \frac{\pi_{\mathrm{posterior}}(D_0|T^-)/\pi_{\mathrm{posterior}}(D_1|T^-)}{\pi_{\mathrm{prior}}(D_0)/\pi_{\mathrm{prior}}(D_1)} \\
&= \frac{\mathrm{specificity}\cdot\pi_{\mathrm{prior}}(D_0)/(1-\mathrm{sensitivity})\cdot\pi_{\mathrm{prior}}(D_1)}{\pi_{\mathrm{prior}}(D_0)/\pi_{\mathrm{prior}}(D_1)} \\
&= \frac{\mathrm{specificity}}{1-\mathrm{sensitivity}}.
\end{aligned}$$

Similarly, it can be shown that

$$\begin{aligned}
\mathrm{BF}(D_1/D_0|T^-) &= \frac{(1-\mathrm{sensitivity})}{\mathrm{specificity}}, \\
\mathrm{BF}(D_0/D_1|T^+) &= \frac{(1-\mathrm{specificity})}{\mathrm{sensitivity}}.
\end{aligned}$$

Although natural resource scientists may be wondering how this medical example relates to applications in their disciplines, analogous examples may be found in natural resource applications. Habitat selection modeling with logistic regression provides one such example. An important diagnostic in examining the comparative goodness of fit of logistic regression models is the receiver operating curve (ROC) c statistic, which measures the area under the sensitivity/(1 − specificity) curve for various cutoff points (see Chapter 6). Hence, we see that this comparative statistic measures the Bayes factors for various cutoff points of logistic regression models.

The model with the highest Bayes factors overall has the highest c statistic and is the best-fitting overall model.

The original case study problem of deciding whether a critical threshold for a parameter has been exceeded that was described in Section 1.1 can also be couched in the same framework as this medical problem. The two states for the parameter space of exceeding the threshold or not exceeding the threshold are analogous to the disease state D_1 and the nondisease state D_0. Sensitivity is the probability of deciding from the monitoring (i.e., the "test") that the threshold has been exceeded (i.e., D_1, the "diseased" state). Specificity is the probability of deciding from the monitoring that the threshold has not been exceeded. Sensitivity and specificity depend on the sample sizes used for the monitoring and can be determined by the probabilities calculated from the binomial model and the prior distribution for the parameter p. A particular example of this application will be illustrated later in the problem assignments at the end of this chapter.

3.2 BAYESIAN DECISION THEORY

An additional advantage to Bayesian statistical inference is that it has been fully integrated with decision theory to provide a completely coherent theory, **Bayesian decision theory**. This theory incorporates **losses**

$$l(\theta,a) = l(\theta,d(y))$$

that are a function of parameters θ and actions $a = d(y)$ resulting from **decisions** d that are based on data y. A loss may be interpreted as an expense or opportunity cost or some other liability. A gain or benefit or profit from the decision is considered a negative loss. This results in a **risk function** that is the expected loss

$$R(\theta,d) = E[l(\theta,d(y))] = \begin{cases} \sum_i l(\theta,d(y_i)) \cdot f(y_i|\theta) & \text{(discrete case)} \\ \int l(\theta,d(y_i)) \cdot f(y|\theta) \cdot dy & \text{(continuous case)} \end{cases},$$

a function of the parameter θ and the decision d. The risk function is the average loss, the expectation weighted with respect to the conditional probability distribution function $f(y|\theta)$ of the data y, given parameter value θ. The **Bayes risk** $B(d)$ of a decision d is the average risk over the distribution function of the parameter

$$B(d) = E[R(\theta,d)] = \begin{cases} \sum_i R(\theta_i,d) \cdot \pi(\theta_i) & \text{(discrete case)} \\ \int R(\theta,d) \cdot \pi(\theta) \cdot d\theta & \text{(continuous case)} \end{cases}$$

The **Bayes rule** is the principle of choosing the decision d^* that minimizes the Bayes risk $B(d)$. The idea is simple in concept; look at the risk functions of the decision options and select the decision that minimizes your risk. In practice, however, this may be difficult to apply. In particular, it may be difficult to assign losses with sufficient precision to apply the theory with practical value. Even if that is the case, however, Bayesian decision theory is still useful because it points out the important role that costs and benefits play, along with probabilities, in the decision-making process.

Bayes risk and the Bayes rule are formulated in terms of a distribution function of a parameter, say, a prior distribution. Suppose you collect data y and calculate the **posterior risk** of an action $a = d(y)$ as the expected loss where the expectation is taken with respect to the posterior distribution of the parameter θ. Note that the posterior risk is defined solely in terms of a sample dataset and not the entire population, unlike the Bayes risk, which requires the entire population of data. It can be shown that a decision $d = d(y)$ that minimizes the posterior risk is a Bayes rule. So Bayesian decision theory permits a choice of decisions to be based on sample data, rather than the entire population of data. The interested reader may seek further details in Rice (1995, Chapter 15).

Let's illustrate Bayesian decision theory with the medical diagnostics example in Section 3.1.2. Let's consider a decision process d_1 where a treatment is applied if the test is positive T^+ and not applied if the test is negative T^-. We specify the following losses for the data, the test results, for this decision d_1 with the two parameter values, D_0 and D_1. If $\theta = D_0$, we will assign losses

$$\begin{aligned} l(\theta,a) &= l(D_0,d_1(T^-)) = 0 \\ &= l(D_0,d_1(T^+)) = L_0{}^+ > 0. \end{aligned}$$

If $\theta = \mathrm{D}_1$, we will assign losses

$$\begin{aligned} l(\theta,a) &= l(D_1,d_1(T^+)) = 0 \\ &= l(D_1,d_1(T^-)) = L_1{}^- > 0. \end{aligned}$$

If the patient does not have the disease, we are assigning a zero loss for a negative test result and a positive loss $L_0{}^+$ for a positive test result and the potentially resulting unnecessary treatment. If the patient does have the disease, we are assigning a zero loss for a positive test result and a positive loss $L_1{}^-$ for a negative test result and the potentially resulting failure to apply treatment for the disease. The risk function for this decision for the two parameter values is given by

$$\begin{aligned} R(D_0,d_1) &= l(D_0,d_1(T^-)) \cdot p(T^-|D_0) + l(D_0, d_1(T^+)) \cdot p(T^+|D_0) \\ &= 0 \cdot \text{specificity} + L_0{}^+ \cdot (1 - \text{specificity}) \\ &= L_0{}^+ \cdot (1 - \text{specificity}) \end{aligned}$$

and

$$\begin{aligned} R(D_1,d_1) &= \mathrm{l}(D_1,d_1(T^-)) \cdot p(T^-|D_1) + \mathrm{l}(D_1,d_1(T^+)) \cdot p(T^+|D_1) \\ &= L_1^- \cdot (1 - \text{sensitivity}) + 0 \cdot \text{sensitivity} \\ &= L_0^+ \cdot (1 - \text{sensitivity}). \end{aligned}$$

The Bayes risk is the average of this risk over the parameter space

$$\begin{aligned} B(d_1) &= R(D_0,d_1) \cdot \pi(D_0) + R(D_1,d_1) \cdot \pi(D_1) \\ &= L_0^+ \cdot (1 - \text{specificity}) \cdot \pi_{\text{prior}}(D_0) \\ &\quad + L_1^- \cdot (1 - \text{sensitivity}) \cdot \pi_{\text{prior}}(D_1). \end{aligned}$$

So the Bayes risk is affected by the losses, the probabilities arising from incorrect decisions, and the probabilities of having the disease or not, as we would expect. This risk can be compared with Bayes risks for other alternative decisions using the Bayes rule. For example, it is interesting to compare this decision process with the alternative decisions of "no treatment regardless of test results" or "treatment regardless of test results." This question will be examined in more detail in the problems at the end of this chapter.

3.3 PREVIEW: MORE ADVANCED METHODS OF BAYESIAN STATISTICAL ANALYSIS—MARKOV CHAIN MONTE CARLO (MCMC) ALGORITHMS AND WinBUGS SOFTWARE

Datasets with conjugate priors, such as continuous, count, and binary datasets with normal, Poisson, and binomial models, respectively, are important datasets commonly sampled by natural resource scientists. These datasets can be analyzed with closed-form mathematical solutions, as was illustrated in Chapter 2. However, many other important natural resource datasets require more sophisticated models that are not necessarily solvable with closed-form mathematical solutions. The analysis of many natural resource datasets requires the use of more complicated models such as multiple linear regression models and generalized linear models. Binary and count datasets may be overdispersed and require models such as hierarchical models or mixed-effects models with random effects. In Chapter 4, we will discuss ways of solving more complicated problems with Bayesian statistical analysis using Markov Chain Monte Carlo (MCMC) algorithms. These algorithms, developed since the mid-1950s, sample parameter values from posterior distributions that provide solutions to general Bayesian problems. The MCMC algorithms generate Markov chains of values for iterative solutions to the Bayesian analysis. The MCMC algorithms have now been implemented in WinBUGS software that can be downloaded from the Web and is free and relatively easy to use for the natural resource scientist.

3.4 SUMMARY

In this chapter we presented an introduction to Bayesian hypothesis testing of multiple hypotheses using Bayes factors. Bayes factors are odds ratios representing the change in probability of hypotheses from the prior to posterior distributions. We illustrated these concepts with an example based on binomial data. We also introduced Bayesian decision theory using a medical example for illustration. The medical example can be extrapolated to the case study 1 example of Chapter 1 of determining whether a parameter has exceeded a critical threshold. We concluded by briefly previewing the ideas that will be presented in the next chapter: Bayesian statistical analysis of more complex natural resource datasets using MCMC algorithms that are implemented with WinBUGS software.

PROBLEMS

3.1 A logging company is initiating a habitat restoration plan for their timberlands and monitoring the progress of an endangered species as they implement the plan. The monitoring procedure calls for annual surveys of presence–absence data using simple random sampling to estimate the proportion of the habitat occupied by the endangered species. The objective of the restoration plan is to restore the presence of the species to at least 75% of the timberlands. The first 2 years of survey data resulted in the species being present in 60 and 74 of 100 sites, respectively, independently sampled each year. Assuming a uniform beta prior for the first year in a sequential Bayesian statistical analysis, what are the posterior distributions for each of the first two years of analysis? What are the posterior probabilities of exceeding the 75% threshold each year? What are the Bayes factors for the analysis in each of the 2 years? On the basis of the posterior probabilities of exceeding the threshold and the Bayes factors, what are your conclusions about the status and progress of the owl under the habitat restoration plan?

3.2 A watershed study of the viability of a bird population includes the monitoring of trends in its abundance with bird count data collected independently annually on 50 randomly selected sites. The first 3 years of data collection yield total counts of $\Sigma_i y_i = 30$, 48, and 55, respectively. Assuming certainty of detection and random spatial distribution of the birds, analyze the datasets for each of the 3 years, using gamma conjugate priors with an initial Jeffreys gamma prior $G(\lambda;\alpha_{\text{prior}} = 0.001,\ \beta_{\text{prior}} = 0.001)$ for the first year's data and a sequential Bayesian statistical analysis. Examine the Bayes factors for the first, second, and third year's analyses. What is the degree of evidence in support of the hypothesis H_1 that the mean plot density of the birds is exceeding 1

$$H_1 : \lambda > 1?$$

3.3 A patient has symptoms indicating a 5% chance of having of a malevolent form of cancer. Her physician proposes a DNA scan testing procedure with a sensitivity of 80% and specificity 95%. If the patient takes the test and it is positive, how would that increase the assessment of the patient's odds of having the cancer, if she has the cancer and if she doesn't have the cancer? If the test is negative, how would that decrease the assessment of her odds of having the cancer, if she has the cancer and if she doesn't have the cancer?

3.4 Consider the general medical diagnostics decision problem of Section 3.2 with decision d_1 having Bayes risk $B(d_1) = L_0{}^{+} \cdot (1 - \text{specificity}) \cdot \pi_{\text{prior}}(D_0) + L_1{}^{-} \cdot (1 - \text{sensitivity}) \cdot \pi_{\text{prior}}(D_1)$. Consider the alternative decisions d_2 of no treatment without the test with loss $L = L_1{}^{-}$ if the patient has the disease and 0 otherwise, and d_3 of treatment without the test with loss $L = L_0{}^{+}$ if the patient does not have the disease and 0 otherwise. What are the Bayes risks for these decisions d_2 and d_3? Use Bayes' rule to determine when each decisions is optimal in minimizing risk. What does the Bayes rule suggest about the need for testing under certain circumstances?

3.5 In the medical example above in Problems 3.3 and 3.4, suppose that the cost of the treatment and its side effects is estimated to be $L_0{}^{+} = \$3000$. Suppose also that the opportunity cost of the delay caused by waiting until other symptoms appear if the patient does have cancer is estimated to be $L_1{}^{-} = \$20{,}000$. What is the Bayes risk of the decision d_1 to take the DNA scan and apply the treatment to correct the disease depending on the results of the test? How does this compare with the Bayes risk for the decision d_2 of not taking the test and not applying the corrective treatment? How does this compare with the Bayes risk for the decision d_3 of not taking the test but applying the treatment to correct the disease without taking the test? Use the Bayes rule to determine which decision d_1, d_2, and d_3 is optimal.

3.6 Extrapolate the concepts of Section 3.3 and Problem 3.4 to case study 1 of Chapter 1 illustrated by the critical threshold problem in Section 3.1.1 for estimating the of proportion of nesting Northern Spotted Owl pairs. What are sensitivity and specificity in this context? What are the meanings of the losses $L_0{}^{+}$ and $L_1{}^{-}$ in terms of the value of the restoration of the viability of the owls and the opportunity costs of the harvesting of the timber? What does the Bayes rule suggest about the need for monitoring of the owls under certain circumstances?

4 Bayesian Statistical Inference III: MCMC Algorithms and WinBUGS Software Applications

Machines should work;
people should think.

—IBM Pollyanna Principle

In this chapter we introduce the methods of Markov chain Monte Carlo (MCMC) simulation and provide an overview of the Markov chain theory necessary to appreciate the properties of the MCMC algorithms. We describe the leading MCMC algorithms: Gibbs sampling and the Metropolis–Hastings algorithm, and conclude by presenting WinBUGS applications of the MCMC algorithms for several examples. The examples include the normal model for continuous data and the linear regression model for paired data. For these examples, we compare the Bayesian results with noninformative priors with frequentist results. We also include several other examples, for count datasets using the Poisson model, for count datasets with overdispersion using the negative binomial and mixed-effects models, and for datasets with hierarchical models.

4.1 INTRODUCTION

In Chapters 2 and 3 we introduced some of the principal ideas of Bayesian statistical analysis and described some of the advantages of using this approach for statistical inference with natural resource datasets. We included in this discussion the topic of conjugate closed-form Bayesian analysis solutions for many important natural resource datasets. However, many other important natural resource datasets require models that are more complex, with hierarchical structures, random effects, and

Contemporary Bayesian and Frequentist Statistical Research Methods for Natural Resource Scientists. By Howard B. Stauffer

large numbers of parameters with and without constraints. Closed-form Bayesian analysis solutions for these models may be difficult if not intractable to derive mathematically and may not be available or accessible to the natural resource scientist. Bayesian statistical inference has been significantly limited historically because of this lack of solution to many practical problems.

All this has changed, however, since the mid-1950s with the development of MCMC algorithms and their recent implementation with the Bayesian MCMC software WinBUGS. Bayesian statistical analysis solutions using these algorithms are now generally available and accessible for the natural resource scientist. We will describe the most important of these MCMC algorithms in this chapter and discuss their statistical properties, illustrating with examples in WinBUGS.

4.2 MARKOV CHAIN THEORY

Let's begin by presenting an overview of the theory behind the MCMC methods that provide general Bayesian analysis solutions for statistical models of natural resource datasets. The mathematical details of this theory are beyond the scope of this book. The interested reader, however, is referred to the many excellent references providing more details and examples of these methods, including Metropolis et al. (1953), Hastings (1970), Gemen and Geman (1984), Gelfand and Smith (1990), Gelman (1992), Draper (2000), Carlin and Louis (2000), Congdon (2001), and Gill (2002).

One of the leading problems historically with the Bayesian approach to statistical analysis has been the apparent requirement for the computation of complex integrals that may be intractable. Recall that the formulation of the Bayesian solution in Bayes' Theorem requires computation of the product of the prior times the likelihood of a dataset, based on a model, for the numerator. This part of the computation presents no problem. However, Bayes' Theorem also requires, in the continuous case, computation of the integral of this product with respect to the parameter in the denominator. This integral may be very complex, multidimensional, and intractable to mathematical closed-form solution. Additionally, for multidimensional models with many parameters, the marginal posterior distribution for each parameter must be calculated.

There are many important cases where these integrals can be calculated. Chapter 2 has revealed that the forms of the likelihoods of some models for datasets lend themselves to the use of particular families of prior distributions called **conjugate priors**. Conjugate prior families of distributions provide closed-form solutions with posteriors from the same family of distributions. Furthermore, the posterior solutions are functions of the parameter values used for the priors and statistics computed from the sample dataset. Conjugate priors for several important examples of natural resource datasets were examined, including normal conjugate priors for the normal model with continuous data, gamma conjugate priors for the Poisson model with count data, and beta conjugate priors for the binomial model with binary data. However, models for natural resource datasets are often far more complicated: multidimensional, hierarchical, and intractable to closed-form solution. With the advent of computers, it might seem that a Monte Carlo simulation approach, using computer iteration to approximate solutions by sampling from the posterior distributions, would provide an effective solution. What is particularly interesting about the MCMC simulation approach, however, is that the Monte Carlo samples are not completely independent, but rather are dependent in a Markov chain.

The MCMC methods represent a collection of algorithms prescribing the sampling of parameter values from posterior distributions that circumvent the need to calculate the integrals required in Bayes' Theorem. An MCMC algorithm prescribes a **stochastic process** that samples random variable values from a **state space** (here, a parameter space). The sampling consists of a sequence, or "chain," of values. The stochastic process consists of a **Markov chain**: each value $\theta_i = \theta_i|\theta_{i-1}$ from the parameter space that is sampled depends conditionally only on the previous value θ_{i-1}. In other words, any specified state in this chain is conditionally independent of all previous values except the last one. The process that defines the probabilities of moving from one state to the next one in the chain is called the **transition kernel**. If the state space has only a finite number of values, then the transition kernel can be described by a matrix $[p_{ij}]$, where p_{ij} = the probability of moving from state i to state j in the chain. Typically, however, the parameter spaces will be infinite and continuous, so a specified rule will be required.

Furthermore, the Markov chain must satisfy certain properties in order to converge to a stationary distribution where it will representatively sample the values from the posterior distribution. The Markov chain must be **ergodic**; that is, it must be **irreducible**, **positive Harris recurrent**, and **aperiodic**. Let's describe each of these properties.

A Markov chain is **irreducible** for a subspace $A \subseteq S$ of the state space S if every point or collection of points in the subspace can be reached from every other point or collection of points in the subspace. The idea is to continue sampling every point in the subspace while moving through the chain. For the Bayesian solution, the subspace of importance consists of the **support** of the posterior distribution, the set of parameter values in the parameter space with nonzero posterior probability.

An irreducible Markov chain is **recurrent** for a subspace $A \subseteq S$ if the probability that the chain occupies A infinitely often over unbounded time is nonzero. An irreducible recurrent Markov chain is **positive recurrent** if the mean amount of time [i.e., time between points θ_i and $\theta_j = t(\theta_i, \theta_j)$] required to return to A is bounded. The idea is to be able to continue to sample every point not only in the subspace, the support of the posterior distribution, but also on average within a finite bounded number of iterations. With unbounded continuous state spaces, we need a slightly stricter definition of recurrence, **Harris recurrence**, which guarantees that the probability of visiting A infinitely often in the limit is 1. We refer the reader to the more detailed technical references given earlier for further explanation.

The last requirement is that the irreducible, positive Harris recurrent Markov chain be **aperiodic**. The **period** of a Markov chain is the length of time required to repeat an identical cycle of chain values. The requirement is that the Markov chain be aperiodic, where the only length of time for which the chain repeats some cycle of values is the trivial case of cycle length 1. The requirement ensures that the Markov chain reaches a state where it will sample representatively from the posterior distribution and not just repeat some cycle, even if that cycle includes every value from the subspace.

In summary, the MCMC algorithms produce ergodic Markov chains that are irreducible, positive Harris recurrent, and aperiodic. An ergodic Markov chain converges to a subspace of the parameter space where it proportionally samples the posterior distribution with a stationary distribution. A Bayesian statistical analysis solution can be provided, hence, by utilizing this property and simulating the sampling iteratively on the computer, as specified by the MCMC algorithm. The sampling will require a "burn-in period" until it converges to the stationary distribution in the Markov chain. Sample values in the stationary chain can then be collected for a subset of sufficient size to represent the posterior distribution. Statistics and graphs of the samples provide an empirical summary of the posterior distribution of the parameter space. The statistics include means, medians, modes, percentiles, and credible interval limits. No computation of the integral in Bayes' Theorem is required, and marginal distributions of the parameters can be obtained from the samples. Let's look at the most important MCMC algorithms.

4.3 MCMC ALGORITHMS

Many classical Monte Carlo methods provide Bayesian analysis solutions, such as Monte Carlo integration, rejection sampling, classical numerical integration, and importance sampling (Gill 2002). These methods produce sets of independent simulated values from a desired probability distribution.

The MCMC algorithms, on the other hand, produce chains of simulated values that are mildly dependent. Each simulated value is dependent upon the previous value in the chain. The chains are ergodic, converging to the desired posterior distributions of interest. Summary statistics and graphs of the convergent stationary sampled values provide descriptions of the posterior distributions. The most important MCMC algorithms are Gibbs sampling and the Metropolis–Hastings algorithm, which are described next.

4.3.1 Gibbs Sampling

The Gibbs sampler (Geman and Geman 1984, Gelfand and Smith 1990) is the most widely used MCMC algorithm. Although it is less flexible than the Metropolis–Hastings algorithm, it is conceptually simpler. Given the parameter vector $\theta = (\theta_1, \theta_2, \ldots, \theta_k)$ and the data $y = \{y_1, y_2, \ldots, y_n\}$, it assumes that the posterior $\pi(\theta|y)$ can be characterized by its complete set of full conditional distributions $\pi(\theta_i|y, \theta_{-i})$, for $i = 1, \ldots, k$, where θ_{-i} refers to the vector θ without the θ_i component. The conditional distributions $\pi(\theta_i|y, \theta_{-i})$ will often be easier to express in closed form than will either $\pi(\theta|y)$ or $\pi(\theta_i|y)$. The **Gibbs sampling algorithm** proceeds as follows:

1. Choose starting values $\theta^{[0]} = (\theta_1^{[0]}, \theta_2^{[0]}, \ldots, \theta_k^{[0]})$.

2. At the ith iteration, complete the single cycle by drawing values from the k distributions

$$\begin{aligned}
\theta_1^{[i]} &\sim \pi\left(\theta_1|\theta_2^{[i-1]}, \theta_3^{[i-1]}, \ldots, \theta_{k-1}^{[i-1]}, \theta_k^{[i-1]}\right), \\
\theta_2^{[i]} &\sim \pi\left(\theta_2|\theta_1^{[i]}, \theta_3^{[i-1]}, \ldots, \theta_{k-1}^{[i-1]}, \theta_k^{[i-1]}\right), \\
\theta_3^{[i]} &\sim \pi\left([\theta_3|\theta_1^{[i]}, \theta_2^{[i]}, \theta_4^{[i-1]}, \ldots, \theta_{k-1}^{[i-1]}, \theta_k^{[i-1]}\right), \\
&\vdots \\
\theta_{k-1}^{[i]} &\sim \pi\left(\theta_{k-1}|\theta_1^{[i]}, \theta_2^{[i]}, \ldots, \theta_{k-2}^{[i]}, \theta_k^{[i-1]}\right), \\
\theta_k^{[i]} &\sim \pi\left(\theta_k|\theta_1^{[i]}, \theta_2^{[i]}, \ldots, \theta_{k-2}^{[i]}, \theta_{k-1}^{[i]}\right).
\end{aligned}$$

3. Increment i and repeat until convergence. It can be shown that Gibbs sampling produces an ergodic Markov chain (i.e., irreducible, positive Harris recurrent, and aperiodic) that converges to a limiting distribution that is the posterior distribution $\pi(\theta|y)$ of the parameter vector. The objective of Gibbs sampling is to move toward and then around this distribution, providing representative samples throughout the posterior distribution.

Although the full set of conditional distributions is required in order to run the Gibbs sampling algorithm, many examples lend themselves to this approach. Hierarchical models, for instance, have natural conditioning structures in their formulations. The interested reader may find examples in Gill (2002, pp. 26–29, 313–317) and Congdon (2001, pp. 216–217). Casella and George (1992) provide a very clear basic introduction to Gibbs sampling in a paper originally entitled "Gibbs for kids." It can be shown that Gibbs sampling is actually a special case of the Metropolis–Hastings algorithm.

Let's conclude this section with an example of Gibbs sampling. We consider the problem of estimating population size of an animal population with mark–recapture estimation using Gibbs sampling as illustrated in Manly (1997). Assume that the population is **closed** during a short sampling period, without births and deaths, and without immigration or emigration. Suppose that $M = 50$ animals are marked and released into the population and allowed to mix freely with U unmarked animals. A random sample is then taken, assuming a fixed probability p of each animal being captured, with $m = 25$ marked animals and $u = 10$ unmarked animals. We will estimate U and p using Gibbs sampling, assuming that U can vary between 10 and 100 and that p can vary between 0.0 and 1.0 in increments of 0.01. We also assume that the priors for U and p are noninformative uniform distributions, hence constant

$$\pi_{\text{prior}}(U, p) = C.$$

Since $m \sim B(m; p, M)$ and $u \sim B(u; p, U)$ are both binomially distributed, we have a probability distribution P for m and u, conditional to U and p, given by

$$\begin{aligned} P(m, u; U, p) &= B(m; p, M) \cdot B(u; p, U) \\ &= {}^{M}C_{m} \cdot p^{m} \cdot (1-p)^{M-m} \cdot {}^{U}C_{u} \cdot p^{u} \cdot (1-p)^{U-u} \end{aligned}$$

and the likelihood–prior function for U and p given by

$$P(U, p;\ m, u) = C \cdot {}^{M}C_{m} \cdot p^{m} \cdot (1-p)^{M-m} \cdot {}^{U}C_{u} \cdot p^{u} \cdot (1-p)^{U-u}.$$

For Gibbs sampling, therefore, the conditional distribution of U, given p, is

$$P(U; p, m, u) \propto {}^{U}C_{u} \cdot (1-p)^{U},$$

and the conditional distribution of p, given U, is

$$P(p; U, m, u) \propto p^{m+u} \cdot (1-p)^{M-m+U-u}.$$

The S-Plus or R code for the Gibbs sampling uses the standardized forms of these expressions in an iterative process with n iterations, starting with initial values $U_1 = 55$ and $p_1 = 0.50$, as indicated in Fig. 4.1a. Figures 4.1b and 4.1c illustrate the trajectories of (U, p) for the first 100 and 1000 iterations. The means of the samples for the first 1000 iterations are $\bar{U} = 21.931$ and $\bar{p} = 0.491$.

4.3.2 The Metropolis–Hastings Algorithm

The groundbreaking event in the development of the MCMC algorithms occurred in 1953 when Nicholas Metropolis and his colleagues Arianna Rosenbluth, Marshall Rosenbluth, Augusta Teller, and Edward Teller (developer of the hydrogen bomb) published their algorithm in the *Journal of Chemical Physics*. Metropolis et al. (1953) were interested in the potential positions of all molecules in an enclosure. Their idea was, rather than try to track and calculate future positions, to establish a molecule movement criterion and simulate a series of potential positions, to determine probabilistically where the molecules were. Their interest was focused at that time primarily on nuclear physics. However, their algorithm was to have more far-reaching application. Concurrently, throughout the 1950s and 1960s, mathematicians were developing the foundations of ergodic Markov chain theory that explain the properties of the simulated series of positions in the Metropolis algorithm.

In 1970 Hastings refined and generalized the Metropolis algorithm, showing that a reversibility condition could be substituted for a symmetry requirement. Hastings' paper in *Biometrika* (Hastings 1970), along with a paper by Peskun (1973), helped introduce these ideas to the statistics community. Two papers by Geman and Geman (1984) and Gelfand and Smith (1990) later clarified the usefulness of

Figure 4.1. Trajectories of the parameters U and p from Gibbs sampling for mark–recapture analysis of a closed animal population with $M = 50$ marked animals, U unmarked animals, $m = 25$ recaptured marked animals, and $u = 10$ captured marked animals. (**a**) Gibbs sampling program code in S Plus or R. (**b**) First 100 iterations. (**c**) First 1000 iterations.

(**a**)

```
function(n)
{
u <- 10; m <- 25; M <- 50
U. <- 10:100; p. <- seq(0,1,0.01)
U <- rep(NA,n); U[1] <- 55
p <- rep(NA,n); p[1] <- 0.50
for (i in 1:(n-1))
   {
   prob1 <- dbinom(u,U.,p[i])*(1-p[i])^u/p[i]^u
   prob1 <- prob1/sum(prob1)
   U[i+1] <- sample(U.,size=1,prob=prob1)
   prob2 <- p.^(m+u)*(1-p.)^(M-m+U[i+1]-u)
   prob2 <- prob2/sum(prob2)
   p[i+1] <- sample(p.,size=1,prob=prob2)
   }
return(U,p)
}
```

(**b**)

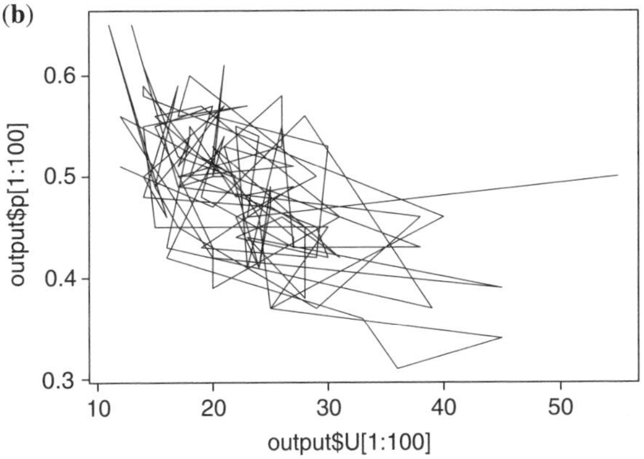

(**c**)

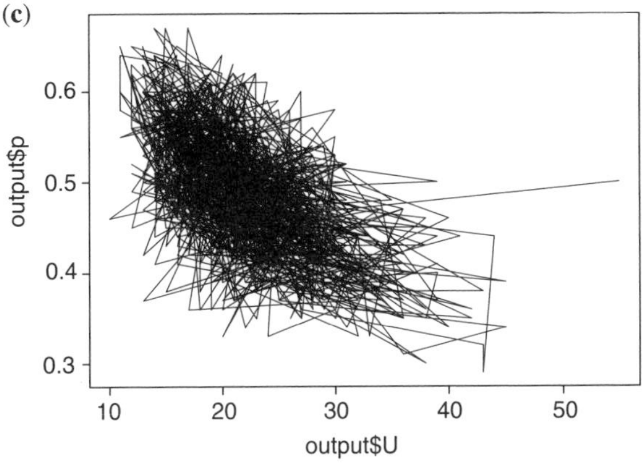

Gibbs sampling for formulating Markov chains to estimate posterior distributions. Gibbs sampling is a special case of Metropolis–Hastings, and Metropolis is a special case of Hastings. These algorithms, along with the explosion of computing power in the 1990s, provide the practical tools for solving realistically formulated complex general problems in natural resource science with Bayesian statistical analysis.

In simplest form, for a parameter θ with posterior $\pi(\theta)$ and data $y = \{y_1, y_2, \ldots, y_n\}$, the **Metropolis–Hastings algorithm** is as follows:

1. Choose a starting value θ_0.
2. At the ith iteration, given θ_{i-1},
   a. choose a candidate value θ_i' from a **proposal distribution** $q_i(\theta_i'|\theta_{i-1})$. This proposal distribution q_i must have a support that includes the support of the posterior distribution $\pi(\theta)$.
   b. Define the acceptance ratio

$$r(\theta_i'|\theta_{i-1}) = \frac{\pi(\theta_i')/q_i(\theta_i'|\theta_{i-1})}{\pi(\theta_{i-1})/q_i(\theta_{i-1}|\theta_i')}$$

   c. Set

$$\theta_i = \begin{cases} \theta_i' & \text{with probability } \min(r(\theta_i'|\theta_{i-1}), 1) \\ \theta_{i-1} & \text{with probability } 1 - \min(r(\theta_i'|\theta_{i-1}, 1) \end{cases}$$

3. Increment i, go to step 2, and repeat until convergence occurs.

Note that the Metropolis–Hastings algorithm does not require movement for each iteration as does Gibbs sampling. The posterior distribution $\pi(\theta)$ is required only up to a proportionality constant since the constant would be canceled in the ratio shown above. So the Bayesian prior–likelihood function whose scaled value equals the posterior may be used in place of the posterior, hence circumventing the problem of scaling with integration in the denominator in Bayes' Theorem. The proposal distribution must cover the nonzero support portion of the domain of the posterior distribution. The divisors in the numerator and denominator of the ratio in the Metropolis–Hastings algorithm adjust for the probabilities of selection from the proposal distribution. The original theorem of Metropolis et al. (1953) imposed symmetry constraints, requiring the two conditionals to be equal: $q_i(\theta_i'|\theta_{i-1}) = q_i(\theta_{i-1}|\theta_i')$. Hastings determined later that this symmetry constraint was unnecessary using the acceptance ratio in step 2b above that could be simplified to the ratio

$$r(\theta_i'|\theta_{i-1}) = \frac{\pi(\theta_i')}{\pi(\theta_{i-1})}$$

in the Metropolis algorithm with the symmetry condition. A convenient symmetric proposal distribution to use is the normal distribution with center at the current value θ_{i-1} and a variance $s^2_{\theta_{i-1}}$ adjusted periodically using the rejection rates of θ_i' determined from previous iterations (with $s^2_{\theta_0} = 1$). So, to apply the special case of the Metropolis algorithm, use a symmetric proposal distribution to generate candidate values for the posterior distribution and the ratio of the posterior distribution values to decide whether to accept candidates θ_i'. Note that it is important to give candidates with low posterior distribution values at least a chance to be selected, to sample representatively from the entire posterior distribution, driven by the probabilities.

As with Gibbs sampling, Metropolis–Hastings sampling produces an ergodic Markov chain that converges to a limiting distribution that is the posterior distribution $\pi(\theta|y)$ of the parameter vector. Metropolis–Hastings sampling moves toward and then around the posterior distribution, providing samples with probabilities proportional to that distribution. For an example of Metropolis–Hastings sampling (actually Metropolis sampling), consider the case of normally distributed data x, with mean mu and known standard deviation `sigma.model`. We will assume a normally distributed prior with mean `mu.prior` and standard deviation `sigma.prior`, and a normally distributed proposal distribution with standard deviation `sigma.proposal`. An analysis of the model parameter `mu`, with initial value `mu.start`, with Metropolis–Hastings samples from `begin` to `end`, is illustrated with S-Plus or R code in Fig. 4.2.

4.4 WinBUGS APPLICATIONS

Although MCMC algorithms provide the required procedure for Bayesian analysis of complex statistical models, computer software furnishes the key to practical application for the natural resource scientist. WinBUGS can be downloaded as freeware from the Web at http://www.mrc-bsu.cam.ac.uk/bugs/. It is reasonably easy to use and comes with a manual and many helpful examples. WinBUGS, the Windows version of BUGS (Bayesian inference Using Gibbs Sampling), was developed as part of a statistical research project at the Medical Research Unit (MRC), Biostatistics Unit of Britain, the United Kingdom Government, starting in 1989, and is now developed jointly with the Imperial College School of Medicine at St. Mary's, London (Spiegelhalter et al. 2001). The Website includes the following: the earlier version, Classic BUGS; a spatial statistics version, GeoBUGS; a convergence output and diagnostic analysis module, CODA; and an e-mail discussion list address. The Classic BUGS program uses a command-line interface with an S-Plus or R text-based model description, whereas WinBUGS has the added option of a graphical user interface called DoodleBUGS.

The novice user can become oriented to WinBUGS by downloading the software, reading the abbreviated manual, and trying some of the examples. The user should be aware of the possible dangers of misleading output from MCMC sampling due to lack of convergence and check the error diagnostics using CODA. Multiple starting values for the iterated chains are recommended, as well as a burn-in period to discard initial chain values that may not have converged to the posterior distribution.

The general procedure to use with WinBUGS programming is as follows:

1. Create a WinBUGS program consisting of three sections, with program code, data, and initial values.

Figure 4.2. Example of Metropolis–Hasting sampling of the normal distribution parameter mu with normal data $x \sim$ N(mu, sigma.model), prior distribution N(mu.prior, sigma.prior), proposal distribution N(, sigma.proposal), starting value mu.start, and posterior samples from begin to end. (**a**) S-Plus or R program code. (**b**) Histogram of posterior distribution samples for mu with normally distributed data $x \sim$ N(mu = 15, sigma.model = 2) and sample size $n = 30$, prior distribution N(mu.prior = 10, sigma.prior = 2), proposal distribution N(,sigma.proposal = 4), starting value mu.start = 10, and posterior samples from begin = 500 to end = 10000, with 10,000 interations.

(**a**)

```
function(mu.prior,sigma.prior,sigma.model,x,sigma.proposal,
   mu.start,begin,end) # MCMC algorithm / Metropolis–Hastings
{
mu <- rep(NA,end); mu[1] <- mu.start
j <- 1; exit <- end-1
for(i in 1:exit)
   {
   likelihood.mu <- 1
   for(k in 1:length(x))
      {likelihood.mu<-likelihood.mu*(1/(sigma.model*sqrt(2*pi)))*
         exp(-(x[k]-mu[i])^2/(2*sigma.model^2))}
   prior.mu <- dnorm(mu[i],mu.prior,sigma.prior)
   p.mu <- likelihood.mu*prior.mu
   j <- j + 1
   repeat
      {
      candidate <- rnorm(1,mu[i],sigma.proposal)
      likelihood.candidate <- 1
      for(k in 1:length(x))
         {likelihood.candidate <-likelihood.candidate*
            (1/(sigma.model*sqrt(2*pi)))*
            exp(-(x[k]-candidate)^2/(2*sigma.model^2))}
      prior.candidate <-dnorm(candidate,mu.prior,sigma.prior)
      p.candidate <- likelihood.candidate*prior.candidate
      r <- p.candidate/p.mu
      if(r >= 1)
         {
         mu[i+1] <- candidate
         break
         }
      if(r < 1 & runif(1,0,1) <= r)
         {
         mu[i+1] <- candidate
         break
         }
      j <- j+1
      next
      }
   rejection.rate <- 1-end/j
   }
return(mu[begin:end],rejection.rate)
}
```

Figure 4.2. *Continued.*

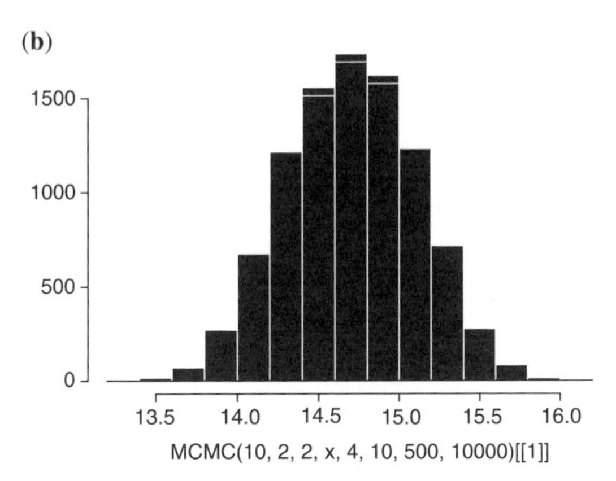

2. Use the `>Model >Specification Tool` menu option (Fig. 4.3) to (a) check the model, (b) load the data, (c) specify the number of chains and compile, and (d) load and generate initial values for the specified number of chains. Using these operations, WinBUGS checks on the syntactical correctness of the program code, data, and initial values. Multiple chains can be generated using initial values spread throughout the parameter range to check for convergence. To execute each step in this process, begin by blocking the first string of characters, or word, in the appropriate location in the program, the model program code, the data, and the initial values sections, respectively. After executing each step, look at the lower left corner of the window for a message indicating the success or failure of the step.
3. Use the `>Model >Update Tool` menu option (Fig. 4.4) to indicate a specified number of posterior sample updates, that are refreshed, thinned, overrelaxed, and adapted, as requested. Refreshing displays the samples on a trace graph periodically as requested. Refreshing every sample (i.e., `1`) provides a complete picture but slows down the processing. Refreshing every 10th sample (i.e., `10`) or every 100th sample (i.e., `100`) is more common. Thinning stores every kth sample only, as specified, to reduce the autocorrelation between the samples and reduce the storage requirements. Overrelaxing produces multiple samples per iteration and chooses the one that is most negatively correlated with the current value, to reduce the autocorrelation between the selected samples. Adapting activates an initial adaptive phase for the tuning of some of the parameters used by Metropolis sampling, such as the standard deviation of a normal proposal distribution. The summary statistics do not include samples collected during this initial adaptive phase. The update button can be turned on and off at the user's discretion, to begin, pause, and resume the MCMC sampling process.

4. Use the `>Inference >Sample Monitor Tool` menu option (Fig. 4.5) to specify the nodes of interest, sampled with specified beginning and end, and thinned to select every *k*th sample for output summary. Select a burn-in period at the beginning to allow the MCMC process to converge before collecting samples. Before executing the program, use the asterisk * to select all specified nodes, and select trace to view dynamical graphical output. At the end of the simulation run, output of the simulated MCMC parameter sample values

Figure 4.3. WinBUGS implementation: program code and `>Model >Specification Tool` windows. (**a**) Program code with `>Model >Specification Tool` dialog box. (**b**) `>Model >Specification Tool` dialog box.

(**a**)

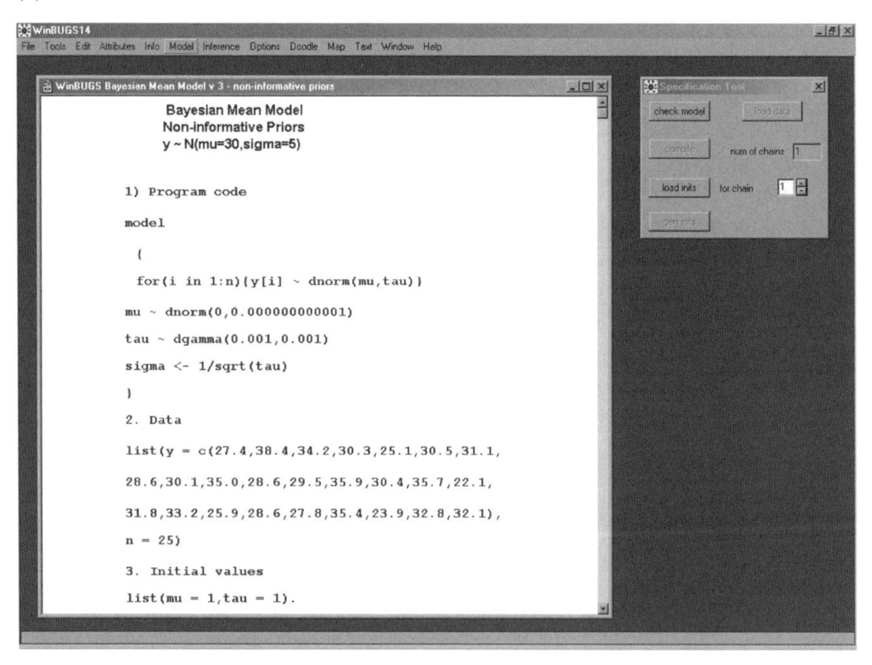

(**b**)

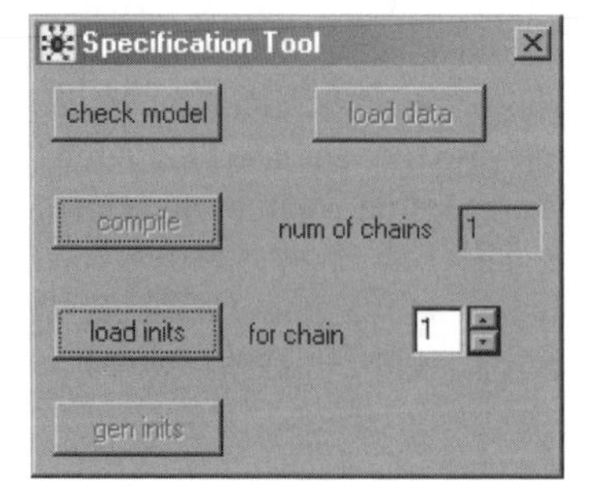

can be displayed as follows: (a) density, (b) statistics, (c) quantiles, (d) history, (e) autocorrelations, (f) Brooks–Gelman–Rubin diagnostic convergence statistics, and (g) output for CODA.

5. Another important option to activate in the menu options is the `>DIC` (Fig. 4.6), which calculates the deviance information criterion, to compare models with the Bayesian equivalent to Akaike's information criterion (AIC) (see Chapter 1). Set the DIC, once it activates at the beginning of the MCMC run. The model with the lowest DIC value is the model that best fits the sample dataset.

Figure 4.4. WinBUGS implementation: program code, `>Model >specification Tool`, and `>Model >Update Tool` windows. (**a**) Program code with `Model >Specification Tool` and `>Model >Update Tool` dialog boxes. (**b**) `>Model >Update Tool` dialog boxes.

(**a**)

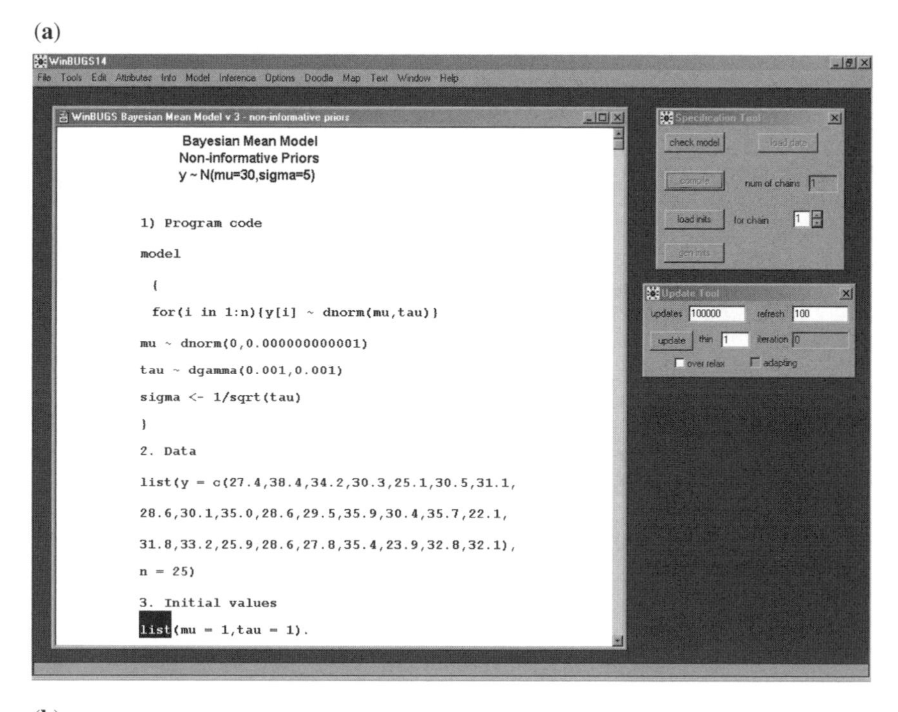

(**b**)

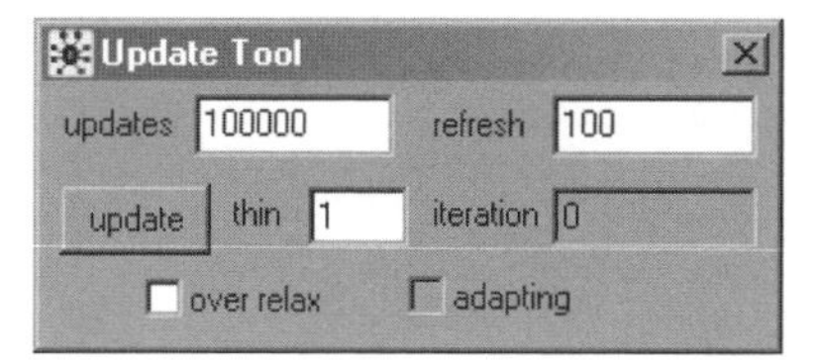

Figure 4.5. WinBUGS implementation: program code, `>Model >Specification Tool`, `>Model >Update Tool`, and `>Inference >Sample Monitor Tool` windows. **(a)** Program code with `>Model >Specification Tool`, `>Model >Update Tool`, and `>Inference >Sample Monitor Tool` dialog boxes. **(b)** `>Inference >Sample Monitor Tool` dialog box.

(a)

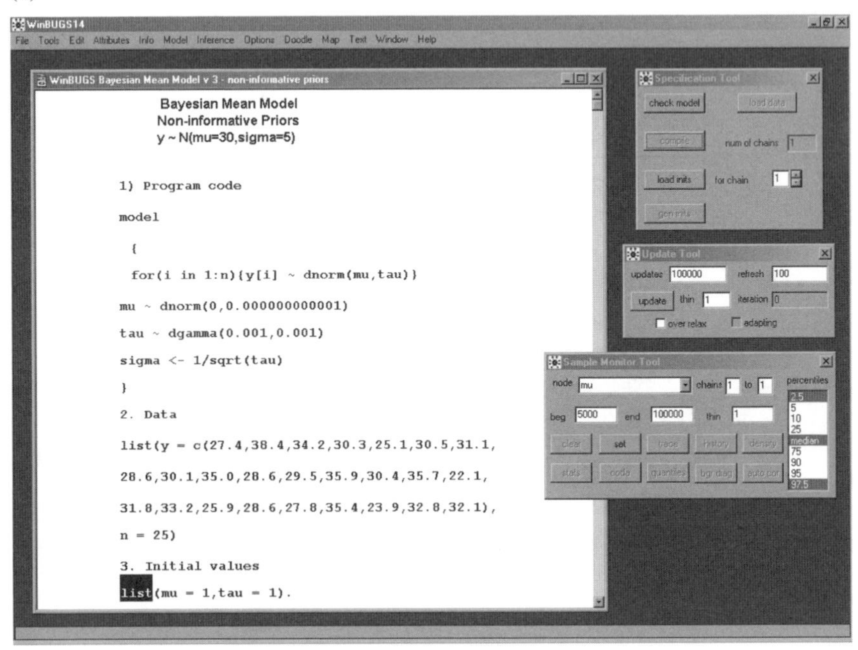

(b)

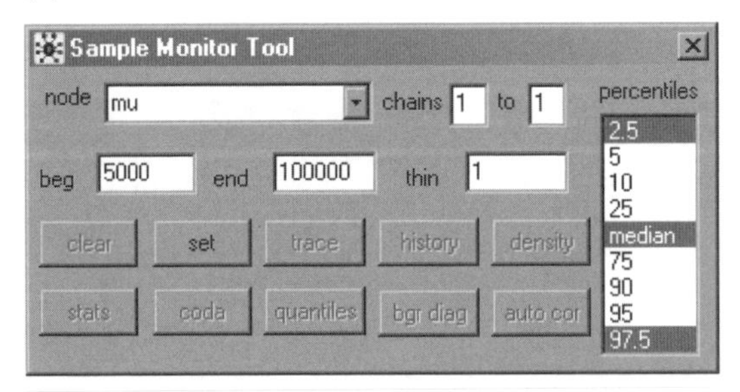

We will illustrate the use of WinBUGS with some introductory examples.

4.4.1 The Normal Mean Model for Continuous Data

Let's conduct a Bayesian statistical analysis with WinBUGS of a simple normal mean model for continuous data. The sample dataset consists of $y \sim N(30, 5)$ from a normal

Figure 4.6. WinBUGS implementation: program code, `>Model >specification Tool`, `>Model >Update Tool`, `>Inference >Sample Monitor Tool`, and `>Inference >DIC` windows. (**a**) Program code with `>Model >Specification Tool`, `>Model >Update Tool`, `>Inference >Sample Monitor Tool`, and `>Inference >DIC Tool` dialog boxes. (**b**) `Inference >DIC Tool` dialog box.

(**a**)

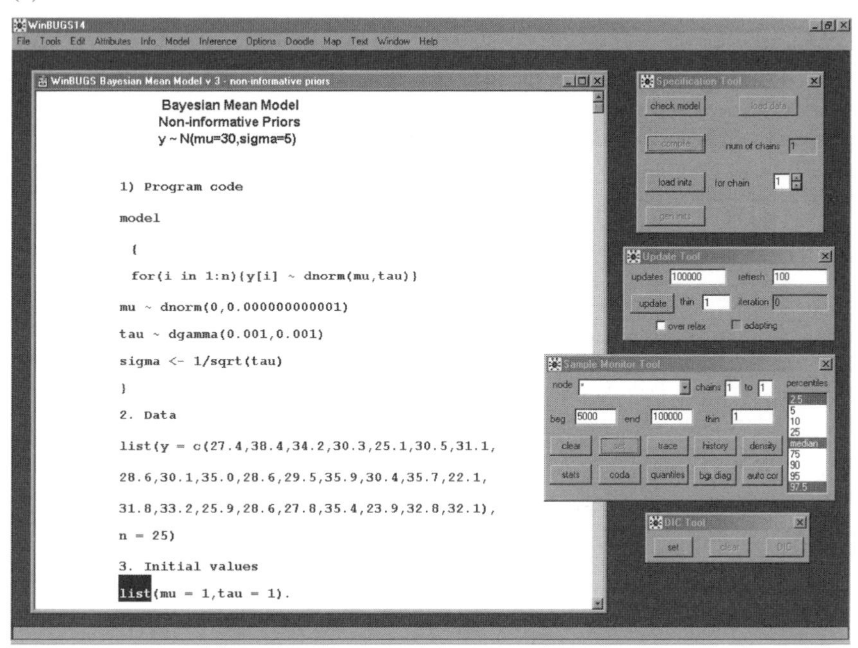

(**b**)

population with mean $\mu = 30$ and standard deviation $\sigma = 5$, of sample size $n = 25$. The WinBUGS code is given in Fig. 4.7a. The program consists of program code, data, and initial values. WinBUGS requires use of the precision parameter tau $=$ $\tau = 1/\sigma^2$ in the program code as a parameter for the normal distribution. Recall that precision is the inverse of the variance. The derived parameter sigma $= \sigma = 1/\tau^{1/2}$ is calculated for each MCMC iteration. Noninformative, vague, approximately "flat" priors are used for the parameters, a normal distribution for the mean parameter μ with precision 10^{-12} (hence standard deviation $\sigma = 10^6$), and the approximate Jeffreys transformation-invariant gamma prior $G(0.001, 0.001)$ for the nonnegative precision parameter τ (Fig. 4.7b). The parameters are initialized at $\mu = 1$ and $\tau = 1$. We blocked the words `model` in the program section, `list` in the data section, and `list` in the initial values section to run the (1) `check`, (2) `load data` and `compile`, and (3) `load initial values` steps in the model specification phase of the run. We requested 10,000 iterations, with a burn-in period of 1000 iterations. Recommended burn-in periods range from 500 to 5000 iterations, and total numbers of iterations range from 5000 to 100,000, depending on the complexity of the model.

Figure 4.7. Graphs and tables for the Bayesian statistical analysis of the normal mean model for the dataset $y \sim N(30, 5)$ with $n = 25$ using WinBUGS. (**a**) Program code for the normal mean model and dataset y. (**b**) The Jeffreys noninformative gamma prior $G(0.001,0.001)$ for τ. (**c**) Dynamic trace output of initial posterior samples for the parameters. (**d**) Posterior distribution density functions for the parameters, from MCMC samples. (**e**) Posterior distribution statistics for the parameters, estimated from MCMC samples. (**f**) DIC statistics for the normal mean model.

(**a**)

```
1) Program code
model
{
for(i in 1:n){y[i] ~ dnorm(mu,tau)}
mu ~ dnorm(0,0.000000000001)
tau ~ dgamma(0.001,0.001)
sigma <- 1/sqrt(tau)
}

2. Data
list(y = c(27.4,38.4,34.2,30.3,25.1,30.5,31.1,
28.6,30.1,35.0,28.6,29.5,35.9,30.4,35.7,22.1,
31.8,33.2,25.9,28.6,27.8,35.4,23.9,32.8,32.1),
n = 25)

3. Initial values
list(mu = 1,tau = 1)
```

(**b**)

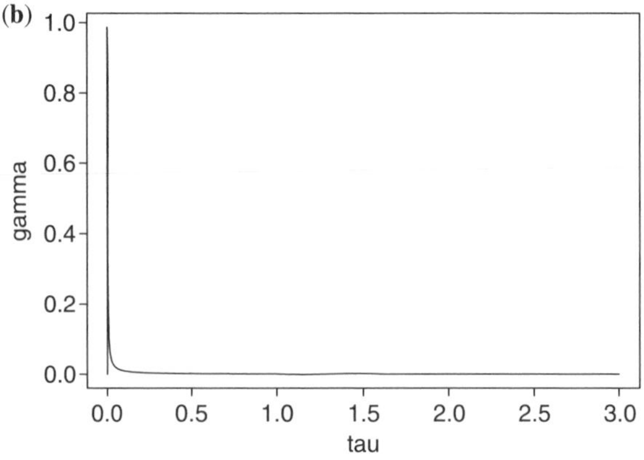

Figure 4.7. *Continued.*

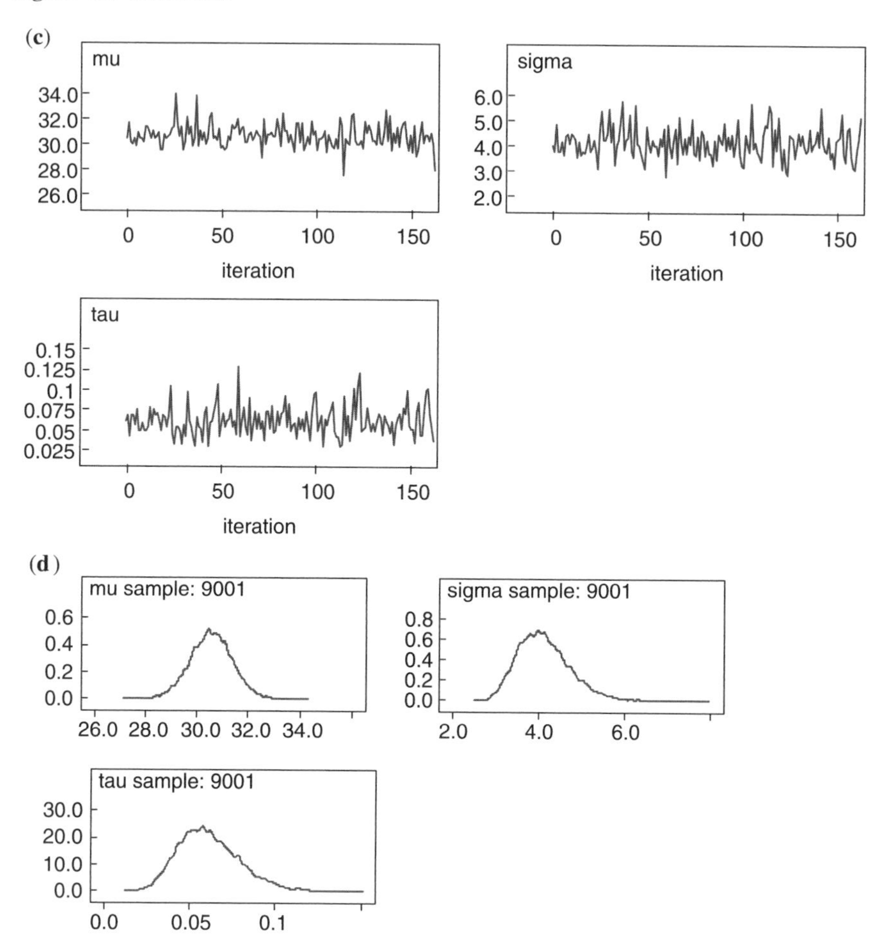

(e)

node	mean	sd	MC error	2.5%	median	97.5%	start	sample
mu	30.57	0.819	0.008845	28.97	30.57	32.19	1000	9001
sigma	4.14	0.6239	0.006665	3.14	4.074	5.55	1000	9001
tau	0.062	0.0179	1.907E-4	0.0324	0.0602	0.1016	1000	9001

(f) null standardized deviance: using means

	Dbar	Dhat	DIC	pD
y	141.468	139.454	143.482	2.014
total	141.468	139.454	143.482	2.014

The dynamic trace output at the beginning of the simulation reveals that MCMC immediately jumped to the range of the parameter samples around $\mu = 30$, $\sigma = 5$, and $\tau = 1/\sigma^2 = 1/5^2 = 0.04$, away from the initial values $\mu = 1$ and $\tau = 1$ (Fig. 4.7c). The posterior distribution density functions for μ and σ, graphed from the sampled MCMC simulated values between iterations 1000 and 10,000, along with the output statistics, are illustrated in Figs. 4.7d–4.7f. The statistics include mean, median, and 95% credible interval estimates (with limits at the 2.5% and 97.5% percentile points) of 30.57, 30.57, and $[28.97, 32.19]$ for the μ mean parameter samples and 4.14, 4.074, and $[3.14, 5.55]$ for the σ sigma parameter samples, respectively. The DIC = 143.482 could be used to compare this normal mean model with other models analyzed with this dataset.

The frequentist statistics estimates from S-Plus and R for this sample dataset are $\hat{\mu} = 30.576, \widehat{\text{median}} = 30.4, \hat{\sigma} = 4.01552$, and 95% confidence interval (CI) = $[28.91847, 32.23353]$ for the mean, in close agreement with the Bayesian statistics for the model with noninformative priors, as we expect. Although the estimates using both approaches to statistical analysis are in close agreement, the two interpretations are, of course, quite different. We could have used informative priors on the basis of previous analysis posterior results if available, or use these posterior results as priors for analysis of additional sample datasets in the future.

4.4.2 Models for Count Data: The Poisson Model, Poisson–Gamma Negative Binomial Model, and Overdispersed Mixed-Effects Poisson Model

Suppose that the sample consists of count data y for a randomly dispersed population. The normal mean model of Section 4.4.1 can be readily adapted to a Poisson model, $y \sim \text{Pois}(\lambda)$, with the nonnegative mean parameter $\lambda > 0$, using the WinBUGS program code given in Fig. 4.8a. For the Poisson model, with count data that are randomly dispersed, the variance parameter is equal to the mean parameter: $\sigma^2 = \mu$.

For count data that are overdispersed, describing an aggregated or clustered population, the variance parameter is greater than the mean parameter: $\sigma^2 > \mu$. An overdispersed population can be modeled by using either a negative binomial model or a Poisson model with dispersion incorporated as a random effect. The negative binomial model is a hierarchical model, a Poisson–gamma distribution of the random variable $y \sim \text{Pois}(\lambda)$ that is Poisson-distributed with a mean $\lambda \sim G(\alpha, \beta)$ that is gamma-distributed with parameters $\alpha > 0$ and $\beta > 0$ (Hilborn and Mangel 1997). The WinBUGS program code for this model is given in Fig. 4.8b. The parameters of this hierarchical model are the shape and scale **hyperparameters** α and β that are "estimated" with posteriors, using Jeffreys noninformative gamma priors. It can be shown that the mean and variance parameters of the negative binomial distribution are given by $\mu = \alpha/\beta$ and $\sigma^2 = (\alpha/\beta + \alpha/\beta^2)$, so these parameters can also be programmed in the WinBUGS code as derived parameters whose posteriors can be sampled.

The WinBUGS code for a model incorporating overdispersion as a random effect $e \sim N(0, \sigma = 1/\sqrt{\tau})$ in a mixed-effects Poisson model is given in Fig. 4.8c. Note that

Figure 4.8. Program code for the Poisson and negative binomial models using WinBUGS. (**a**) Program code for the Poisson model. (**b**) Program code for the negative binomial model using the Poisson–gamma distrubition. (**c**) Program code for the negative binomial model using random effects. (**d**) Program code for the negative binomial model with multiplicative random effects.

(**a**)
```
model
{
for(i in 1:n){y[i] ~ dpois(lambda)}
lambda ~ dgamma(0.001,0.001)
}
```
(**b**)
```
model
{
for(i in 1:n)
{
y[i] ~ dpois(lambda[i])
lambda[i] ~ dgamma(alpha,beta)
}
alpha ~ dgamma(0.001,0.001)
beta ~ dgamma(0.001,0.001)
mu <- alpha/beta
variance <- alpha/beta+alpha/(beta*beta)
}
```
(**c**)
```
model
{
for(i in 1:n)
{
y[i] ~ dpois(lambda[i])
lambda[i] <- max(0.000001,mu+e[i])
e[i] ~ dnorm(0,tau)
}
mu ~ dgamma(0.001,0.001)
tau ~ dgamma(0.001,0.001)
sigma <- 1/sqrt(tau)
}
```
(**d**)
```
model
{
for(i in 1:n)
{
y[i] ~ dpois(lambda[i])
log(lambda[i]) <- mu+e[i]
e[i] ~ dnorm(0,tau)
}
mu ~ dgamma(0.001,0.001)
tau ~ dgamma(0.001,0.001)
sigma <- 1/sqrt(tau)
}
```

the mean λ_i of the count data must be forced to be positive in the WinBUGS code to enable the program to execute.

There are other ways of modeling overdispersion in a population in WinBUGS, such as with a multiplicative random effect and a log transformation (see Fig. 4.8d). WinBUGS now also includes the negative binomial model for data $y \sim$ dnegbin (p, r) with parameters p and r as one of its model options. We will leave it to interested readers to explore these alternative approaches (Hilborn and Mangel 1997).

4.4.3 The Linear Regression Model

Next, let's conduct a Bayesian statistical analysis with WinBUGS of a linear regression model using simulated sample data $x \sim \mathrm{Unif}(0, 200)$, $y \sim 100 + 1.2^{*}x + N(0, 40)$, with sample size $n = 30$ (Fig. 4.9a). The linear regression model "reality" is given by

$$y_i = 100 + 1.2 * x_i + e_i$$

with residual errors $e_i \sim N(0, 40)$ that are normally distributed with mean $\mu = 0$ and standard deviation $\sigma = 40$. For the simulated population, $\beta_0 = 100$, $\beta_1 = 1.2$, and $\sigma = 40$. The WinBUGS code is given in Fig. 4.9b. We use noninformative priors for the parameters, "flat" normal distributions for β_0 and β_1 with precision 10^{-12} and standard deviation $= 10^6$, and the approximate Jeffreys gamma prior $G(0.001, 0.001)$ for the nonnegative precision parameter τ. The dynamic trace output at the beginning of the simulation reveals that MCMC quickly jumps to the realistic range for the parameter samples centered around $\beta_0 = 100$, $\beta_1 = 1.2$, and $\sigma = 40$, far from the initial values $\beta_0 = 0$, $\beta_1 = 0$, and $\sigma = 1$ (Fig. 4.9c). The posterior distribution density functions for β_0, β_1, and σ, graphed from the simulated MCMC samples between iterations 1000 and 10,000, along with the output statistics, are illustrated in Figs. 4.9d–4.9f. The parameter β_0, β_1, and σ posterior sample statistics are as follows: means (and standard errors) 94.56 (14.1), 1.264 (0.12), and 36.64 (5.17); medians 94.62, 1.26, and 36.08; and 95% credible intervals [66.02, 122.5], [1.02, 1.51], and [28.25, 48.38], respectively. The DIC = 303.818 could be compared with other models analyzed with this dataset.

The S-Plus and R frequentist estimates are reasonably similar to the Bayesian output for the linear regression model with noninformative priors, as expected: $\hat{\beta}_0 = 94.7946$, $\hat{\beta}_1 = 1.2626$, and $\hat{\sigma} = 35.68$.

These examples illustrate the power of the WinBUGS approach to statistical modeling. Data can be easily and accurately modeled, using realistic assumptions. The focus for the natural resource scientist can be on the model specifications rather than the development of methods for analytical solution. WinBUGS is particularly suited for hierarchical structures that are so often present in biological systems, as illustrated by the overdispersed Poisson–gamma model in Section 4.4.2. WinBUGS has a large collection of examples in its `>help` menu that illustrate a wide range of important models for data analysis, including generalized linear models (GLMs) and mixed-effects models. The interested reader is encouraged to

Figure 4.9. Graphs and tables for the Bayesian statistical analysis of the linear regression model for the dataset (x,y), with $x \sim$ Unif (0, 200) and $y \sim 100 + 1.2^{*}x + N(0, 40)$ with $n = 30$ for the linear regression model using WinBUGS. (**a**) Scatterplot of the dataset (x, y). (**b**) Program code. (**c**) Initial dynamic trace output of posterior samples for the parameters. (**d**) Posterior distribution density functions for the parameters, estimated from MCMC samples. (**e**) Posterior distribution statistics for the parameters, estimated from MCMC samples. (**f**) Statistics for linear regression conditional means, μ[i] terms estimated from MCMC samples. (**g**) DIC statistics for the linear regression model, estimated from MCMC.

(**a**)

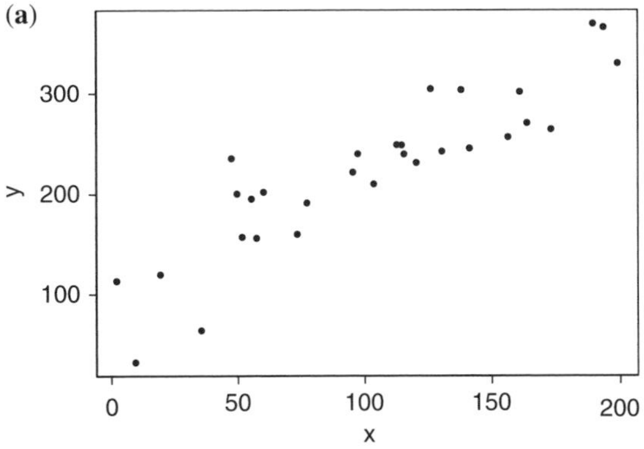

(**b**)

```
1) Program code
model
{
for(i in 1:n)
{
y[i] ~ dnorm(mu[i],tau)
mu[i] <- beta0 + beta1*x[i]
}
beta0 ~ dnorm(0,0.000000000001)
beta1 ~ dnorm(0,0.000000000001)
tau ~ dgamma(0.001,0.001)
sigma <- 1/sqrt(tau)
}

2. Data
list(x = c(47.6,103.7,57.4,19.2,163.9,9.4,35.7,
114.7,189.5,55.3,60.1,156.4,161.0,115.7,95.5,
120.5,138.2,97.4,126.1,73.4,51.7,193.5,130.6,
112.8,141.4, 2.1,173.1,199.0,49.8,77.3),
y = c(235.3,210.0,156.1,119.6,270.6,32.3,64.1,
248.6,369.8,195.1,202.0,256.6,302.0,239.5,
221.7,230.9,303.9,239.8,305.0,160.1,157.2,
366.2,242.4,248.8,245.3,113.3,264.3,330.4,
200.3,191.3),
n = 30)

3. Initial values
list(beta0 = 0, beta1 = 0, tau = 1)
```

Figure 4.9. *Continued.*

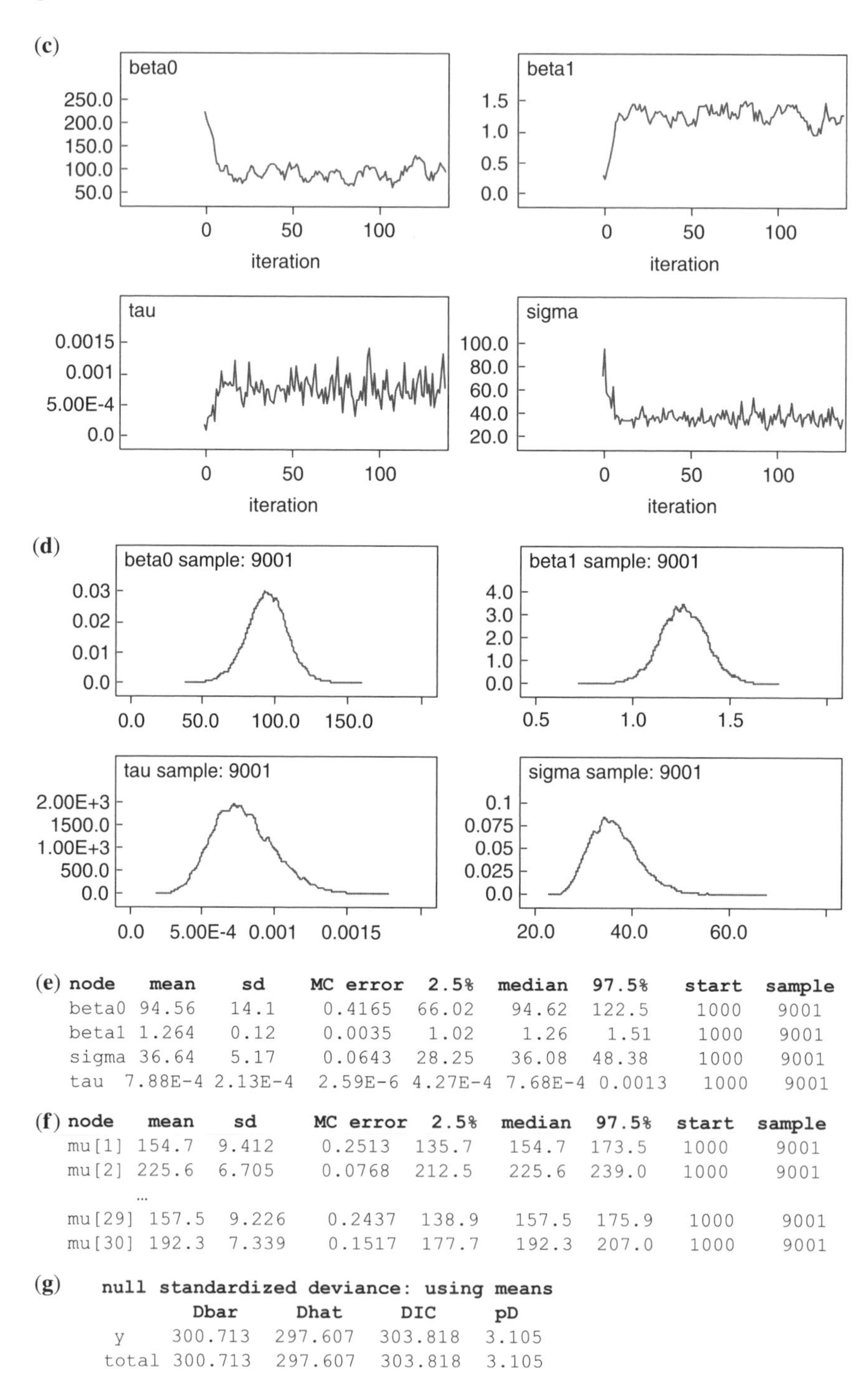

(e)

node	mean	sd	MC error	2.5%	median	97.5%	start	sample
beta0	94.56	14.1	0.4165	66.02	94.62	122.5	1000	9001
beta1	1.264	0.12	0.0035	1.02	1.26	1.51	1000	9001
sigma	36.64	5.17	0.0643	28.25	36.08	48.38	1000	9001
tau	7.88E-4	2.13E-4	2.59E-6	4.27E-4	7.68E-4	0.0013	1000	9001

(f)

node	mean	sd	MC error	2.5%	median	97.5%	start	sample
mu[1]	154.7	9.412	0.2513	135.7	154.7	173.5	1000	9001
mu[2]	225.6	6.705	0.0768	212.5	225.6	239.0	1000	9001
...								
mu[29]	157.5	9.226	0.2437	138.9	157.5	175.9	1000	9001
mu[30]	192.3	7.339	0.1517	177.7	192.3	207.0	1000	9001

(g) **null standardized deviance: using means**

	Dbar	Dhat	DIC	pD
y	300.713	297.607	303.818	3.105
total	300.713	297.607	303.818	3.105

analyze these examples in WinBUGS. We will continue to illustrate WinBUGS Bayesian statistics MCMC solutions to natural resource problems throughout the remainder of this book, analyzing problems of increasing complexity, and comparing their results to frequentist results. As always, the ultimate advantage of the Bayesian approach to statistical analysis for the natural resource scientist and manager is its direct interpretation of probabilistic inferences to parameters, rather than the indirect interpretation of probabilistic inferences to sample datasets, with repeated surveys or experiments, which is provided by frequentist statistical analysis.

4.5 SUMMARY

In this chapter, we introduced the topic of MCMC simulation, providing an overview of the MCMC algorithms and the Markov chain theory necessary to appreciate their ergodic properties. We described the most important MCMC algorithms, Gibbs sampling, and the Metropolis–Hastings algorithm, and presented several WinBUGS applications of these algorithms that provided Bayesian statistical analysis for model examples. We included examples of Bayesian MCMC statistical analysis in WinBUGS for normal, Poisson, negative binomial, and linear regression models and compared the Bayesian and frequentist results. Markov chain Monte Carlo algorithms, and their implementation in WinBUGS software, provide general Bayesian statistical analysis solutions for natural resource datasets. The solutions provide the Bayesian statistical inferences, based on posterior distributions for the parameters, obtained from datasets, models, and prior distributions, which can be particularly useful for natural resource scientists and management decisionmakers to provide answers to adaptive management problems.

PROBLEMS

4.1 WinBUGS example. Implement the models discussed in Section 4.4.2 in WinBUGS for count data, the Poisson model, the Poisson–gamma negative binomial model, and the overdispersed mixed-effects Poisson model, with noninformative priors, as follows. Analyze the simulated Poisson dataset with mean $\lambda = 3.0$ and sample size $n = 30$: data $= c(4,2,2,4,5,1,2,2,2,4,5,6,0,0,1,3,3,4,3,2,3,3,4,1,4,6,4,4,1,3)$. Draw inferences on the meaning of the parameters of the models in light of the fact that this Poisson dataset is not overdispersed (i.e., examine the relationship between the derived parameters mean $= \mu$ and variance $= \sigma^2$ in the negative binomial Poisson–gamma model and the significance of the parameter sigma $= \sigma$ in the mixed-effects Poisson model). Your results provide solution options for case study 2 presented in Section 1.2.2.

4.2 WinBUGS example. Implement a negative binomial multiplicative mixed-effects model for count data in WinBUGS using the Poisson count dataset in

Figure 4.10. Normally distributed sample data $x \sim N(\mu = 15, \sigma = 2)$ with sample size $n = 100$.

x
16.86
19.59
16.18
16.44
15.13
17.15
15.70
17.20
11.28
16.94
11.79
14.35
17.00
14.02
14.81
15.39
18.02
15.60
18.16
11.72
18.65
12.77
12.47
12.93
16.92
16.97
14.89
15.96
15.83
16.54
12.82
14.13
16.09
16.81
14.45
15.05
19.18
14.68
14.63
14.89
13.81
14.69
15.47
12.76
16.55
15.45
16.96
14.67
14.28
16.84
19.84

Figure 4.10. *Continued.*

```
14.97
14.31
17.30
17.13
14.31
10.30
14.04
15.19
11.12
13.54
17.66
17.89
16.38
12.75
16.72
14.50
16.17
14.70
12.93
19.98
13.56
19.33
13.71
13.58
13.31
13.08
14.91
15.16
13.50
12.91
12.29
14.46
12.53
14.45
12.51
11.96
12.53
19.57
14.66
13.36
15.85
15.97
14.94
12.78
12.97
14.27
14.45
12.60
14.90
```

Problem 4.1. Use a multiplicative random effect and a log transformation for the mixed-effects model. Use noninformative priors. Draw inferences on the meaning of the parameters in the model in light of the fact that this Poisson dataset is not overdispersed (i.e., examine the significance of the parameter sigma $= \sigma$ in this multiplicative mixed-effects model). Your results provide an additional solution option for case study 2 presented in Section 1.2.2.

4.3 Gibbs sampling program. Write a program in S-Plus or R to implement Gibbs sampling on Poisson data $y = \{y_A = y_1, y_2, \ldots, y_k; y_B = y_{k+1}, y_{k+2}, \ldots, y_n\}$, with $k = 20$ and sample size $n > k$, that is, randomly sampled from two strata, A and B, with $y_A \sim \text{Pois}(\lambda_A)$ and $y_B \sim \text{Pois}(\lambda_B)$. Use approximate Jeffreys gamma priors $G(\alpha, \beta)$ with $\alpha = \beta = 0.001$ for λ_A and $G(\chi, \delta)$ with $\chi = \delta = 0.001$ for λ_B, with starting values $\lambda_{A_0} = 1.0$ and $\lambda_{B_0} = 1.0$. You can use conjugacy properties of the gamma prior for the mean parameter λ of the Poisson model to encode the program (Gelfand and Smith 1990; Gill 2002, pp. 313–317, 348–349). Implement the program on the dataset in Problem 4.1. Interpret your results in light of the fact that $\lambda_A = \lambda_B = \lambda = 3.0$ for the simulated sample dataset.

4.4 Metropolis–Hastings iterations. Use the Metropolis–Hastings algorithm to generate the first three sample values p_1, p_2, and p_3, in the Markov chain for the posterior distribution of the parameter p for the binary sample dataset data $= c(1,0,1)$ with the binomial model. Use a uniform beta prior $\text{BE}(p; \alpha = 1, \beta = 1)$, with starting value $p_0 = 0$, and candidate values $p_1{}' = 0.34$, $p_2{}' = 0.67$, and $p_3{}' = 0.52$ obtained from a proposal distribution. Use random numbers 0.91, 0.17, 0.45 $\sim$ Unif(0,1), as many as are required for the ratios $r_i < 1$, $i = 1 - 3$.

4.5 Metropolis–Hastings program. Utilize the S-Plus and R program in Fig. 4.2a to implement the Metropolis–Hastings algorithm for normally distributed measurements $y \sim N(\mu, \sigma)$ with mean μ and known standard deviation $\sigma = 2.0$. The program generates an ergodic Markov chain of samples $\{\mu_0, \mu_1, \mu_2, \ldots\}$ converging to the posterior distribution of μ. Use a normally distributed proposal distribution for $m \sim N(\mu_{\text{proposal}}, \sigma_{\text{proposal}})$ with mean $\mu_{\text{proposal}} = \mu_i$ and standard deviation σ_{proposal}. Apply the program to the simulated sample dataset in Fig. 4.10 with mean $\mu = 15.0$. Use a normally distributed prior for $\mu \sim N(\mu_{\text{prior}} = 10.0, \sigma_{\text{prior}} = 2.0)$ with mean $\mu_{\text{prior}} = 10.0$ and standard deviation $\sigma_{\text{prior}} = 2.0$, and an initial value for the Markov chain $\mu_0 = 15.0$ (close to the mean of the sample data y). Use a standard deviation $\sigma_{\text{proposal}} = 0.24$ for the normally distributed proposal distribution that provides a rejection rate of approximately 33%. Run 500,000 iterations with burn-in of 5000 samples. Interpret the results of the simulation, describing the relationship of the precision and mean of the posterior to the precision and means of the prior and sample dataset (see Section 2.4.1).

4.6 Analyze the sample dataset in Fig. 4.10 again, using WinBUGS. Use the same initial conditions as in the MCMC simulation above in Problem 4.5: sample dataset $y \sim N(\mu, \sigma)$ with mean μ and known standard deviation $\sigma = 2.0$, normally distributed prior for $\mu \sim N(\mu_{prior} = 10.0, \sigma_{prior} = 2.0)$ with mean $\mu_{prior} = 10.0$ and standard deviation $\sigma_{prior} = 2.0$, an initial value for the Markov chain of $\mu_0 = 15.0$, and 500,000 iterations with a burn-in of 5000 samples. Compare the posterior results with those obtained from the MCMC simulation sampling in Problem 4.5 and with the theoretically expected results (see Section 2.4.1).

5 Alternative Strategies for Model Selection and Inference Using Information-Theoretic Criteria

In this chapter we describe two contrasting strategies for model selection and inference, a descriptive strategy consisting of a posteriori exploratory model selection and inference and a predictive strategy consisting of a priori parsimonious model selection and inference using information-theoretic criteria. We will first discuss the descriptive strategy of a posteriori exploratory model selection and inference, sometimes known pejoratively as "data dredging," and illustrate with examples. We will then discuss an alternative predictive strategy of a priori parsimonious model selection and inference using information-theoretic criteria, such as Akaike's information criterion (AIC). We will include in the discussion a review of the standard methods of least squares, maximum likelihood, and Bayesian fit for the estimation of model parameters. We can use these two strategies of model selection and inference with either a frequentist or a Bayesian approach to statistical analysis and inference. We will also review some standard methods of evaluation of fit, or goodness of fit, based on cross-validation techniques and test datasets. We conclude the chapter with a discussion of model averaging, for prediction, coefficient estimates and error, and importance of covariates.

5.1 ALTERNATIVE STRATEGIES FOR MODEL SELECTION AND INFERENCE: DESCRIPTIVE AND PREDICTIVE MODEL SELECTION

5.1.1 Introduction

In this section we describe two alternative strategies for statistical model selection and inference. We will be able to use either of these two strategies with either a frequentist or Bayesian approach to statistical analysis and inference. Let's assume that we are

Contemporary Bayesian and Frequentist Statistical Research Methods for Natural Resource Scientists. By Howard B. Stauffer

interested in formulating, fitting, and evaluating statistical models of a dependent response that is a function of a collection of covariates. We restrict ourselves to numerically valued covariates, although categorically valued treatment factors may be incorporated as well. The model could be a habitat selection model (Manly et al. 1995, 2004) that expresses the preference of a wildlife species for habitat expressed as a function of its attributes. For example, the wildlife species might be a bird or mammal species threatened with extinction, and the habitat attributes might be late-seral-stage, moisture, temperature, and proximity-to-water conditions particularly suitable to the animal. More generally, the habitat attributes may include vegetation, geologic, and climatic attributes. The habitat attributes, obtained from computerized GIS (geographic information system) information, may be more easily sampled than the wildlife species itself. The dependent variable y, the response, would measure the abundance, or presence–absence, of the wildlife species. The "independent," "predictor," variables or covariates, $x_1, x_2, \ldots, x_k$, would measure habitat attributes. The relationship can be expressed by the function f given by $y = f(x_1, x_2, \ldots, x_k)$, which may be linear or otherwise. We will use multiple linear regression throughout this chapter to illustrate.

The first of the two model selection strategies is most appropriate for populations where the relationship between the independent covariates and dependent response variable is not well understood. For this strategy, sample datasets consist of measurements of the response and candidate covariates that are believed to be related to the response variable. The model selection is determined a posteriori after data collection. After the sample data have been collected, graphs and statistics are examined to evaluate the apparent relationship of the covariates with the response. Tests of the relationships such as correlation tests and linear regression modeling can be conducted to determine the statistical significance of subsets of the covariates with the response. It is permissible to examine and compare the graphs and test statistics for any subset of covariates. However, because of the multiplicity of tests, compounded type I–type II error may occur. Also, because of the possible overreliance of the results on the particular sample dataset, models may tend to overfit the data. Hence the statistical results must be regarded as preliminary, exploratory, and tentative, descriptive of the sample dataset rather than predictive of the population in general. The results of analysis using this strategy can be useful in formulating hypotheses and selecting important covariates for further study. Although this strategy is sometimes pejoratively called "data dredging" because of its propensity for error, we wish to emphasize that it does have its place as a statistical strategy for data analysis and inference, particularly for natural resource populations that are not well understood.

An alternative strategy is more appropriate for populations that are better understood. This strategy, a priori parsimonious model selection and inference (Burnham and Anderson 1998), has come into favor in more recent years with ecology and wildlife management scientists and managers to address the problems associated with compounding of error and overfitting of sample data associated with the first strategy. For this alternative strategy, a collection of candidate models must be specified prior to data collection. The collection of models must be

parsimonious; the number of models in the collection must not be too large, and each model must not contain too many covariates. Typically, the collection must be limited to 10–30 models, although in practical application, the collection may contain up to several hundred models. The proposed candidate models may contain covariates (and factors) x_i of various specified forms such as linear x_i, quadratic x_i^2, and pseudothreshold (i.e., logarithmic) $\log(x_i)$, and their lower-order interactions such as the second-order $x_i \cdot x_j$. The candidate models are hypothesized on the basis of a biological assessment of those relatively independent covariates that are most likely in combination to efficiently describe the response. The results of previous studies, collective current scientific wisdom, and cumulative personal experience and intuition can be used to identify the candidate models. After the collection of candidate models has been specified, the sample data can be collected and analyzed. The analysis consists of assessing the relative competitiveness of the candidate models in fitting the data. Information-theoretic criteria have proved to be most effective at providing this assessment. For most models of biological populations that are high-dimensional or infinite-dimensional in complexity, with frequentist statistical analysis and inference, the corrected AIC (AIC_c) (Akaike 1973, 1974; Burnham and Anderson 1998) is a particularly effective criterion. The best fitting models are those with the lowest AIC_c. The AIC_c criterion measures relative error between the comparative models, and not absolute error, so additional methods must be used to examine goodness of fit of the best-fitting models. With Bayesian statistical analysis and inference, the deviance information criterion DIC (Carlin and Louis 2000, Spiegelhalter et al. 2001) is an analogous relative measure of goodness of fit, particularly with hierarchical models. This strategy circumvents the problems with compounding of error and overfitting of data that are inherent in the first strategy. It is more appropriate for populations that are better understood, however, where the collection of candidate models is likely to include good-fitting models. If the population is not well understood, the analyst runs the risk with this second strategy of not choosing good-fitting models in the collection of candidate models. So this strategy should be used with a degree of caution. In successful application, however, with populations that are better understood, and with proper testing of goodness of fit, this approach to statistical modeling, predictive a priori parsimonious model selection and inference using AIC_c, provides a reliable strategy for predictive model selection and inference of natural resource populations.

5.1.2 The Metaphor of the Race

The advantages and disadvantages of the two model selection and inference strategies are illustrated with a metaphor of the race. The question is as follows:

> How might we decide with one race who is the best automobile racing car driver in the world?

One strategy would be to conduct an open racing event. All racers in the world would be invited to participate. The number of entrants might be huge, with many candidates of varying levels of skill participating. Many unlikely and uncontrollable events might occur during the course of the race. Leading contenders might fall by the wayside because of the crowded conditions and collisions with less skilled and amateurish opponents. Local racers accustomed to conditions of the track might have an advantage and rise to the lead. Although all candidates could participate, with conditions of apparent fairness and equality, the results would be unpredictable and subject to a significant and indeterminable amount of error. Leading racers of the world might very well lose the race. Winners of the race would need to be further tested to determine whether their performance could withstand the test of time.

An alternative strategy for determining the best racer in the world would be to choose the initial contestants according to their previous track records. A small number of finalists would be chosen among those who had previously proved themselves by consistently winning races. The race would be a final contest, among the leading racers of the world, to determine the ultimate winner. This strategy would be less subject to error than the first strategy because conditions would be better controlled. Although some uncertainty would still exist, this final race would be better suited to determining the best overall racer in the world. If, however, not much is known about the leading racers in the world prior to the contest, if other preliminary races had not occurred to provide a screening process, this preselection process might not work out well, failing to identify the leading contestants for the final race. So the key to making this strategy work well is a prior knowledge of the leading contestants for the final race.

5.2 DESCRIPTIVE MODEL SELECTION: A POSTERIORI EXPLORATORY MODEL SELECTION AND INFERENCE

We discuss here in greater detail the first strategy, the descriptive a posteriori exploratory approach to model selection and inference based on "data dredging." This strategy provides a large degree of flexibility in exploring graphical and analysis methods to assess models fitting the sample data. However, this flexibility in examining and assessing models fitting the sample dataset must be provided in tandem with increased amounts of risk of compounded error and the overfitting of models to the dataset.

Numerous techniques may be used with this strategy once the sample dataset has been collected. Graphs such as scatterplots may be examined to look at possible relationships between the covariates and response, as well as relationships among the covariates. Correlations may be examined, between the covariates and response, and among the covariates. The general idea is to choose covariates for a model that are highly correlated with the response but not with each other. The ideal design is an orthogonal design where the covariates are independent of each other. One approach is to group the covariates by attribute type, such as vegetation, geologic, and climatic attributes. Covariate attributes can also be grouped by scale, such as microhabitat, macrohabitat, and landscape scales. The idea is to arrange the covariates into attribute type and scale groups such that covariates within groups are correlated with each other and covariates between groups are not correlated with each other. The leading attribute or attributes within each group that is correlated with the response can then be selected for inclusion in the best-fitting models. Numerous multivariate statistical methods exist to assess the relationships among covariates, such as discriminant analysis, principal-components analysis, factor analysis, and canonical correlation analysis.

Stepwise and best-subsets selection methods can then be used, with statistical modeling such as multiple linear regression, to assess the collection of all possible models of linear combinations of covariates for their relationship with the response. These standard methods with frequentist analysis are well known and are discussed in detail in many leading references (Seber 1977, Draper and Smith 1981, Manly 1994, Ryan 1997, Cook and Weisberg 1999).

With descriptive a posteriori exploratory model selection strategy, all plausible models may be examined in the quest to find the best-fitting model. The danger with this strategy, however, is the potential for compounding type I and II errors from multiple tests, of choosing covariates that shouldn't be in leading models, and failing to choose covariates that should be in leading models. In the "fishing expedition" that can ensue with abuse of this strategy, specious covariates can enter into models that are only fortuitously related in sample datasets by random chance, covariates that are not biologically related to the response in the population in general.

Care must be exercised to avoid overfitting of models too closely to sample datasets. It is the **signal** or **information** in a sample dataset that we wish to model, and not the **noise**. Burnham and Anderson (1998) suggest the analogy of drawing a profile of an elephant for the fitting of a model to a reality. The profile of an elephant based on one sample elephant should not be overdrawn, capturing every detail, every mole and whisker, every idiosyncrasy of the sample elephant. The profile should rather be an attempt to capture the general characterizes of all elephants, such as the large trunk and the lumbering feet.

Figure 5.1 illustrates this point. The scatterplot graph of a randomly sampled dataset of small sample size $n_1 = 3$ from a larger population is illustrated in Fig. 5.1a. Figure 5.1b presents an apparently reasonably fitting quadratic model to this sample dataset that is curvilinear in shape. This sample size, although extremely small, is illustrative of the small sizes of many sample datasets currently used in natural resource studies. It is quite common, for instance, for natural resource research studies to be based on sample datasets with 100 or fewer observations and 30 or more covariates, with three samples per covariate.

A larger randomly sampled dataset from the same population reveals, however, that a linear model, not a quadratic model, provides a much better fit (Figs. 5.1c and 5.1d). We were apparently misled into overfitting a model to the smaller sample dataset, overcharacterizing that particular sample dataset, its "whiskers and moles," with our model. The quadratic model for the smaller sample dataset required, however, estimates of three parameters, a constant, linear coefficient, and

Figure 5.1. Schematic representations of overfitting a model to a sample dataset. (**a**) Sample dataset of size $n_1 = 3$ from a population and (**b**) Quadratic model overfitting the sample dataset of size $n_1 = 3$. (**c**) Larger sample dataset of size $n_2 = 50$ from the same population, with the smaller sample of size $n_1 = 3$. (**d**) Larger sample dataset of size $n_2 = 50$ from the same population, with the smaller sample of size $n_1 = 3$, with the overfit model.

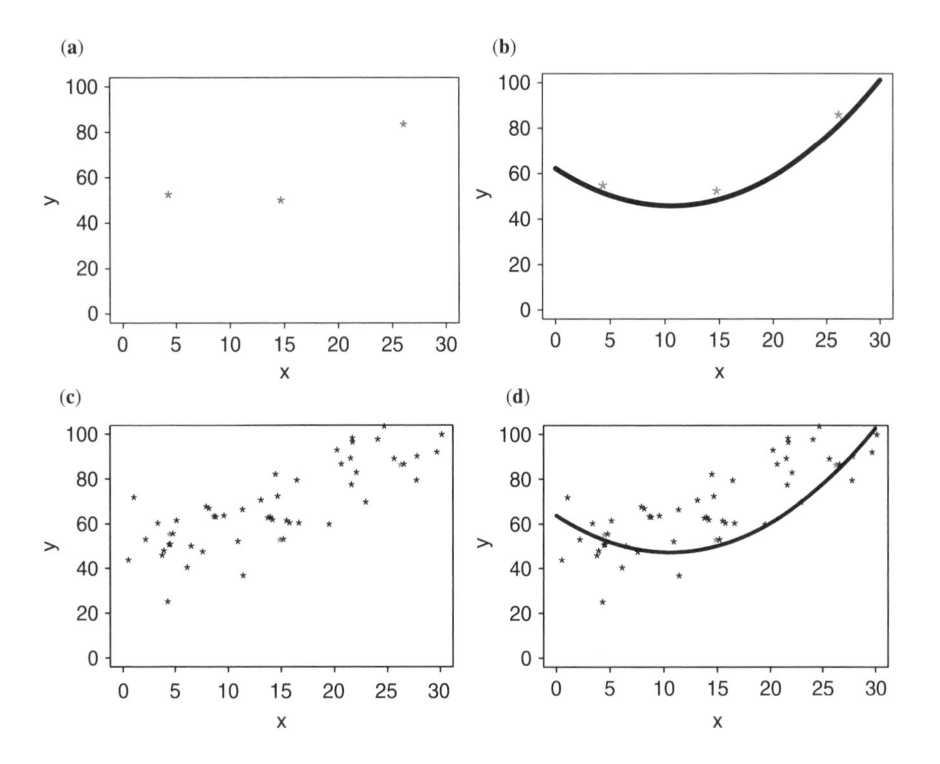

second-degree coefficient, with just three sample data points: 1 sample per parameter. The linear model, on the other hand, would require estimates of just two parameters, a constant and linear coefficient: 1.5 samples per parameter. The precision of the estimates for the linear model would generally be higher than those for the overfit quadratic model, with more samples per parameter. A model that is well-fitting must balance the requirements of minimizing the bias or error of the fit, using more parameters, with maximizing the precision of the parameter estimates, using fewer parameters.

Therefore, modeling results based on this strategy of a posteriori exploratory model selection and inference should be viewed as preliminary and tentative, descriptive of sample datasets, but subject to further study, analysis, and confirmation before predictive inferences can be made with assurance about populations.

This strategy is best applied to populations that are not well understood. Results can then be used to formulate hypotheses for more definitive studies with additional sample datasets.

We next discuss an alternative strategy of model selection: a priori parsimonious model selection and inference. The principle of parsimony will be discussed in greater detail and an information-theoretic criterion, AIC_c, will be described for identifying the best-fitting model. The best-fitting model will have the lowest AIC_c value. Furthermore, AIC_c values can provide Akaike weights that estimate the probabilities of the models being the best-fitting to the population, based on the sample dataset. This criterion, coupled with the principle of parsimony, provides an alternative strategy for model selection and inference that is a highly effective way of avoiding the overfitting of models to sample datasets.

5.3 PREDICTIVE MODEL SELECTION: A PRIORI PARSIMONIOUS MODEL SELECTION AND INFERENCE USING INFORMATION-THEORETIC CRITERIA

The exploratory data analysis strategy described in the previous section provides an effective approach for the descriptive modeling of a sample dataset. Powerful and easy-to-use statistical software is available to implement this strategy, such as with stepwise and best-subsets selection for multiple linear regression modeling. This strategy has the advantage of considering a wide range of models, even after data have been collected. It is exploratory and open to the possibility that unknown relationships between covariates and response may be uncovered from the sample dataset. It is particularly effective with populations that are little understood. However, as we have pointed out, the descriptive modeling strategy does have its limitations. It is subject to compounded error and the overfitting of data. Specious variables may enter into the best-fitting models from type I error, variables that are not biologically significantly related to the response within the population in general but statistically significant because of the vagaries of the sample dataset. Other variables that are biologically significantly related to the response, but not statistically significant, may be overlooked with the analysis from type II error. This a posteriori strategy places a premium on the analysis phase of a study, after data are collected, rather than on careful reflection and biological thinking at the beginning of a study, before data are collected.

For populations that are better understood, with biological understanding that has cumulated from prior studies, there is an alternative strategy to model selection and inference that avoids compounded error and the overfitting of sample data. This alternative strategy, a priori parsimonious model selection and inference, using AIC_c for frequentist statistical analysis and inference or DIC for Bayesian statistical analysis and inference, provides more reliable predictive results. Burnham and Anderson (1998) proposed this statistical data analysis strategy for model selection and inference to address the problems of compounded error and overfitting of natural resource sample datasets. Their strategy consists of several components:

1. Hypothesize a relatively small collection of candidate models for analysis, a priori to data collection.
2. Choose a parsimonious number of candidate models for the collection and a parsimonious number of covariates for each of the candidate models.
3. Use AIC_c with frequentist statistical analysis or DIC with Bayesian statistical analysis as a criterion for selection of the best-fitting models; these are the models with the lowest AIC_c or DIC.
4. Use AIC_c or DIC weights for assessment of the relative competitiveness of the models and for the model averaging of the estimates of covariate coefficients and error, predictive values, and the importance of covariates.

The parsimonious a priori model selection–inference strategy advocated by Burnham and Anderson provides a way of avoiding compounded type I and type II error and the overfitting of sample data. Compound error is avoided by using information-theoretic criteria such as AIC_c, rather than multiple tests of hypotheses. Overfitting is avoided by limiting the number of models and their numbers of covariates. This strategy, combined with goodness of fit tests, provides a more rigorous method for obtaining reliable predictive models in contrast to the descriptive strategy of a posteriori exploratory model selection. On the other hand, this strategy can miss important models with covariates that are overlooked in the initial selection of the candidate models with populations that are inadequately understood. So it is important to use the descriptive a posteriori model selection strategy for populations that are poorly understood and the predictive a priori model selection strategy for populations that are better understood.

5.4 METHODS OF FIT

Model selection and inference presupposes the use of appropriate methods of model fit. We described the three most common methods of fitting statistical models to sample datasets in Chapter 2: least-squares (LS) fit, maximum-likelihood (ML) fit, and Bayesian fit. The two alternative strategies of model selection and inference described in this chapter can be used with any of these three methods of model fit, whether frequentist LS, frequentist ML, or Bayesian. Least-squares fit, based on the minimization of the sums of the squared residuals, provides frequentist estimators of parameter covariate coefficients with multiple linear regression that are BLUE (best linear unbiased estimators), with estimators that are unbiased with minimum variance. We introduced the analysis of linear regression models in Chapter 1. We will illustrate multiple linear regression analysis using both strategies for model selection and inference later in this chapter. See Appendix A for a review of linear regression and multiple linear regression analysis. Maximum-likelihood fit, based on the maximization of the likelihood function, provides frequentist estimators that are asymptotically (i.e., as sample size $n \rightarrow \infty$) unbiased of minimum variance. We will illustrate maximum likelihood estimation with the two model selection and inference strategies with generalized linear modeling (GLM) in Chapter 6. Bayesian estimators are based on the posterior distributions for parameters that are obtained from Bayes' Theorem expressing the conditional probability for parameters, given the sample dataset. The posterior distributions are the scaled products of priors and likelihoods. The posteriors can be obtained for some datasets and models by using conjugate priors or more generally by Markov Chain Monte Carlo (MCMC) simulation. We introduced Bayesian methods in Chapters 2–4. We will continue to illustrate the use of Bayesian statistical analysis with multiple linear regression in this chapter, generalized linear modeling (GLM) in Chapter 6, and mixed-effects modeling in Chapter 7.

5.5 EVALUATION OF FIT: GOODNESS OF FIT

Regardless of which strategy is used to evaluate the best-fitting models, whether it is a posteriori descriptive modeling with exploratory data analysis, or a priori parsimonious predictive modeling using $\mathrm{AIC_c}$ or DIC, it is still necessary to evaluate goodness of fit of the leading models. In certain cases all models under consideration, even the most competitive models with lowest $\mathrm{AIC_c}$ or DIC, may provide poor fit to the population. In other cases, all of the most competitive models may provide reasonable fit. Remember that $\mathrm{AIC_c}$ and DIC provide an information-theoretic criterion (see Section 1.3.5) that measures the relative goodness of fit among a collection of competing models. All the statistics used to evaluate the best-fitting models that we have mentioned previously, including $\mathrm{AIC_c}$ and DIC, do not adequately address the issue of absolute goodness of fit. It is therefore critical, particularly with a priori modeling where the objective is predictive results, to follow an analysis of the best-fitting models with a goodness of fit analysis. A review of research articles in leading biological and natural resource journals revealed that only 7% of studies included a goodness of fit analysis, 5% with cross-validation and the remaining 2% with test datasets (Stauffer 1999). Hence we will outline here goodness of fit approaches that should be an integral part of any analysis. We will use multiple linear regression modeling as as example.

Assume, first, that the statistics for the leading models have been examined for statistical significance, say, at the 95% probability level. With frequentist (or Bayesian) multiple linear regression analysis, for instance, check that the following conditions generally apply:

1. The estimates of the parameters are significant, with confidence intervals (credible intervals) that do not include 0, or equivalently with p values for t tests or F tests that do not exceed α (i.e., t or F statistics are in the rejection region).

2. The R^2 statistic is acceptably large.
3. The residual standard error $s_{y|x}$ is acceptably small.
4. The F test is significant.
5. The AIC_c (DIC) is relatively small.
6. Other statistics such as the adjusted R^2 (and Mallows' C_p) are acceptably large (small).

Next, conduct a residual analysis to assess whether the residual errors between the observed and predicted response values are acceptably small and satisfy the assumptions of the model. For multiple linear regression, the residual errors must be independent, homoscedastic, and normally distributed.

Finally, check the leading models for predictive accuracy. There are two primary ways to accomplish this: with cross-validation and with test datasets. With multiple linear regression, the model prediction interval can be calculated and checked for its predictive accuracy with test datasets. A prediction interval with a 95% confidence level should contain on average 19 of 20 points in the test dataset. It is better to use independent test datasets rather than the original developmental dataset that was used to estimate and assess the competitiveness of the models. The predictive accuracy will tend to be overly optimistic on the developmental dataset used to estimate the models. If test datasets are not available, divide the sample dataset into two subsamples, if it is not too small, before analysis, with one used for development and the other for testing. If this is not possible, and test datasets are not available, cross-validation should be used as the next best option.

With cross-validation, individual points are omitted from the developmental sample dataset, the model is estimated on the remaining subsample dataset, and the omitted point is then checked to determine whether it falls within the prediction interval. This process is repeated for each point in the sample dataset. The predictive accuracy of the model is the proportion of the individually omitted points that fell within the prediction interval of the model fit to the remaining subsample dataset. This proportion can then be compared with the confidence-level standard that is used for the predictive accuracy of the model prediction intervals.

With test datasets, the prediction interval for the model can be estimated from the developmental sample dataset. The proportion of points in the test sample dataset that fall within the prediction interval can then be compared with the confidence-level standard that is used for the predictive accuracy of the model predictive intervals to assess the goodness of fit of the model. With rigorous studies that have the objective of developing reliable predictive models for populations, multiple test datasets are highly recommended. It should be borne in mind that populations are dynamic and may change with respect to location and time. Hence, degraded prediction accuracies of best-fitting models with test datasets may be due to changes in a dynamic population with respect to location and time, as well as to the poor fit of the models to a static population.

With other forms of modeling, such as logistic regression, which predicts the probabilities of success for data consisting of binary responses of "success" (1) or "failure" (0), a classification analysis is recommended. With logistic regression, optimal cutoff points can be determined, based on correct classification, sensitivity, and specificity rates for the developmental dataset, and applied to test datasets or to deleted points with cross-validation, to determine the predictive accuracy of the best-fitting models. This process will be described in greater detail in Chapter 6.

5.6 MODEL AVERAGING

One advantage of the Burnham and Anderson (1998, 2002) strategy for frequentist model selection and inference with the use of $\mathrm{AIC_c}$ is that Akaike weights can be calculated for each model in a collection of the m candidate models $\{M_1, M_2, \ldots, M_m\}$ that are being evaluated as best-fitting models using $\mathrm{AIC_c}$. The Akaike weights are calculated as

$$w_i = \frac{\exp[-(\mathrm{AIC}_{\mathrm{c}_i} - \min_{k=1,2,\ldots,m}(\mathrm{AIC}_{\mathrm{c}_k}))/2]}{\sum_{j=1}^{m} \exp[-(\mathrm{AIC}_{\mathrm{c}_j} - \min_{k=1,2,\ldots,m}(\mathrm{AIC}_{\mathrm{c}_k}))/2]}, i = 1, 2, \ldots, m.$$

The Akaike weights w_i, which sum to 1, are the scaled likelihoods for the models and can be interpreted as the probability of the ith model being the best-fitting model, given the collection of candidate models and the sample dataset. These weights not only provide a probability for the relative competitiveness of the candidate models but also may be used as weights for model averaging. Similar ideas may sometimes be used for Bayesian statistical analysis and inference, with DIC weights instead of $\mathrm{AIC_c}$ weights. There are three primary ways in which these weights may be utilized for model averaging.

5.6.1 Unconditional Estimators for Parameters: Covariate Coefficient Estimators, Errors, and Confidence Intervals

One way to utilize model averaging is to obtain unconditional estimators for parameters such as covariate coefficient estimators, errors, and confidence intervals (CIs). An **unconditional shrinkage estimator** $\hat{\theta}_{\mathrm{un}_1}$ (and its **unconditional standard error** $\mathrm{se}_{\hat{\theta}_{\mathrm{un}_1}}$) for a parameter θ can be obtained by taking the average of the candidate model M_j conditional estimators $\hat{\theta}_j$ (and their conditional standard errors $\mathrm{se}_{\hat{\theta}_j}$) for θ, weighted by the Akaike weights w_j

$$\hat{\theta}_{\mathrm{un}_1} = \sum_{j=1}^{m} w_j \cdot \hat{\theta}_j \left(\mathrm{se}_{\hat{\theta}_{\mathrm{un}_1}} = \sum_{j=1}^{m} w_j \cdot \mathrm{se}_{\hat{\theta}_j} \right).$$

This summation for $j = 1, 2, \ldots, m$ is over all m candidate models. If the parameter θ_j is not in model M_j, then assume $\theta_j = 0$ and $\hat{\theta}_j = 0$ ($\text{se}_{\hat{\theta}_j} = 0$).

If the summation is restricted only to those models with the parameter and the weights w_j adjusted accordingly to $w_j{}'$ that sum to one, we obtain another **unconditional estimator** (and **unconditional standard error**)

$$\hat{\theta}_{\text{un}_2} = \sum_{\substack{j'=1; \\ \theta_{j'} \text{ in } M_{j'}}}^{m} w_{j'} \cdot \hat{\theta}_{j'}$$

$$\text{se}_{\hat{\theta}_{\text{un}_2}} = \sum_{\substack{j'=1; \\ \theta_{j'} \text{ in } M_{j'}}}^{m} w_{j'} \cdot \text{se}_{\hat{\theta}_{j'}}.$$

The estimator $\hat{\theta}_j$ for parameter θ_j and its error are calculated conditional to the model M_j being selected. The first of these two unconditional estimators is sometimes called a **shrinkage estimator** because it is smaller than the second unconditional estimator and "shrunk" toward zero.

For example, with multiple linear regression models, we can obtain unconditional estimator $\hat{\beta}_{\text{u}}$ for the covariate coefficient parameter β, along with its unconditional standard error $\text{se}_{\hat{\beta}_{\text{u}}}$, sampling error $E_{\hat{\beta}_{\text{u}}}$, and confidence interval $\text{CI}_{\hat{\beta}_{\text{u}}}$, as follows

$$\hat{\beta}_{\text{u}} = \sum_{j=1}^{m} w_j \cdot \hat{\beta}_j,$$

$$\text{se}_{\hat{\beta}_{\text{u}}} = \sum_{j=1}^{m} w_j \cdot \text{se}_{\hat{\beta}_j},$$

$$E_{\hat{\beta}_{\text{u}}} = t_{n-p-1} \cdot \text{se}_{\hat{\beta}_{\text{u}}},$$

$$\text{CI}_{\hat{\beta}_{\text{u}}} = \hat{\beta}_{\text{u}} \pm E_{\hat{\beta}_{\text{u}}} = [\hat{\beta}_{\text{u}} - E_{\hat{\beta}_{\text{u}}}, \hat{\beta}_{\text{u}} + E_{\hat{\beta}_{\text{u}}}],$$

where $\hat{\beta}_j$ is the conditional estimator of β in M_j, $\text{se}_{\hat{\beta}_j}$ is its conditional standard error, and t_{n-p-1} is the critical t value with the appropriate degrees of freedom $(n-p-1) = (n-k)$ for the multiple linear regression model with p covariates and $k = p+1$ parameters (including the constant). Burnham and Anderson (1998, 2002) recommend using the estimated mean-square error ($\widehat{\text{MSE}}$) for the unconditional standard error, given by

$$\widehat{\text{MSE}} = \widehat{\text{var}}\,\hat{\beta}_{\text{u}} = \left[\sum_{j=1}^{m} w_j \cdot \sqrt{\widehat{\text{var}}\,\hat{\beta}_j + (\hat{\beta}_j - \hat{\beta}_{\text{u}})^2}\right]^2$$

$$= \left[\sum_{j=1}^{m} w_j \cdot \sqrt{(\text{se})^2_{\hat{\beta}_j} + (\hat{\beta}_j - \hat{\beta}_{\text{u}})^2}\right]^2$$

with

$$\text{se}_{\hat{\beta}_u} = \sqrt{\hat{\text{var}}\,\hat{\beta}_u} = \sum_{j=1}^{m} w_j \cdot \sqrt{\hat{\text{var}}\hat{\beta}_j + (\hat{\beta}_j - \hat{\beta}_u)^2}.$$

The MSE incorporates both error and bias in its expression.

5.6.2 Unconditional Estimators for Prediction

Model averaging can also be used for prediction. Each model function f provides a predictive value at each p-dimensional covariate point $x_i = (x_{1_i}, x_{2_i}, \ldots, x_{p_i})$

$$\hat{y}_i = f(x_{1_i}, x_{2_i}, \ldots, x_{p_i})$$

that estimates the response value at that point, along with confidence intervals $\text{CI}_{\hat{y}_i}$ and prediction intervals $\text{PI}_{\hat{y}_i}$ for specified levels of confidence. For multiple linear regression, $\hat{y}_i$ is the estimated mean response at x_i. For logistic regression, $\hat{y}_i$ is the estimated probability at x_i. These predictive values for covariate points are conditional on the "correctness" of the model. **Unconditional model averaged prediction values** can be estimated from a collection of models $\{M_1, M_2, \ldots, M_m\}$, using Akaike weights w_j for each of the M_j model prediction functions $\hat{y}_j, j = 1, 2, \ldots, m$:

$$\hat{y}_u = \sum_{j=1}^{m} w_j \cdot \hat{y}_j.$$

The unconditional predictive value at covariate point $x_i = (x_{1_i}, x_{2_i}, \ldots, x_{p_i})$ is then given by

$$\hat{y}_{u,i} = \sum_{j=1}^{m} w_j \cdot \hat{y}_j(x_{1_i}, x_{2_i}, \ldots, x_{p_i}).$$

Unconditional confidence and prediction intervals can be obtained by taking weighted averages of the model conditional standard errors to calculate unconditional standard errors, sampling errors, and confidence and prediction intervals.

5.6.3 Importance of Covariates

Model averaging using Akaike weights also provides a method for assessing the **relative importance of the covariates**, the "independent" predictor variables x_k, in the candidate models. The importance of each covariate x_k is given by the following

$$\text{Importance}\,(x_k) = \sum_{x_k \text{ is in } M_j} w_j.$$

Traditionally, a covariate was deemed "important" when it occurred in a "best-fitting model." Alternatively, the "importance" of a covariate has been measured by the number of models in an analysis that included it. However, Akaike weights provide a more sensitive way of weighing these numbers, incorporating the probabilities estimating the relative competitiveness of the models into a cumulative sum of the weights of the models containing the covariate.

5.7 APPLICATIONS: FREQUENTIST STATISTICAL ANALYSIS IN S-Plus AND R; BAYESIAN STATISTICAL ANALYSIS IN WINBUGS

We now will illustrate the ideas of this chapter by analyzing a sample dataset with habitat selection modeling (Manly et al. 1995, 2004) using multiple linear regression, employing both strategies, predictive a priori parsimonious model selection and inference with AIC and DIC, and descriptive a posteriori model selection and inference. The sample dataset is simulated from a known "reality" representing the response of a wildlife species to habitat selection. The habitat is described by seven covariates distributed as follows for measurements on 200-ha sites:

1. aspect: binomial with probabilities = 0.5, for south- (0) and north- (1) facing sites
2. species: binomial with probabilities = 0.2 and 0.8, respectively, for RW (redwood) (1) and Other (0).
3. old.growth: uniform (0, 200)
4. rock: binomial with probabilities = 0.8 and 0.2, respectively, for absence (0) and presence (1)
5. moss: binomial with probabilities = 0.6 and 0.4, respectively for absence (0) and presence (1)
6. temp: uniform (15, 30)
7. moist: uniform (0, 100)

Note that aspect, species, rock, and moss are represented by so-called design or dummy variables. The wildlife biomass response is described by the "reality"

```
response = 5.0+2.0* rock+3.0* moss+0.5* temp+0.1* moist
           +error
```

where error $\sim N(0, 3)$ is normally distributed with mean $\mu = 0$ and standard deviation $\sigma = 3$. The S-Plus and R code used to generate the sample with a sample size of $n =$ 50, along with the simulated sample dataset `data1`, is presented in Fig. 5.2.

Figure 5.2. The S-Plus and R code used to generate the simulated habitat selection dataset *data1*.

```
> aspect <- sample(c(0,1),size=50,prob=c(0.5,0.5),replace=T)
> species1 <-
sample(c("RW","Other"),size=50,prob=c(0.2,0.8),replace=T)
> species <- rep(NA,50)
> for (i in 1:50) if (species1[i]=="RW") species[i]<-1 else
species[i]<-0
> old.growth <- runif(50,0,200)
> old.growth <- round(old.growth,1)
> rock <- sample(c(0,1),size=50,prob=c(0.8,0.2),replace=T)
> moss <- sample(c(0,1),size=50,prob=c(0.6,0.4),replace=T)
> temp <- runif(50,15,30)
> temp <- round(temp,1)
> moist <- runif(50,0,100)
> moist <- round(moist,1)
> response <-
5+2*rock+3*moss+0.5*temp+0.1*moist+rnorm(50,0.0,3.0)
> response <- round(response,1)
> data <-
data.frame(aspect,species,old.growth,rock,moss,temp,moist,response)
> data
   aspect species old.growth rock moss temp moist response
 1      0       0       30.0    0    1 25.9  35.5     20.8
 2      1       0      151.5    0    0 21.9  88.8     22.8
 3      0       0       61.6    1    1 17.4  37.6     24.7
 4      1       0       26.4    1    0 27.0  79.3     28.4
 5      1       0      121.1    0    0 23.0  24.1     19.9
 6      0       0      122.9    1    0 23.1  33.1     23.1
 7      1       0      161.2    0    1 23.6  69.5     26.2
 8      1       0      134.6    0    0 28.6  46.6     22.0
 9      0       0       44.5    1    0 27.9  64.6     30.4
10      0       0        7.7    0    0 22.7  11.6     17.6
11      1       0      198.1    0    1 26.2  38.7     24.7
12      0       0      134.9    0    1 15.6  26.7     17.8
13      1       0       97.3    0    0 22.3  76.5     21.4
14      0       0       51.6    0    0 19.9  10.2     14.5
15      1       0       62.3    0    1 22.9  67.4     25.2
16      0       1       54.6    0    0 18.2  50.8     17.1
17      1       0      151.7    0    0 26.3  99.5     26.5
18      0       1       84.7    0    0 27.0  85.2     25.3
19      1       0       54.6    1    0 15.8  64.9     19.9
20      0       0       31.0    0    0 26.8  93.1     28.2
21      0       1      122.7    1    1 22.3  12.3     21.1
22      0       0       25.1    0    1 23.4  23.1     25.3
23      1       0      143.7    0    0 18.0  17.1     16.1
24      0       0       30.4    1    0 27.8  36.7     24.3
25      1       0      186.0    0    0 22.6   9.1     18.4
26      0       0       14.5    0    1 21.8  54.4     25.7
27      0       0       84.7    0    1 22.6  59.2     25.6
28      1       0        8.2    0    0 22.5  24.5     18.8
```

Figure 5.2. *Continued.*

```
29    1    0    177.8   0    0 26.1  75.4    27.4
30    0    1     55.2   0    0 26.7  26.5    19.5
31    1    0     37.9   1    0 20.2  74.6    23.8
32    1    0    130.2   0    1 19.2   4.4    20.8
33    0    1    103.5   0    0 22.3  65.7    26.2
34    0    1     99.2   0    0 24.0  72.1    24.7
35    1    0    121.7   0    0 21.4   2.9    13.2
36    1    0    198.5   0    1 22.0  25.6    23.1
37    1    1    174.0   0    1 28.1  42.5    24.5
38    1    1    124.3   1    0 24.1  19.4    23.5
39    0    0     23.2   0    1 15.1  63.2    19.5
40    1    0     81.5   1    0 17.5  97.9    28.1
41    1    0    165.0   0    0 27.3  65.1    26.4
42    1    1      7.6   1    0 19.1  93.7    27.8
43    1    0    108.7   1    0 23.1  40.1    21.6
44    1    1     83.6   0    0 23.7  41.0    17.6
45    0    0     81.2   0    0 23.2  22.4    18.1
46    1    0     47.5   1    1 17.2  72.0    25.3
47    0    0    121.6   0    0 29.9  35.9    18.1
48    1    0    118.5   0    1 29.7  88.4    29.7
49    1    0      7.9   0    0 22.2  17.2    21.2
50    1    1    122.3   1    0 29.8  13.2    22.1
> names(data)
[1] "aspect"  "species"  "old.growth"  "rock"  "moss"  "temp"
"moist"  "response"
```

5.7.1 Frequentist Statistical Analysis in S-Plus and R: Predictive A Priori Parsimonious Model Selection and Inference Using the Akaike Information Criterion (AIC)

For the a priori parsimonious model selection and inference strategy applied to the sample dataset `data1`, we consider a collection of 10 models, hypothesized from biological considerations, with the following covariates:

1. {aspect,species,old.growth}
2. {aspect,species,moss}
3. {aspect,species,temp}
4. {aspect,species,moist}
5. {aspect,old.growth,temp}
6. {species,old.growth,temp}
7. {temp,moist}
8. {rock,moss,temp,moist} (the correct model)
9. {1} (the null model)
10. {aspect,species,old.growth,rock,moss,temp,moist} (the full model)

The frequentist statistical analysis results of the LS fit of the 10 multiple linear regression models with the collections of covariates to the sample data listed above are presented in Fig. 5.3. Note that, for the most part, the coefficient estimates $\hat{\beta}_i$ for the covariates x_i are statistically insignificant at the 95% confidence level (i.e., we cannot reject the null hypothesis H_0: $\beta_i = 0$ for the covariate x_i, with 95% confidence) for the biologically insignificant covariates aspect, species, and old.growth (i.e., $\beta_i = 0$), and statistically significant at the 95% confidence level (i.e., we can reject the null hypothesis H_0: $\beta_i = 0$ for the covariate x_i, with 95% confidence) for the biologically significant covariates rock, moss, temp, and moist (i.e., $b_i \neq 0$). There are two interesting exceptions, however, for the correct model 8, where the coefficient estimate for rock is statistically insignificant at the 95% confidence level even though the covariate is biologically significant, and for model 2, where the coefficient estimate for moss is statistically insignificant at the 95% confidence level even though the covariate is biologically significant.

Figure 5.4 summarizes the statistics from the analysis of `data1`. Note that the residual standard error $S_{y|x}$ usually decreases and the coefficient of determination R^2 increases as nested models contain increasing numbers of covariates. The F-test statistic is statistically significant at the 5% level of significance (i.e., 95% confidence level) for only some of the models containing the biologically significant covariates rock, moss, temp, and moist: models 4, 7, 8, and 10. Akaike's information criterion correctly identifies model 8, with the biologically significant covariates, as the best-fitting model, with minimal AIC of 218.437 and maximal Akaike weight of 94.8%. The full model 10 is the only slight competitor, with AIC of 224.246 and Akaike weight of 5.2%. The null model 9 and models 1 and 2 with specious covariates only, except for model 2's moss covariate, are the poorest-fitting, with the highest AIC values. In conclusion, the a priori model selection–inference strategy correctly identifies model 8 as the best-fitting model.

5.7.2 Frequentist Statistical Analysis in S-Plus and R: Descriptive A Posteriori Model Selection and Inference

We perform descriptive a posteriori model selection and inference in S-Plus and R, beginning with the command `pairs(data1)` applied to the data frame `data1` of covariates and response to examine the scatterplots of the paired variables (Fig. 5.5a). We use the S-Plus and R commands with `type <- "forward"`, `type <- "backward"`, or the default `type <- "both"` for stepwise multiple linear regression (Figs. 5.5b and 5.5c). The initial model formula should be that of the null model `response~1` for forward stepwise multiple regression and the full model `response~aspect+species+old.growth+rock+moss+temp+moist` for backward stepwise multiple regression. The option `both` is either forward and backward or backward and forward depending on the initial model. The addition or deletion of covariates is based on minimal Mallows C_p (S-Plus and R call it AIC).

For best subsets multiple linear regression in S-Plus and R, we construct the vector of strings `covariates` (Fig. 5.5d) for the models consisting of all linear

Figure 5.3. The S-Plus and R frequentist statistical analysis results of dataset `data1`, using predictive a priori parsimonious model selection and inference with AIC, described in Section 5.7.1.

```
> output1 <- lm(response~aspect+species+old.growth,data=data1)
> summary(output1)
...
Coefficients:
               Value Std. Error  t value Pr(>|t|)
(Intercept)  22.5121    1.2350   18.2288   0.0000
     aspect   0.9245    1.3123    0.7045   0.4847
    species   0.1507    1.4348    0.1050   0.9168
 old.growth  -0.0044    0.0116   -0.3765   0.7083

Residual standard error: 4.151 on 46 degrees of freedom
Multiple R-Squared: 0.01089
F-statistic: 0.1688 on 3 and 46 degrees of freedom, the p-value
is 0.9169
...
> output2 <- lm(response~aspect+species+moss,data=data1)
> summary(output2)
...
Coefficients:
              Value Std. Error t value Pr(>|t|)
(Intercept) 21.4713   1.1205   19.1627  0.0000
     aspect  0.9477   1.1888    0.7972  0.4294
    species  0.4528   1.4253    0.3177  0.7522
       moss  1.7483   1.2620    1.3853  0.1726

Residual standard error: 4.073 on 46 degrees of freedom
Multiple R-Squared: 0.04758

F-statistic: 0.766 on 3 and 46 degrees of freedom, the p-value is
0.519
...
> output3 <- lm(response~aspect+species+temp,data=data1)
> summary(output3)
...
Coefficients:
               Value Std. Error  t value Pr(>|t|)
(Intercept)  14.0796   3.4341     4.1000  0.0002
     aspect   0.6253   1.1304     0.5532  0.5828
    species  -0.3790   1.3603    -0.2786  0.7818
       temp   0.3602   0.1459     2.4684  0.0173

Residual standard error: 3.906 on 46 degrees of freedom
Multiple R-Squared: 0.1239
F-statistic: 2.168 on 3 and 46 degrees of freedom, the p-value is
0.1046
```

Figure 5.3. *Continued.*

```
...
> output4 <- lm(response~aspect+species+moist,data=data1)
> summary(output4)
...
Coefficients:
                Value Std. Error  t value Pr(>|t|)
(Intercept)  17.7958    0.9624    18.4902   0.0000
     aspect  -0.0147    0.8578    -0.0171   0.9864
    species   0.0529    1.0138     0.0522   0.9586
      moist   0.1017    0.0150     6.7698   0.0000

Residual standard error: 2.942 on 46 degrees of freedom
Multiple R-Squared: 0.503

F-statistic: 15.52 on 3 and 46 degrees of freedom, the p-value is
4.129e-007
...
> output5 <- lm(response~aspect+old.growth+temp,data=data1)
> summary(output5)
...
Coefficients:
                Value Std. Error  t value Pr(>|t|)
(Intercept)  14.0391    3.3988     4.1306   0.0002
     aspect   1.1641    1.2118     0.9607   0.3417
 old.growth  -0.0112    0.0111    -1.0099   0.3178
       temp   0.3894    0.1471     2.6472   0.0111

Residual standard error: 3.867 on 46 degrees of freedom
Multiple R-Squared: 0.1414
F-statistic: 2.526 on 3 and 46 degrees of freedom, the p-value is
0.06909
...
> output6 <- lm(response~species+old.growth+temp,data=data1)
> summary(output6)
...
Coefficients:
                Value Std. Error  t value Pr(>|t|)
(Intercept)  14.5186    3.3843     4.2900   0.0001
    species  -0.4947    1.3450    -0.3678   0.7147
 old.growth  -0.0069    0.0102    -0.6781   0.5011
       temp   0.3856    0.1493     2.5822   0.0131

Residual standard error: 3.9 on 46 degrees of freedom
Multiple R-Squared: 0.1268
```

Figure 5.3. *Continued.*

```
F-statistic: 2.226 on 3 and 46 degrees of freedom, the p-value is
0.09776
...
> output7 <- lm(response~temp+moist,data=data1)
> summary(output7)
...
Coefficients:
              Value Std. Error t value Pr(>|t|)
(Intercept) 10.8971  2.3457     4.6456  0.0000
       temp  0.3050  0.0982     3.1042  0.0032
      moist  0.0987  0.0135     7.3332  0.0000

Residual standard error: 2.652 on 47 degrees of freedom
Multiple R-Squared: 0.5875
F-statistic: 33.47 on 2 and 47 degrees of freedom, the p-value is
9.154e-010
...
> output8 <- lm(response~rock+moss+temp+moist,data=data1)
> summary(output8)
...
Coefficients:
             Value Std. Error t value Pr(>|t|)
(Intercept) 6.6079 1.9155     3.4497  0.0012
       rock 2.8965 0.6518     4.4440  0.0001
       moss 3.0248 0.6299     4.8024  0.0000
       temp 0.4187 0.0768     5.4518  0.0000
      moist 0.0963 0.0103     9.3599  0.0000

Residual standard error: 2.01 on 45 degrees of freedom
Multiple R-Squared: 0.7731
F-statistic: 38.33 on 4 and 45 degrees of freedom, the p-value is
5.906e-014
...
> output9 <- lm(response~1,data=data1)
> summary(output9)
...
Coefficients:
              Value Std. Error t value Pr(>|t|)
(Intercept) 22.6800  0.5719    39.6598  0.0000

Residual standard error: 4.044 on 49 degrees of freedom
Multiple R-Squared: 1.235e-029
F-statistic: Inf on 0 and 49 degrees of freedom, the p-value is NA
...
> output10 <-
lm(response~aspect+species+old.growth+rock+moss+temp+moist,
data=data1)
> summary(output10)
...
```

Figure 5.3. *Continued.*

```
Coefficients:
                Value Std. Error t value Pr(>|t|)
(Intercept)    6.5631  2.0027     3.2771  0.0021
     aspect    0.0185  0.6949     0.0267  0.9788
    species   -0.2810  0.7409    -0.3793  0.7064
 old.growth    0.0004  0.0064     0.0689  0.9454
       rock    2.9321  0.6998     4.1902  0.0001
       moss    2.9913  0.6703     4.4627  0.0001
       temp    0.4214  0.0825     5.1112  0.0000
      moist    0.0962  0.0108     8.8728  0.0000

Residual standard error: 2.077 on 42 degrees of freedom
Multiple R-Squared: 0.774
F-statistic: 20.54 on 7 and 42 degrees of freedom, the p-value is
1.107e-011
...
> AIC(output1,output2,output3,output4,output5,output6,output7,
output8,output9,output10)
          df      AIC
 output1   5 290.0520
 output2   5 288.1621
 output3   5 283.9863
 output4   5 255.6408
 output5   5 282.9741
 output6   5 283.8205
 output7   4 244.3194
 output8   6 218.4372
 output9   2 284.5995
 output10  9 224.2456
```

combinations of the covariates under consideration and perform multiple linear regression on all its entries (Fig. 5.5e).

The stepwise regression method chose the model with covariates {moist, temp, moss, rock}, selected in that order, based on lowest Mallows C_p. The best-subsets selection method also chose the model with covariates {aspect, moss, temp, moist} as best-fitting, using lowest AIC for the selection criterion. We conclude that Mallows C_p and AIC successfully identified the best-fitting model 8, using stepwise and best-subsets selection a posteriori methods, respectively.

Generally, however, stepwise and best-subsets selection a posteriori methods do tend to overfit models and compound error. Regardless of strategy, AIC is the most appropriate indicator for model selection and inference for infinite-dimensional realities. We used the defaulted AIC function that is available in both S-Plus and R. However, we could obtain the most accurate results by using the corrected AIC_c in this example, particularly since the sample size was relatively small at $n = 50$.

Figure 5.4. Statistics for 10 models: a priori parsimonious model selection and inference using AIC and Akaike weights.

Model	Covariates	Fitted Model	k	$s_{y\|x}$	R^2	*F*-Test *p* Value	AIC	Akaike Weights
1	aspect, species, old.growth	y = 22.512+0.9245*aspect+0.1507*species-0.0044*old.growth (aspect, species, and old.growth are statistically insignificant)	5	4.151	1.089%	0.9169	290.052	<0.00005
2	aspect, species, moss	y = 21.4713+0.9477*aspect+0.4528*species+1.7483*moss (aspect, species, and moss are statistically insignificant)	5	4.073	4.758%	0.5190	288.162	<0.00005
3	aspect, species, temp	y = 14.0796+0.6253*aspect-0.3790*species+0.3602*temp (aspect and species are statistically insignificant)	5	3.906	12.390%	0.1046	283.986	<0.00005
4	aspect, species, moist	y = 17.7958-0.0147*aspect+0.0529*species+0.1017*moist (aspect, and species are statistically insignificant)	5	2.942	50.300%	<0.00005	255.641	<0.00005
5	aspect, old.growth, temp	y = 14.0391+1.1641*aspect-0.0112*old.growth+0.3894*temp (aspect, and old.growth are statistically insignificant)	5	3.867	14.140%	0.0691	282.974	<0.00005
6	species, old.growth, temp	y = 14.5186-0.4947*species-0.0069*old.growth+0.3856*temp (species and old.growth are statistically insignificant)	5	3.900	12.680%	0.0978	283.821	<0.00005
7	temp, moist	y = 10.8971+0.3050*temp+0.0987*moist	4	2.652	58.750%	<0.0005	244.319	0.0002%
8	rock, moss, temp, moist	y = 6.6079+2.8965*rock+3.0248*moss+0.4187*temp+0.0963*	6	2.010	77.310%	<0.0005	218.437	94.8052%
9	none	y = 22.6800	2	4.044	<0.0005%	NA	284.600	<0.00005
10	aspect, species, old.growth, rock, moss, temp, moist	y = 6.5631+0.0185*aspect-0.2810*species+0.0004*old.growth +2.9321*rock +2.9913*moss+0.4214*temp+0.0962*moist (aspect, species, and old,growth are statistically insignificant)	9	2.077	77.400%	<0.0005	224.246	5.1946%
							Total:	100.0000%

Note: k = number of parameters, $s_{y|x}$ = residual standard error, R^2 = coefficient of determination, confidence level = 95%.

Figure 5.5. S-Plus and R results for exploratory data analysis using stepwise and best-subsets selection of multiple linear regression. (**a**) Pairwise scatterplots. (**b**) Stepwise selection method code. (**c**) Stepwise selection method results. (**d**) Best-subsets selection code. (**e**) Best subsets selection results.

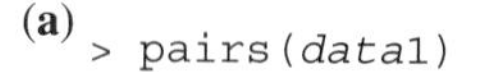

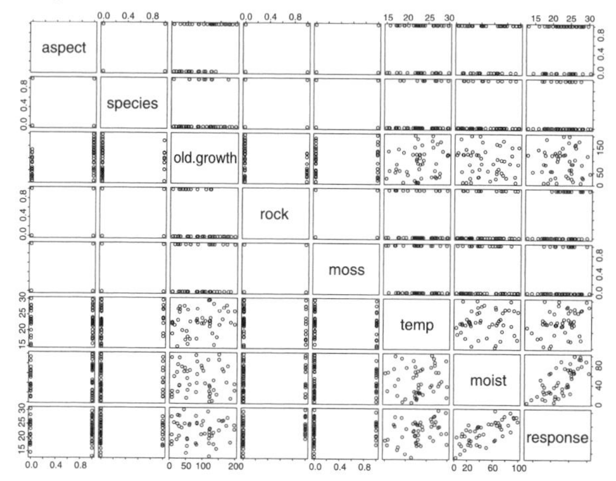

(**b**)

```
> object0 <- lm(initial model formula,data1)
> scope <- ".~.+aspect+species+old.growth+rock+moss+temp+moist"
> step(object0,scope,direction=type)
```

(**c**)

```
> step(lm(response~1,data1),scope=.~.+aspect+species+old.growth
+rock+moss+temp+moist,direction="forward")

Start:  AIC= 833.9229
 response ~1
Single term additions
Model:
response ~ 1
...
            Df Sum of Sq      RSS       Cp
    <none>                801.2200 833.9229
    aspect   1    6.1857  795.0343 860.4400
   species   1    0.0007  801.2193 866.6250
old.growth   1    0.1734  801.0466 866.4523
      rock   1   70.0889  731.1311 796.5368
      moss   1   26.9388  774.2812 839.6869
      temp   1   92.6286  708.5914 773.9972
     moist   1  402.9880  398.2320 463.6377

Step:  AIC= 463.6377
 response ~ moist

Single term additions

Model:
response ~ moist
...
```

Figure 5.5. *Continued.*

```
               Df Sum of Sq      RSS       Cp
    <none>                  398.2320 463.6377
    aspect  1    0.00513 398.2269 496.3354
   species  1    0.02616 398.2058 496.3144
old.growth  1    2.41694 395.8151 493.9236
      rock  1   38.98145 359.2505 457.3591
      moss  1   44.42449 353.8075 451.9161
      temp  1   67.75629 330.4757 428.5843

Step:  AIC= 428.5843
 response ~ moist + temp

Single term additions

Model:
response ~ moist + temp
...
               Df Sum of Sq      RSS       Cp
    <none>                  330.4757 428.5843
    aspect  1    0.01668 330.4590 461.2704
   species  1    1.04099 329.4347 460.2461
old.growth  1    0.12947 330.3462 461.1577
      rock  1   55.50659 274.9691 405.7805
      moss  1   68.89202 261.5837 392.3951

Step:  AIC= 392.3951
 response ~ moist + temp + moss

Single term additions

Model:
response ~ moist + temp + moss
...
               Df Sum of Sq      RSS       Cp
    <none>                  261.5837 392.3951
    aspect  1    0.53863 261.0450 424.5593
   species  1    0.01234 261.5713 425.0856
old.growth  1    2.19419 259.3895 422.9038
      rock  1   79.78681 181.7969 345.3111

Step:  AIC= 345.3111
 response ~ moist + temp + moss + rock

Single term additions

Model:
response ~ moist + temp + moss + rock
...
```

Figure 5.5. *Continued.*

```
              Df Sum of Sq       RSS       Cp
    <none>                  181.7969 345.3111
    aspect   1 0.0716715 181.7252 377.9423
   species   1 0.6551721 181.1417 377.3588
old.growth   1 0.0301936 181.7667 377.9838

Call:
lm(formula = response ~ moist + temp + moss + rock, data1)

Coefficients:
 (Intercept)     moist      temp     moss     rock
    6.607909 0.0962838 0.4187392 3.024792 2.896458

Degrees of freedom: 50 total; 45 residual
Residual standard error (on weighted scale): 2.009958
```

(d)

```
> covariates <- c("aspect","species",…,"aspect+species",…,
"aspect+species+old.growth+rock+moss+temp+moist")
> models <- list(rep(NA,length(covariates)))
> for (i in 1:length(covariates)) {models[[i]] <- lm(paste(
"response~",covariates[i]),data1)}
> models
> aic <- rep(NA,length(covariates))
> for (i in 1:length(covariates)) {aic[i] <- AIC(lm(paste(
"response~",covariates[i],data1))}
> min(aic)
> for (i in 1:length(covariates)) {if (aic[i]==min(aic)) k<-i}
> k
> covariates[k]
> summary(models[[k]])
```

(e)

```
> covariates <- c("aspect","species",…,"aspect+species",…,
    "aspect+species+old.growth+rock+moss+temp+moist")
> for (i in 1:length(covariates)) {models[[i]] <-
    lm(paste("response~",covariates[i]),data1)}
> models
[[1]]:
…
Coefficients:
               Value Std. Error t value Pr(>|t|)
(Intercept) 22.2667  0.8881    25.0722  0.0000
     aspect  0.7126  1.1661     0.6111  0.5440

Residual standard error: 4.07 on 48 degrees of freedom
Multiple R-Squared: 0.00772
F-statistic: 0.3735 on 1 and 48 degrees of freedom, the
p-value is 0.544
…
[[2]]:
…
```

Figure 5.5. *Continued.*

```
Coefficients:
                 Value Std. Error  t value Pr(>|t|)
(Intercept)    22.6821    0.6542   34.6705   0.0000
    species    -0.0093    1.3948   -0.0067   0.9947

Residual standard error: 4.086 on 48 degrees of freedom
Multiple R-Squared: 9.31e-007
F-statistic: 0.00004469 on 1 and 48 degrees of freedom,
the p-value is 0.9947

...
...
[[127]]:
...
Coefficients:
                Value Std. Error t value Pr(>|t|)
(Intercept)   6.5631  2.0027      3.2771  0.0021
     aspect   0.0185  0.6949      0.0267  0.9788
    species  -0.2810  0.7409     -0.3793  0.7064
 old.growth   0.0004  0.0064      0.0689  0.9454
       rock   2.9321  0.6998      4.1902  0.0001
       moss   2.9913  0.6703      4.4627  0.0001
       temp   0.4214  0.0825      5.1112  0.0000
      moist   0.0962  0.0108      8.8728  0.0000

Residual standard error: 2.077 on 42 degrees of freedom
Multiple R-Squared: 0.774
F-statistic: 20.54 on 7 and 42 degrees of freedom, the
p-value is 1.107e-011
```

5.7.3 Bayesian Statistical Analysis in WinBUGS: A Priori Parsimonious Model Selection and Inference Using the Deviance Information Criterion (DIC)

The Bayesian statistical analysis of `data1` in WinBUGS, using the parsimonious a priori model selection and inference strategy, is presented in Figs. 5.6 and 5.7. Figure 5.6 provides the WinBUGS code used for model 8, the "correct model." Note the straightforward specification of the multiple linear regression model with normal errors in the program code, and the noninformative, approximately flat, normally distributed priors with precision $\tau = 10^{-12}$ and standard deviation $\sigma = 10^6$. Figure 5.7 presents the WinBUGS DIC and DIC weight statistics for the Bayesian MCMC analysis, along with the AIC and AIC weights for the frequentist analysis. The Bayesian DIC and DIC weights are very similar to the frequentist AIC and AIC weights, with rankings and weights of the models preserved, regardless of frequentist or Bayesian approaches.

Although the statistics are similar for the frequentist approach and Bayesian approach with noninformative priors, the advantages and disadvantages of each should be considered by the natural resource scientist when deciding which to use in the planning phase of a data collection project. The frequentist approach to multiple

Figure 5.6. Bayesian statistical analysis WinBUGS program code and output results for the "correct model" 8. (**a**) Program code. (**b**) Output results: kernel density graphys of the posteriors. (**c**) Output results: sample statistics. (**d**) Output results: DIC.

(**a**)

```
        Bayesian Multiple Linear Regression Model:
              Model #8 - Correct Model with
                 {rock,moss,temp,moist}
                    Howard Stauffer

                       Reality:
response = 5+2*rock+3*moss+0.5*temp+0.10*moist+rnorm(50,0,3)
-------------------------------------------------------------
1. Program code
model
{
for(i in 1:n)
  {
response[i] ~ dnorm(mu[i],tau)
mu[i] <- beta0 + beta4*rock[i] + beta5*moss[i]
+ beta6*temp[i] + beta7*moist[i]  }

beta0 ~ dnorm(0,0.000000000001)
beta4 ~ dnorm(0,0.000000000001)
beta5 ~ dnorm(0,0.000000000001)
beta6 ~ dnorm(0,0.000000000001)
beta7 ~ dnorm(0,0.000000000001)
tau ~ dgamma(0.001,0.001)
sigma <- 1/sqrt(tau)
}

2. Data
list(rock=c(0,0,1,1,0,1,0,0,1,0,0,0,0,0,0,0,0,0,1,0,1,0,
0,1,0,0,0,0,0,0,1,0,0,0,0,0,0,1,0,1,0,1,1,0,0,1,0,0,0,1),

moss=c(1,0,1,0,0,0,1,0,0,0,1,1,0,0,1,0,0,0,0,0,1,1,0,0,
0,1,1,0,0,0,0,1,0,0,0,1,1,0,1,0,0,0,0,0,0,1,0,1,0,0),

temp=c(25.9,21.9,17.4,27.0,23.0,23.1,23.6,28.6,27.9,22.7,
26.2,15.6,22.3,19.9,22.9,18.2,26.3,27.0,15.8,26.8,22.3,
23.4,18.0,27.8,22.6,21.8,22.6,22.5,26.1,26.7,20.2,19.2,
22.3,24.0,21.4,22.0,28.1,24.1,15.1,17.5,27.3,19.1,23.1,
23.7,23.2,17.2,29.9,29.7,22.2,29.8),

moist=c(35.5,88.8,37.6,79.3,24.1,33.1,69.5,46.6,64.6,
11.6,38.7,26.7,76.5,10.2,67.4,50.8,99.5,85.2,64.9,93.1,
12.3,23.1,17.1,36.7,9.1,54.4,59.2,24.5,75.4,26.5,74.6,
4.4,65.7,72.1,2.9,25.6,42.5,19.4,63.2,97.9,65.1,93.7,
40.1,41.0,22.4,72.0,35.9,88.4,17.2,13.2),
```

Figure 5.6. *Continued.*

```
response=c(20.8,22.8,24.7,28.4,19.9,23.1,26.2,22.0,30.4,
17.6,24.7,17.8,21.4,14.5,25.2,17.1,26.5,25.3,19.9,28.2,
21.1,25.3,16.1,24.3,18.4,25.7,25.6,18.8,27.4,19.5,23.8,
20.8,26.2,24.7,13.2,23.1,24.5,23.5,19.5,28.1,26.4,27.8,
21.6,17.6,18.1,25.3,18.1,29.7,21.2,22.1),

n = 50)

3. Initial values

list(beta0 = 1, beta4=1,beta5=1,beta6=1,beta7=1,tau = 1)
```

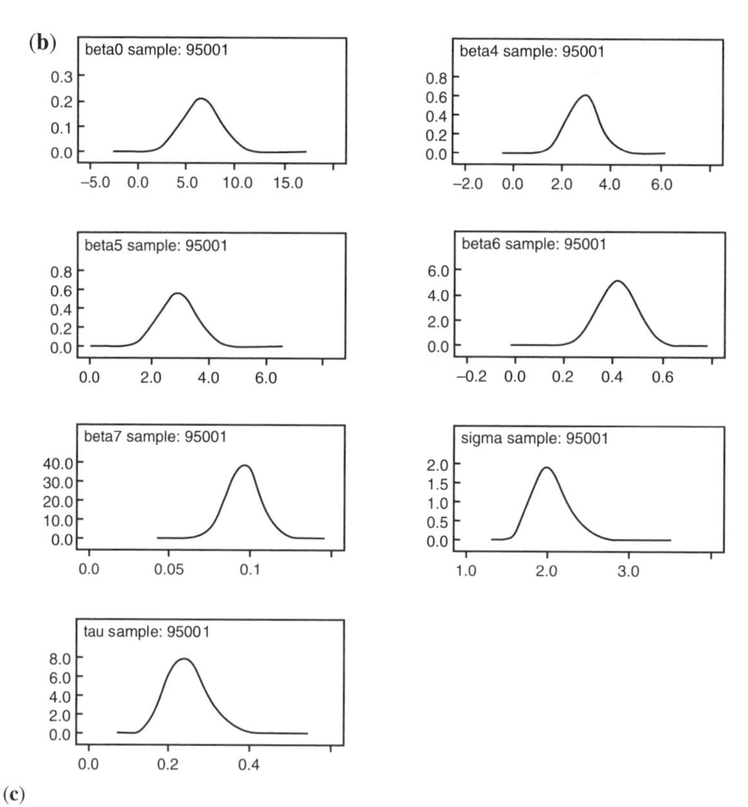

(c)

node	mean	sd	MC error	2.5%	median	97.5%	start	sample
beta0	6.613	1.953	0.006026	2.794	6.608	10.47	5000	95001
beta4	2.898	0.6664	0.002083	1.591	2.9	4.212	5000	95001
beta5	3.022	0.6453	0.002114	1.753	3.02	4.298	5000	95001
beta6	0.4186	0.07833	2.407E-4	0.2645	0.4188	0.5712	5000	95001
beta7	0.09624	0.01054	3.357E-5	0.07551	0.09626	0.1168	5000	95001
sigma	2.045	0.2215	8.021E-4	1.667	2.026	2.535	5000	95001
tau	0.2474	0.05221	1.826E-4	0.1557	0.2437	0.3597	5000	95001

(d)

Dbar = post.mean of -2logL; Dhat = -2LogL at post.mean of stochastic nodes

	Dbar	Dhat	pD	DIC
response	212.829	206.709	6.121	218.950
total	212.829	206.709	6.121	218.950

Figure 5.7. Statistics for 10 models: AIC and Akaike weights from frequentist statistical analysis; DIC and DIC weights from Bayesian statistical analysis.

Model	Covariates	k	p_D	AIC	Akaike Weights	DIC	DIC Weights
1	aspect, species, old.growth	5	5.079	290.052	<0.00005%	290.379	<0.00005%
2	aspect, species, moss	5	5.079	288.162	<0.00005%	288.489	<0.00005%
3	aspect, species, temp	5	5.079	283.986	<0.00005%	284.313	<0.00005%
4	aspect, species, moist	5	5.079	255.641	<0.00005%	255.968	<0.00005%
5	aspect, old.growth, temp	5	5.079	282.974	<0.00005%	283.301	<0.00005%
6	species, old.growth, temp	5	5.079	283.821	<0.00005%	284.147	<0.00005%
7	temp, moist	4	4.066	244.319	0.0002%	244.546	0.0003%
8	rock, moss, temp, moist	6	6.121	218.437	94.8052%	218.950	96.0015%
9	none	2	2.027	284.600	<0.00005%	284.664	<0.00005%
10	aspect, species, old.growth, rock, moss, temp, moist	9	9.172	224.246	5.1946%	225.307	3.9983%
				Total:	100.0000%	*Total:*	100.0000%

Note: k = number of parameters
p_D = Bayesian number of parameters

linear regression analysis is reliable and well known, with standards that are well established. Alternatively, the Bayesian approach to multiple linear regression incorporates previous information with priors that can be generalized and provides direct probability inferences for parameters, of interest to natural resource managers. Each has its place in the toolbox of contemporary statistical methods available to the natural resource scientist.

Similarly, the natural resource scientist should consider which model selection and inference strategy to employ during the planning phase of a data collection project. The a posteriori strategy provides descriptive models for populations that are not well understood. Alternatively, a priori strategy provides predictive models for populations that are better understood. Each also has its place in the toolbox of contemporary model selection and inference strategies available to the natural resource scientist.

5.8 SUMMARY

In this chapter we described two contrasting strategies for model selection and inference and illustrated with a multiple linear regression example. The two strategies are a descriptive strategy of a posteriori exploratory data analysis, sometimes pejoratively called "data dredging," and a predictive strategy of a priori parsimonious model selection and inference using information-theoretic criteria such as AIC and DIC.

The descriptive strategy is particularly appropriate for populations that are poorly understood. This strategy has the advantage of examining models selected a posteriori to data collection. However, models fit to sample datasets using this approach may experience compounded error and overfitting. Hence, this strategy provides descriptive results that should be viewed as preliminary, tentative, and hypothetical. Predictive conclusions based on this approach should be inferred with extreme caution. Additional studies and analyses, using independent datasets, may be necessary in order to ensure reliable predictive results.

We contrasted this approach with an alternative predictive strategy of a priori parsimonious model selection and inference using information-theoretic criteria, such as AIC and AIC_c with frequentist statistical analysis and DIC with Bayesian statistical analysis. This strategy is more appropriate for populations that are better understood. Its results avoid the compounding of error and overfitting of models and, if subjected to additional testing for goodness of fit, are more reliable for prediction.

Each of these strategies can be used with both frequentist and Bayesian statistical analysis. We reviewed in this chapter the standard methods of fit for the estimation of parameters in models: least-squares fit, maximum-likelihood fit, and Bayesian fit. We also reviewed the standard methods of evaluation of fit, using cross-validation and test datasets. Finally, we discussed the use of model averaging with AIC_c and DIC weights, for coefficient estimates and error, prediction, and the importance of covariates. We concluded with a multiple linear regression example using S-Plus and R for frequentist statistical analysis and WinBUGS for Bayesian statistical analysis.

PROBLEMS

5.1 Conduct a frequentist statistical analysis of the sample dataset `data2` (Fig. 5.8) in S-Plus or R, using a predictive a priori parsimonious model selection and inference strategy with AIC_c. The dataset `data2` consists of habitat selection modeling data for salamander biomass response with covariates aspect, species, old.growth, rock, moss, temp, and moist. Use multiple linear regression for a comparative analysis of the following 12 models with linear covariates:

1. {aspect,species,old.growth}
2. {aspect,species,rock}
3. {aspect,species,moss}
4. {aspect,species,temp}
5. {aspect,species,moist}
6. {aspect,old.growth,temp}
7. {species,old.growth,temp}
8. {moss,temp,moist}
9. {rock,temp}
10. {temp,moist}
11. {rock,moss,temp,moist}
12. {aspect,species,old.growth,rock,moss,temp,moist} (full model)

5.2 Conduct a frequentist statistical analysis of the sample dataset `data2` (Fig. 5.8) in S-Plus or R, using a descriptive a posteriori model selection and inference strategy. Use multiple linear regression for a comparative analysis of models, including stepwise and best-subsets selection of the models with the 128 different combinations of covariates. Use the Mallows C_p and AIC as the model selection criteria for the stepwise and best-subsets selection methods, respectively.

5.3 Conduct a Bayesian statistical analysis of the sample dataset `data2` (Fig. 5.8) in WinBUGS, using a predictive a priori parsimonious model selection and inference strategy with DIC. Use multiple linear regression for a comparative analysis of the 12 models with the linear covariates prescribed in Problem 5.1 (above).

5.4 Conduct model averaging on the a priori parsimonious model selection results of Problems 5.1 and 5.3, using AIC and DIC weights. Use the mean and standard deviation of the posteriors with the DIC weights. Calculate the unconditional estimates of the coefficients of temp and moist, with and without shrinkage, along with their unconditional estimates of standard error. Also calculate the importance of the covariates. Are the results very different, with and without shrinkage, and with AIC and DIC weights?

Figure 5.8. Habitat selection modeling dataset `data2` for Problems 5.1–5.6.

sample	aspect	species	old.growth	rock	moss	temp	moist	response
1	0	0	0.812	1	1	22.5	12.1	33.6
2	1	0	0.564	0	0	22.9	35.2	45.5
3	0	0	0.455	0	1	29.7	52.3	67.6
4	1	1	0.808	0	0	15.2	15.2	26.9
5	0	1	0.225	0	0	17.5	38.7	53.1
6	0	0	0.765	0	0	29.8	93.0	105.5
7	1	0	0.059	0	0	17.4	2.0	15.6
8	1	1	0.038	0	0	20.6	80.5	84.3
9	1	0	0.277	0	0	27.9	90.9	102.1
10	1	0	0.417	0	1	16.1	63.6	65.9
11	0	0	0.757	0	0	27.8	51.4	62.5
12	1	0	0.806	0	0	21.9	96.8	108.7
13	1	1	0.015	0	1	21.6	77.9	90.7
14	0	0	0.050	0	1	25.3	3.5	20.3
15	0	0	0.006	0	1	16.8	49.2	50.8
16	0	0	0.646	0	1	15.5	59.6	68.0
17	0	1	0.800	0	1	19.9	40.5	48.3
18	1	1	0.358	1	0	26.7	81.5	90.8
19	1	1	0.151	0	1	18.6	20.3	37.9
20	0	0	0.745	0	0	29.1	11.3	25.2
21	1	0	0.905	0	1	19.7	93.4	100.2
22	0	0	0.206	0	1	16.6	25.0	29.1
23	1	1	0.576	0	0	18.4	34.1	40.4
24	1	0	0.644	0	1	25.4	46.3	57.7
25	1	0	0.501	0	1	22.8	43.0	58.5
26	1	0	0.204	0	0	22.0	96.0	105.6
27	0	0	0.575	0	1	15.0	33.8	37.7
28	1	0	0.777	0	0	25.5	62.5	76.6
29	0	0	0.484	0	1	25.0	57.0	67.3
30	0	0	0.231	0	1	23.9	4.2	12.9
31	1	0	0.489	0	1	20.7	94.5	102.5
32	1	1	0.012	0	1	25.4	36.5	55.3
33	1	1	0.205	0	0	17.6	7.5	25.6
34	0	0	0.238	0	1	22.8	84.6	88.0
35	0	0	0.144	0	0	19.7	22.9	26.8
36	1	0	0.887	0	0	17.6	54.8	62.9
37	0	0	0.001	0	0	15.7	66.8	78.4
38	1	1	0.388	0	0	29.1	42.5	57.6
39	0	0	0.396	0	1	17.7	72.4	77.1
40	0	0	0.751	0	0	18.4	3.8	19.0
41	0	0	0.687	0	1	20.9	14.8	27.6
42	1	1	0.514	0	0	21.9	52.8	60.3
43	1	0	0.371	1	1	27.8	64.2	71.7
44	0	0	0.494	0	0	21.5	92.4	95.1
45	1	0	0.205	1	0	21.5	22.3	36.3
46	0	0	0.919	0	1	29.4	57.9	74.7
47	0	0	0.746	1	0	27.6	90.0	98.7
48	1	0	0.197	0	0	22.1	5.7	21.1
49	1	1	0.655	1	1	26.4	68.1	77.2
50	0	0	0.666	1	0	24.8	29.5	44.6

5.5 Conduct a goodness of fit analysis of the best-fitting model from the a priori model selection and inference results of Problems 5.1 and 5.3, including the following:

(a) Examine the residuals for independence and normality.

(b) Conduct a cross-validation analysis in S-Plus or R, developing and utilizing a program to delete successive points, fit the best-fitting model to the remaining data, and assessing the predicted versus observed fits of the deleted points with respect to independence, normality, and prediction interval.

5.6 Write a 3–5-page report, with appendix, summarizing the results of the analysis in Problems 5.1–5.5. Specify the statistics used for the comparative analysis of the different multiple linear regression models, with figures and tables when helpful. Contrast the results and inferences from the two strategies for model selection. What can be inferred for the best-fitting model

$$\begin{aligned} \text{response} \sim\ & \beta_0 + \beta_1 \cdot \text{aspect} + \beta_2 \cdot \text{species} + \beta_3 \cdot \text{old.growth} \\ & + \beta_4 \cdot \text{rock} + \beta_5 \cdot \text{moss} + \beta_6 \cdot \text{temp} + \beta_7 \cdot \text{moist} + e \end{aligned}$$

where $e \sim N(0,\sigma)$? Interpret the biological meaning of the statistical results from a management perspective. Include in your report an abstract and introduction, problem statement, objectives, methods, results, discussion, and conclusions sections, along with an appendix with figures and tables.

6 An Introduction to Generalized Linear Models: Logistic Regression Models

In this chapter we will provide an introduction and overview of an exciting contemporary frequentist approach to statistical modeling, one that generalizes multiple linear regression modeling: generalized linear modeling (GLM). Generalized linear modeling generalizes multiple linear regression modeling by permitting varying types of error structures and incorporating a link function for the response. It also uses frequentist maximum likelihood (ML) estimation of model parameters rather than least-squares (LS) fit. We will discuss the importance of design for generalized linear modeling and present the analysis, including the assumptions, fit, significant statistics, and selection of models for the exponential family of distributions. We will then concentrate on the most important generalized linear model used in natural resource science, the logistic regression model, including its assumptions, fit, significant statistics, model selection, and goodness of fit. We will briefly describe alternative generalized linear models such as the probit and complementary log–log models for binomial data, the Poisson and negative binomial models for count data, and the log-linear model for categorical data in contingency tables. We will conclude by illustrating the application of frequentist GLM with logistic regression examples in S-Plus and R. We will also illustrate the Bayesian approach to GLM with examples in WinBUGS.

6.1 INTRODUCTION TO GENERALIZED LINEAR MODELS (GLMs)

Although general linear models such as ANOVA and multiple linear regression have traditionally provided useful statistical tools for natural resource scientists and managers, they are based on important assumptions that limit their application; residual errors are required to be normally distributed, and the mean response is required to be a linear function of covariates and factors. Increasingly, natural resource scientists

Contemporary Bayesian and Frequentist Statistical Research Methods for Natural Resource Scientists. By Howard B. Stauffer

and managers require the use of models with error terms that are not normally distributed, for binary datasets that are binomially distributed and count datasets that are Poisson or negative binomially distributed. Additionally, mean responses for natural resource datasets may be probabilities or proportions that are nonlinear and bounded. Because of these limitations, more general forms of models are required such as generalized linear models, which are of increasing importance in natural resource science applications.

Generalized linear models (GLMs) relax the assumptions of normally distributed error and linear mean response. They achieve the latter by requiring the incorporation of a link function of the mean response to a form that is a linear function of covariates and factors. The most important of these models can be described by a general family of distributions, the exponential family of distributions, with mean responses that are nonlinear in form. In this chapter, we introduce this contemporary method, which should become an integral part of the natural resource scientist's statistical analysis toolbox. We will focus on a model for binomial response, the most popular GLM of interest to natural resource scientists, logistic regression modeling. We will also briefly introduce the probit and complementary log–log GLMs for binomial data, and the Poisson and negative binomial regression GLMs for count data. We refer the interested reader to the extensive literature on this subject, for further detail beyond this brief introduction (Dobson 1990, McCullagh and Nelder 1996, Hosmer and Lemeshow 2000, Hardin and Hilbe 2001, McCulloch and Searle 2001). We particularly recommend Dobson (1990) for an elegant, more mathematical, introductory discussion of the topic that is beyond the scope of this book.

6.2 GLM DESIGN

All GLM data should be collected with careful a priori consideration given to design. As with any data used for statistical modeling, the analyst should be cognizant of the distinction between design-based and model-based inference. Design-based inference provides estimators with frequentist properties such as unbiasedness and precision, which are a function of the statistical design for the study and not model assumptions for the distribution of the population. Model-based inference, on the other hand, provides estimators with properties that depend on distributional assumptions for the population that are required by the model. Hence, a study based on a rigorous statistical design can provide a dataset with design-based analytical results that are robust and not dependent on the adherence of the population to the distributional assumptions required by model-based inference. On the other hand, model-based analytical results can benefit from a statistical design that maximizes efficiency in the analysis of a sample survey and power in an experiment.

For example, consider a GLM model examining the linear relationship between an independent variable x and a dependent response variable y. With a model-based approach, where the form of the model is known, independent x values for the dataset should be chosen that are concentrated at important regions of its domain.

With a design-based approach, alternatively, x values should be randomly selected throughout its domain. In both cases, the y values for the fixed x values should be chosen randomly. With a design-based approach, therefore, the x values and y values should both be selected randomly, whereas with a model-based approach, only the y values should be randomly chosen. The danger with a model-based design is that the results provide inferences only for a population that satisfies the distributional assumptions of the model. However, the advantage of a model-based design is that the analysis results can be maximized for efficiency and power.

6.3 GLM ANALYSIS

Traditionally, statistical analyses of continuous data for natural resource scientists and managers have been based on general linear models of the form

$$y_i = \beta_1 \cdot x_{i1} + \beta_2 \cdot x_{i2} + \cdots + \beta_k \cdot x_{ik} + e_i, i = 1, 2, \ldots, n,$$

or in matrix notation

$$\mathbf{y} = \mathbf{x} \cdot \boldsymbol{\beta} + \mathbf{e},$$

where $\mathbf{y}$ is an $n \times 1$ vector, $\mathbf{x}$ is an $n \times k$ design matrix of explanatory covariate or factor measurements, $\boldsymbol{\beta}$ is a $k \times 1$ parameter vector, and $\mathbf{e}$ is an $n \times 1$ error vector of independent, identically, and normally distributed (iid) elements. In this formulation, n is the sample size and k is the number of parameters. If a constant is used in the model, $x_{i1} = 1$ and β_1 is the constant. Both ANOVA and multiple linear regression are described by these models.

More recent advances in statistical theory and computer software, however, provide a generalization of these methods to response variables that are more generally distributed, categorical as well as continuous, from the **exponential family of distributions** (Dobson 1990). Additionally, these methods permit a relationship between the independent explanatory $\mathbf{x}$ and dependent response $\mathbf{y}$ variables to be nonlinear, of the form $\mathbf{y} = f(\mathbf{x} \cdot \boldsymbol{\beta})$, and the estimation procedure for the parameters $\boldsymbol{\beta}$ of the model is based on ML estimation rather than LS fit (see Section 2.2). The ML estimation uses an iterative weighted LS algorithm and is included in many statistical packages. Maximum-likelihood estimation has the important property of asymptotic unbiasedness and efficiency, providing approximately unbiased estimators with minimum error. Least-squares estimation also provides unbiased estimators of maximal efficiency, is computationally simpler, and does not require probability distributional assumptions. However, there is little advantage in using it since probability distributional assumptions are required anyway to ensure the distributional properties of LS estimators.

The exponential family of distributions is of the form

$$\begin{aligned} f(y;\theta) &= s(y)\cdot t(\theta)\cdot e^{a(y)\cdot b(\theta)} \\ &= e^{a(y)\cdot b(\theta)+c(\theta)+d(y)}, \end{aligned}$$

with $s(y)=e^{d(y)}$ and $t(\theta)=e^{c(\theta)}$. Here y is the data and θ is the parameter. If $a(y)=y$, the distribution is said to be in **canonical form**, and $b(\theta)$ is called the **natural parameter** of the distribution. It can be shown that the normal, binomial, Poisson, negative binomial, and exponential distributions all belong to the exponential family of distributions. To illustrate this, the Poisson distribution is in the exponential family since its distribution is given by

$$\begin{aligned} f(y;\lambda) &= \lambda^y\cdot e^{-\lambda}/y! \\ &= e^{y\cdot\log(\lambda)-\lambda-\log(y!)}, \end{aligned}$$

so that $a(y)=y$, $b(\lambda)=\log(\lambda)$, $c(\lambda)=-\lambda$, and $d(y)=-\log(y!)$. It is in the canonical form with natural parameter $\log(\lambda)$. Similarly, it can be shown that the normal, binomial, negative binomial, and exponential distributions are all in canonical form with $a(y)=y$ (Dobson 1990). The natural parameters b, c, and d for these important distributions in the exponential family are indicated as follows:

Distribution	$b(\theta)$	$c(\theta)$	$d(y)$
Normal $f(y;\mu)$	μ/σ^2	$\frac{1}{2}\cdot(\mu^2/\sigma^2-\log(2\cdot\pi\cdot\sigma^2))$	$-\frac{1}{2}\cdot y^2/\sigma^2$
Binomial $f(y;p)$	$\log(p/(1-\pi))$	$n\cdot\log(1-\pi)$	$\log\binom{y}{n}$
Poisson $f(y;\lambda)$	$\log(\lambda)$	$-\lambda$	$-\log(y!)$
Negative binomial $f(y;\theta)$	$\log(1-\theta)$	$r\cdot\log(\theta)$	$\log\binom{y+r-1}{r-1}$
Exponential $f(y;\lambda)$	λ	$\log(\lambda)$	0

It can be shown that the expected value, or mean, and variance of $a(Y)$ are given by the first- and second-derivative expressions (Dobson 1990)

$$\begin{aligned} E[a(Y)] &= -c'(\theta)/b'(\theta), \\ \text{var}[a(Y)] &= [b''(\theta)\cdot c'(\theta)-c''(\theta)\cdot b'(\theta)]/b'(\theta)^3. \end{aligned}$$

Generalized linear models unify statistical methods that incorporate a linear combination of parameters (Nelder and Wedderburn 1972). A GLM consists of a set of independent random response variables $y_1, y_2, \ldots, y_n$, each with the same distribution from the exponential family of canonical form depending on a single parameter $\theta_i, i=1,2,\ldots,n$. Thus the joint probability density function of $y_1, y_2, \ldots, y_n$, is given by

$$f(y_1, y_2, \ldots, y_n; \theta_1, \theta_2, \ldots, \theta_n) = \exp\left[\sum_{i=1}^{n} y_i \cdot b(\theta_i) + \sum_{i=1}^{n} c(\theta_i) + \sum_{i=1}^{n} d(y_i)\right].$$

The parameters θ_i are usually not of direct interest since there is one for each observation, so we consider a smaller set of parameters $\beta_1, \beta_2, \ldots, \beta_k$ where $k < n$ such that

$$g(\mu_i) = \beta_1 \cdot x_{i1} + \beta_2 \cdot x_{i2} + \cdots + \beta_k \cdot x_{ik}, \quad i = 1, 2, \ldots, n,$$

or, in the language of matrices,

$$g(\boldsymbol{\mu}) = \mathbf{x} \cdot \boldsymbol{\beta},$$

where g is a monotone, differentiable function, called the **link function**, of the expected value $\boldsymbol{\mu}_i$ of y_i. Here $\boldsymbol{\mu}$ is the $n \times 1$ vector of mean responses, $\mathbf{x}$ is the $n \times k$ design matrix of explanatory variables (covariates and factors), and $\boldsymbol{\beta}$ is the $k \times 1$ vector of parameters.

Both ANOVA and multiple linear regression are examples of a generalized linear model with a normally distributed response and link function equal to the identity. Logistic regression and Poisson and negative binomial regression are also examples of generalized linear models for binary and count data, respectively.

The method of fit for generalized linear models is ML estimation of the parameters, based on the sample dataset. The parameters are the coefficients of the linear model, related to the mean response by the link function. Maximum-likelihood estimation is based on an iterative weighted LS procedure. Models can be compared using AIC or AIC_c and their Akaike weights. The models can be evaluated for goodness of fit by examining their deviances and comparing them with a χ^2 distribution. Readers interested in further details should consult the appropriate references (Dobson 1990, McCullagh and Nelder 1996). We will illustrate these ideas by focusing on the most commonly used GLM for natural resource applications, logistic regression.

6.4 LOGISTIC REGRESSION ANALYSIS

Logistic regression analysis is based on a generalized linear model for binary response using a logit link function. A binary random variable z has one of two possible values, "success" or "failure," "yes" or "no," "present" or "absent," or 0 or 1, with probability$(z = 1) = p$ and probability$(z = 0) = 1 - p$ with $0 \leq p \leq 1$. If there are n independent random variables $z_1, z_2, \ldots, z_n$ with probability $(z_i = 1) = p_i$, then their joint probability distribution is given by

$$f(z_1, z_2, \ldots, z_n; \theta_1, \theta_2, \ldots, \theta_n) = \prod_{i=1}^{n} p_i^{z_i}(1 - p_i)^{1-z_i}$$

$$= \exp\left[\sum_{i=1}^{n} z_i \cdot \log\left(\frac{p_i}{1 - p_i}\right) + \sum_{i=1}^{n} \log(1 - p_i)\right]$$

and is a member of the exponential family. If the p_i values are all equal and $y = \sum_{i=1}^{n} z_i$, then the random variable y has the binomial distribution

$$B(y; n, p) = \text{probability}(y) = \binom{y}{n} \cdot p^y \cdot (1 - p)^{n-y}.$$

If there are n such independent random variables $y_1, y_2, \ldots, y_n$, the log-likelihood function is given by

$$\ell(p_1, p_2, \ldots, p_n; y_1, y_2, \ldots, y_n)$$
$$= \sum_{i=1}^{n}\left[y_i \cdot \log\left(\frac{p_i}{1 - p_i}\right) + n_i \cdot \log(1 - p_i) + \log\binom{n_i}{y_i}\right].$$

This distribution, if written in terms of the binary variables z_i, can be shown to belong to the exponential family of distributions.

The link function for logistic regression is given by the logit function

$$\text{logit} = g(p) = \log\left(\frac{p}{1 - p}\right),$$

the log of the odds ratio. For the logistic regression GLM with one explanatory variable x, the link function, or logit, is given by

$$\text{logit} = g(p) = \log\left(\frac{p}{1 - p}\right) = \beta_1 + \beta_2 \cdot x.$$

For the more general logistic regression GLM with one or more explanatory variable, the link function is given by

$$\text{logit} = g(p) = \log\left(\frac{p}{1 - p}\right) = \mathbf{x} \cdot \boldsymbol{\beta}$$

$$= \sum_{i=1}^{k} \beta_i x_i = \beta_1 \cdot x_1 + \beta_2 \cdot x_2 + \cdots + \beta_k \cdot x_k,$$

where k is the number of parameters and $x_1 = 1$. The inverse $f = g^{-1}$ of the logit link function g is given by

$$p = f(\mathbf{x}\cdot\boldsymbol{\beta}) = \frac{\exp(\beta_1\cdot x_1 + \beta_2\cdot x_2 + \cdots + \beta_k\cdot x_k)}{1+\exp(\beta_1\cdot x_1 + \beta_2\cdot x_2 + \cdots + \beta_k\cdot x_k)} = \frac{\exp(\mathbf{x}\cdot\boldsymbol{\beta})}{1+\exp(\mathbf{x}\cdot\boldsymbol{\beta})}.$$

Note that the complement of p is given by

$$\begin{aligned}(1-p) &= 1 - \frac{\exp(\beta_1\cdot x_1 + \beta_2\cdot x_2 + \cdots + \beta_k\cdot x_k)}{1+\exp(\beta_1\cdot x_1 + \beta_2\cdot x_2 + \cdots + \beta_k\cdot x_k)}\\ &= \frac{1}{1+\exp(\beta_1\cdot x_1 + \beta_2\cdot x_2 + \cdots + \beta_k\cdot x_k)} = \frac{1}{1+\exp(\mathbf{x}\cdot\boldsymbol{\beta})}.\end{aligned}$$

Two other generalized linear models are sometimes used for binary response data, the probit model and the complementary log–log model. The probit model uses an inverse cumulative normal probability function

$$g(p) = \Phi^{-1}(p) = \mathbf{x}\cdot\boldsymbol{\beta}$$

as its link function, where Φ is the cumulative probability function for the standard normal distribution

$$\Phi\left(\frac{x-\mu}{\sigma}\right) = \frac{1}{\sigma\sqrt{2\pi}}\int_{-\infty}^{x} e^{-\frac{1}{2}[(t-\mu)/\sigma]^2}\,dt.$$

The complementary log–log model uses the complementary log–log function

$$g(p) = \log[-\log(1-p)] = \mathbf{x}\cdot\boldsymbol{\beta}$$

for its link function. Although all three models can be applied to binary data, they differ in shape, particularly in the tails for values of p near 0 and 1. Each model is based on different assumptions for its derivation, but the logistic regression model has a particularly appealing interpretation for its odds ratios, as discussed in Section 6.4.3.

6.4.1 The Link Function and Error Assumptions of the Logistic Regression Model

Recall that the link function for logistic regression is given by the logit function

$$\text{logit} = g(p) = \log\left(\frac{p}{1-p}\right) = \mathbf{x}\cdot\boldsymbol{\beta}$$

with inverse

$$f(\mathbf{x}\cdot\boldsymbol{\beta}) = \frac{\exp(\beta_1\cdot x_1 + \beta_2\cdot x_2 + \cdots + \beta_k\cdot x_k)}{1+\exp(\beta_1\cdot x_1 + \beta_2\cdot x_2 + \cdots + \beta_k\cdot x_k)} = \frac{\exp(\mathbf{x}\cdot\boldsymbol{\beta})}{1+\exp(\mathbf{x}\cdot\boldsymbol{\beta})}.$$

The errors for logistic regression are binomially distributed $B(y_i; n_i, p_i)$.

6.4.2 Maximum-Likelihood (ML) Fit of the Logistic Regression Model

The log-likelihood function for logistic regression is given by

$$\begin{aligned}
&\ell(p_1, p_2, \ldots, p_n; y_1, y_2, \ldots, y_n) \\
&\quad = \sum_{i=1}^{n}\left[y_i\cdot\log\left(\frac{p_i}{1-p_i}\right) + n_i\cdot\log(1-p_i) + \log\binom{n_i}{y_i}\right] \\
&\quad = \sum_{i=1}^{n}\left[y_i\cdot\log(p_i) + (n_i - y_i)\cdot\log(1-p_i) + \log\binom{n_i}{y_i}\right].
\end{aligned}$$

This function is maximized with respect to the parameters in the generalized linear model $\beta_1, \beta_2, \ldots, \beta_k$ by using an iterative weighted LS algorithm (Dobson 1990, pp. 39–41) to derive the GLM ML solution.

6.4.3 Logistic Regression Statistics

The ML solution for the logistic regression GLM results in estimates $\hat{\beta}_1, \hat{\beta}_2, \ldots, \hat{\beta}_k$ for the k parameters $\beta_1, \beta_2, \ldots, \beta_k$ in the linear model, along with estimates of their standard error se $\hat{\beta}_1$, se $\hat{\beta}_2, \ldots,$ se $\hat{\beta}_k$. These estimates can be assessed for statistical significance by checking whether their confidence intervals do not contain zero or equivalently by testing the null hypothesis H_0: $\beta_i = 0$. The estimated model can be evaluated by examining its deviance, R^2, and AIC or AIC_c statistics. These statistics are discussed next.

Confidence intervals for the parameters can be calculated using estimates $\hat{\beta}_1, \hat{\beta}_2, \ldots, \hat{\beta}_k$ for the p parameters $\beta_1, \beta_2, \ldots, \beta_k$ in the linear model, and standard error se $\hat{\beta}_1$, se $\hat{\beta}_2, \ldots,$ se $\hat{\beta}_k$. The confidence intervals are given by

$$\mathrm{CI}_{\beta_i} = \hat{\beta}_i \pm E_{\hat{\beta}_i} = \left[\hat{\beta}_i - E_{\hat{\beta}_i}, \hat{\beta}_i + E_{\hat{\beta}_i}\right], \quad i = 1, 2, \ldots, k,$$

where the sampling error for the estimate of β_i is

$$E_{\hat{\beta}_i} = t_{1-\alpha/2, n-k}\cdot \mathrm{se}_{\hat{\beta}_i}.$$

The estimates $\hat{\beta}_i$ can then be examined for their statistical significance by determining whether each CI_{β_i} is distinct from 0. Alternatively and equivalently, the

Wald test for the null hypothesis H_0: $\beta_i = 0$ with alternative hypothesis H_A: $\beta_i \neq 0$ can be conducted on the basis of the test statistic

$$t_s = \frac{\hat{\beta}_i}{\mathrm{se}_{\hat{\beta}_i}}, \quad i = 1, 2, \ldots, k,$$

which is $t_{n\text{-}k}$-distributed if H_0 is true. With specified type I error α and confidence level of $P = 1 - \alpha$, reject the null hypothesis and conclude that the estimates are statistically significant if the p value satisfies the following Neyman–Pearson condition, equivalent to the test statistic being in the rejection region: $p \leq \alpha$. The comparative capability of the model at fitting the sample dataset can be assessed by using the measurement of the "error," based on the log-likelihood ratio **deviance** statistic

$$D = 2 \cdot \log\left[\frac{\mathcal{L}(\hat{p}_{\max}; y)}{\mathcal{L}(\hat{p}; y)}\right] = 2 \cdot [\ell(\hat{p}_{\max}; y) - \ell(\hat{p}; y)].$$

The log-likelihood ratio is multiplied by 2 because the D statistic under certain conditions is χ^2-distributed. The **maximum model**, or **saturated model**, is the GLM that uses the same distribution and link function as the model of interest and has the number of parameters equal to the number of observations. The maximum model assumes that the p_i terms are the parameters to be estimated, so the ML estimates for this model, obtained by setting the derivatives of the log-likelihood function equal to 0

$$\frac{\partial \ell}{\partial p_i} = \frac{y_i}{p_i} - \frac{n_i - y_i}{1 - p_i} = 0,$$

are given by

$$\hat{p}_{\max_i} = \frac{y_i}{n_i}.$$

Therefore the log-likelihood ratio for the maximum model is given by

$$\ell(\hat{p}_{\max}; y) = \sum_{i=1}^{n} \left[y_i \cdot \log\left(\frac{y_i}{n_i}\right) + (n_i - y_i) \cdot \log\left(1 - \frac{y_i}{n_i}\right) + \log\binom{n_i}{y_i} \right].$$

If the estimates of p for the model of interest, corresponding to the estimates $\hat{\beta}_1, \hat{\beta}_2, \ldots, \hat{\beta}_k$ for the k parameters $\beta_1, \beta_2, \ldots, \beta_k$, are given by $\hat{p}_i$, then the deviance is given by

$$\begin{aligned} D &= 2 \cdot \sum_{i=1}^{n} \left[y_i \cdot \log\left(\frac{y_i}{n_i \cdot \hat{p}_i} \right) + (n_i - y_i) \cdot \log\left(\frac{n_i - y_i}{n_i - n_i \cdot \hat{p}_i} \right) \right] \\ &= 2 \cdot \sum_{i=1}^{n} \left[o_i \cdot \log\left(\frac{o_i}{e_i} \right) \right], \end{aligned}$$

where o_i are the observed frequencies and e_i are the estimated expected frequencies. If the estimated model of interest fits the data well, then its likelihood should be similar to that of the maximum model and therefore the deviance is χ^2-distributed

$$D \sim \chi^2_{n-k}.$$

In other words, if the estimated model is correct, the deviance should be χ^2-distributed with $n-k$ degrees of freedom. In conclusion, the null hypothesis that the model is correct

$$H_0 : M = M_{\max}$$

versus the alternative hypothesis

$$H_{\mathrm{A}} : M \neq M_{\max}$$

can be tested by comparing its deviance statistic with the $\chi^2_{n\text{-}k}$ distribution. The test statistic, on average, should be equal to $n-k$, the mean of this χ^2 distribution, if the null hypothesis H_0 is true, and the p value can be compared with a prescribed type I error α.

Nagelkerke (1991) proposed an **adjusted R^2 coefficient of determination** for logistic regression, comparable to the adjusted R^2 statistic for multiple linear regression, given by

$$R^2_{\mathrm{adj}} = \frac{R^2}{R^2_{\max}},$$

where

$$R^2 = 1 - \left[\frac{\mathcal{L}(0)}{\mathcal{L}(\hat{\beta})} \right]^{2/n}$$

and

$$R^2_{\max} = 1 - [\mathcal{L}(0)]^{2/n}.$$

Here $\mathcal{L}(0)$ is the likelihood of the null model, the intercepts-only model, $\mathcal{L}(\hat{\beta})$ is the likelihood of the specified model, and n is the sample size. The Nagelkerke statistic varies between 0 and 1. The Nagelkerke adjusted R^2 statistic "measures" the proportion of variation in the response explained by the specified model, with consideration given to parsimony.

There are limitations to the use of statistics such as Nagelkerke's R^2 and deviance error D in comparing competitive models for goodness of fit. Although Negalkerke's R^2 measures the proportion of variation of the response explained by the model, it tends to favor models that are overfit, those with "steeper" relationships between the response and explanatory variables. Deviance error D also does not adequately account for the need for parsimony; as the number of covariates and factors in the model increases, the error or "bias" of the model fit becomes less, but the precision of the estimates for the coefficient parameters also becomes less. As the number of covariates and factors in a linear model increases, so does the number of parameters. Since each parameter estimate is based on a dataset with a fixed number of shared samples, the precision of the parameter estimates tends to decline as the number of parameters increases. A balance must be struck between the reduction of error or bias and the reduction of precision that occurs as the number of parameters in the models increases. The statistic that most efficiently separates the noise from the information in the sample dataset, which compares natural resource models for goodness of fit with due consideration given to parsimony, is the AIC criterion (Akaike 1973, 1974).

Akaike's information criterion is given by

$$\text{AIC} = D + 2 \cdot k,$$

where the deviance D for the logistic regression model is

$$D = 2 \cdot \sum_{i=1}^{n} \left[y_i \cdot \log\left(\frac{y_i}{n_i \cdot \hat{p}_i}\right) + (n_i - y_i) \cdot \log\left(\frac{n_i - y_i}{n_i - n_i \cdot \hat{p}_i}\right) \right],$$

the estimates $\hat{p}_i$ of p are derived from the estimates $\hat{\beta}_1, \hat{\beta}_2, \ldots, \hat{\beta}_k$ in the logit and the values for the covariates and factors at the ith covariate pattern, n_i is the number of observations at the ith covariate pattern, and k is the number of parameters in the model. A **covariate pattern** is a single set of values for the covariates in the model; that is, several observations may have the same covariate values, and hence may have a covariate pattern. The AIC criterion is (up to a constant) a first-order Taylor series approximation of the **Kullback–Leibler distance** between the model and the data (Burnham and Anderson 1998). The Kullback–Liebler distance measures the amount of **entropy**, **noise**, or error, separated from the **information**, in the sample dataset. A more precise second-order approximation of the Kullback–Liebler distance is given by

$$\text{AIC}_c = D + 2 \cdot k + \frac{2 \cdot k \cdot (k + 1)}{n - k - 1},$$

where n is the sample size. Since AIC_c is more precise than AIC, we always recommend using AIC_c, particularly for datasets with small sample size.

Models with lower AIC_c are better-fitting models. Because AIC_c is a relative measure of error, correct up to a fixed constant, its value should not be interpreted as an absolute measure of error. Models may therefore be compared using AIC_c but should not be evaluated for goodness of fit solely on the basis of this measurement. Other methods should be used to evaluate goodness of fit, such as the Hosmer–Lemeshow test (Hosemer and Lemeshow 2000), classification methods including ROC, concordance, and residual analysis for logistic regression (see discussion below).

Burnham and Anderson (2002) recommend that models with AIC_c within two units of the best-fitting model with minimum AIC_c be considered highly competitive. Akaike weight may also be used, however, to compare the models. Recall that the **Akaike weight** w_i is given by

$$w_i = \frac{\exp\left\{\left[-(\mathrm{AIC}_{c_i} - \min_{j=1}^{m}(\mathrm{AIC}_{c_j}))\right]/2\right\}}{\sum_{k=1}^{m} \cdot \exp\left\{\left[-(\mathrm{AIC}_{c_k} - \min_{j=1}^{m}(\mathrm{AIC}_{c_j}))\right]/2\right\}},$$

and may be interpreted as the likelihood, or probability, that the model M_i is the best-fitting model of the collection of models M_j, $j = 1, 2, \ldots, m$, under consideration. The leading models with a cumulative Akaike weight of 95% may be interpreted as a 95% **credible interval** of models best fitting the sample dataset, given the collection of models under consideration in the analysis.

Reasons for the importance of the logistic regression GLM among natural resource scientists for the analysis of data with binary response include (1) the shape of the model curve, particularly for p near 0 and 1; and (2) interpretation of the linear coefficient β_i of a covariate x_i as the log of the odds ratio Ψ of the specified linear model in comparison with the null model without the covariate. We will explain this idea next.

For a model M_1 with one dichotomous independent variable x and logit

$$M_1\text{: logit} = g(p) = \beta_0 + \beta_1 \cdot x,$$

we can show that the coefficient β_1 is equal to the log of the ratio Ψ of the odds, with and without the variable x in the model. In other words, if the odds for a model are given by

$$\text{Odds} = \frac{\Pr(y=1)}{\Pr(y=0)} = \frac{\Pr(y=1)}{(1-\Pr(y=1))},$$

where Pr is the probability of the response y, then the odds ratio is given by

$$\Psi = \left(\frac{\Pr_{M_1}(y=1)/1 - \Pr_{M_1}(y=1)}{\Pr_{M_0}(y=1)/1 - \Pr_{M_0}(y=1)}\right) = \frac{[e^{\beta_0+\beta_1}/(1+e^{\beta_0+\beta_1})]/[1/(1+e^{\beta_0+\beta_1})]}{[e^{\beta_0}/(1+e^{\beta_0})]/[1/(1+e^{\beta_0})]}$$

$$= \frac{e^{\beta_0+\beta_1}}{e^{\beta_0}} = e^{\beta_1},$$

where M_0 is the null model without the independent variable

$$M_0 : \text{logit} = g(p) = \beta_0$$

and M_1 is the specified model. Hence

$$\beta_1 = \log(\Psi) = \log\left(\frac{\Pr_{M_1}(y=1)/\Pr_{M_1}(y=0)}{\Pr_{M_0}(y=1)/\Pr_{M_0}(y=0)}\right)$$
$$= \log\left(\frac{\Pr_{M_1}(y=1)/(1-\Pr_{M_1}(y=1))}{\Pr_{M_0}(y=1)/(1-\Pr_{M_0}(y=1))}\right)$$

Alternatively, Ψ can be expressed in terms of the linear coefficient β_1 as $\Psi = e^{\beta_1}$. If the covariate is continuous rather than dichotomous, then this relationship can be interpreted in terms of the addition of one unit to the covariate x, the change from $x = 0$ to $x = 1$ in value. This amount of change in value of x will result in the magnitude of change in the odds given by the odds ratio $\Psi = e^{\beta_1}$. See Hosmer and Lemeshow (2000) for further details.

As with multiple linear regression, exploratory methods for model selection with logistic regression can be utilized such as stepwise and best subsets selection. The selection procedure for the stepwise methods is commonly based on the significance levels of the variables. As usual, be aware of the danger of compounded error and the overfitting of data with these methods. Hosmer and Lemeshow (2000) provide more detailed descriptions of these methods.

6.4.4 Goodness of Fit of the Logistic Regression Model

The statistics for the logistic regression model provide evidence for the comparative capability, or competitiveness, of the models to fit the sample dataset. However, these statistics do not provide definitive evidence for goodness of fit. The following methods can be used to examine the goodness of fit of the most competitive best-fitting models in order to complete the analysis.

The **Hosmer–Lemeshow test** is perhaps the most popular test for goodness of fit of the logistic regression model (SAS Institute Inc. 1995, Hosmer and Lemeshow 2000). The test consists of dividing the data that have been sorted by increasing order of estimated probabilities into m percentile groups of approximately equal size. The test is based on the Pearson χ^2 test statistic obtained from the $2 \times m$

table of observed and expected frequencies, where m is the number of groups and 2 refers to the "yes" and "no" responses. The number m of groups is often set to 10. The statistic is given by

$$X_{\mathrm{HL}}^2 = \sum_{i=1}^{m} \frac{(o_i - n_i \cdot \hat{p}_i)^2}{[n_i \cdot \hat{p}_i \cdot (1 - \hat{p}_i)]}$$

where n_i is the number of observations in the group, o_i is the number of "yes" observations in the group, and $\hat{p}_i$ is the mean estimated probability of an event for the group. If the null hypothesis H_0 is true, that is, that the model provides a good fit to the data, this statistic X_{HL}^2 is χ_{m-2}^2, χ^2-distributed with $(m-2)$ degrees of freedom. The test can be conducted by comparing the p-value, derived from the X_{HL}^2 test statistic, with the prescribed type I error α. The Hosmer–Lemeshow test is a conservative test with low power. It is dependent on how the observations are grouped and the number of groups. Nonetheless, it is easy to apply and is the most common test for goodness of fit applied to the logistic regression model.

Another statistical method for testing goodness of fit is based on **nonparametric bootstrapping**, with computer-intensive Monte Carlo resampling of the original observed dataset. Nonparametric bootstapping provides robust estimates of variance, standard error, and confidence intervals. The method was first described by Bradley Efron (1979) and has prove useful in a wide range of applications (Efron and Tibshirani 1993, Manly 1997). With nonparametric bootstrapping, statistics are generated from the resampled data and used to provide estimates associated with the statistics from the originally observed dataset. In practical application, the original sample dataset is resampled with replacement, with 5000–10,000 samples usually required to obtain satisfactory results. Nonparametric bootstrapping estimates of the standard error of a parameter estimate from a sample dataset can be obtained by resampling the original dataset with the same sample size, with replacement, and calculating the standard deviation of the resampled parameter estimates. The empirically generated sample estimates are therefore used to estimate the standard deviation, or standard error of the original estimate.

Nonparametric bootstrapping can also be used to estimate model selection frequencies and estimation of precision based on model selection uncertainty in a collection of candidate models (Burnham and Anderson 1998). The observed dataset is resampled with replacement. The frequencies of occurrence of the best-fitting models based on Akaike's information criterion (AIC or $\mathrm{AIC_c}$) provide estimates of the model selection frequencies. Estimates of variance and standard deviation based on model selection uncertainty are obtained by calculating the means of the resampled variance and standard deviation parameter estimates of the leading models. The reader is warned that, although nonparametric bootstrapping is generally successful for large sample sizes, it is not always reliable for small sample sizes $n = 5$–20 (Efron and Gong 1983). Furthermore, it is not always successful for model selection (Burnham and Anderson 1998, 2002).

An alternative approach to bootstrapping, **parametric bootstrapping**, does not resample the originally observed sample dataset, but instead generates simulated new sample datasets, based on the model parameter estimates obtained by fitting it to the original observed sample dataset. Statistics are generated iteratively from the simulated new sample datasets. Statistics generated from the original sample dataset are then compared to the statistics generated from the simulated new sample datasets. If the original sample dataset statistics are incongruent with the simulated new dataset statistics, then the null hypothesis H_0 of goodness of fit is rejected. Otherwise, the null hypothesis H_0 is not rejected.

The deviance statistic for logistic regression models can be compared with simulated new sample statistics for goodness of fit with the **deviance test** using parametric bootstrapping. If the deviance of the original sample dataset falls outside the confidence region of the empirical distribution of the simulated sample deviances, then the null hypothesis H_0 of goodness of fit is rejected. For example, suppose that a deviance is estimated from an original sample dataset and this statistic is at the 96.5% percentile among the deviance statistics generated from the Monte Carlo simulation process. Then, with a confidence level of 95% in a one-tailed test, the null hypothesis H_0 of goodness of fit would be rejected. Alternatively, if the deviance statistic were at the 67.2% percentile of the parametrically bootstrapped deviances, then the null hypothesis H_0 would not be rejected. Parametric bootstrapping is a standard method for accessing goodness of fit of logistic regression models.

Another indication of model goodness of fit is provided by the empirical evidence of a model's predictive effectiveness with **classification analysis**. By evaluating the extent of accuracy of the model at predicting the response values for samples from the population, the analyst can assess the reliability of the model as a predictive tool and can also assess accuracy on the original developmental dataset used to fit the model. Better and more rigorously, the analyst can resample the original sample with nonparametric bootstrapping, viewing the new samples as representative of the population, and assess the accuracy of the model on the new samples. With **cross-validation**, the analyst sequentially deletes individual points in the sample and assesses the accuracy of the model on the remaining subsets. Best of all and most rigorously, the analyst can assess the accuracy of the model on new test sample datasets to provide the best indication of the predictive accuracy of the model for a population.

The accuracy of logistic regression models is assessed with classification analysis as follows. A probability cutoff point p_c for the fitted model must first be specified. Observations with predicted response probability values $\hat{p} \geq p_c$ are assigned predictive values of $y_p = 1$. Otherwise, observations with predicted response probability values $\hat{p} < p_c$ are assigned predictive values of $y_p = 0$. An observed–prediction classification table is then constructed that provides the frequency counts of the sample dataset with respect to observed response values $y(x)$ and predicted response values $y_p(x)$ for all observed values x, as follows:

In the table, $\text{frequency}_{0,0}$ is the number of response values whose observed value is 0 and predicted value is 0, $\text{frequency}_{0,1}$ is the number of response values whose observed value is 0 and predicted value is 1, and so forth; $n_{y=0}$ is the number of

Frequency Table

	Predicted = 0	Predicted = 1	
Observed = 0	$\text{frequency}_{0,0}$ $y(x) = 0,\ y_p(x) = 0$	$\text{frequency}_{0,1}$ $y(x) = 0,\ y_p(x) = 1$	$n_{y=0}$
Observed = 1	$\text{frequency}_{1,0}$ $y(x) = 1,\ y_p(x) = 0$	$\text{frequency}_{1,1}$ $y(x) = 1,\ y_p(x) = 1$	$n_{y=1}$
	$n_{y_p=0}$	$n_{y_p=1}$	n

responses with observed values equal to 0 with

$$n_{y=0} = \text{frequency}_{0,\,0} + \text{frequency}_{0,\,1},$$

and $n_{y=1}$ is the number of responses with observed values equal to 1 with

$$n_{y=1} = \text{frequency}_{1,\,0} + \text{frequency}_{1,\,1}.$$

Similarly, $n_{y_p=0}$ is the number of responses with predicted values equal to 0 with

$$n_{y_p=0} = \text{frequency}_{0,\,0} + \text{frequency}_{1,\,0},$$

and $n_{y_p=1}$ is the number of responses with predicted values equal to 1 with

$$n_{y_p=1} = \text{frequency}_{0,\,1} + \text{frequency}_{1,\,1}.$$

For example, consider the sample dataset of (x,y) values given by

$$x = (6.6, 6.1, 4.3, 1.5, 9.6, 6.8, 1.8, 2.3, 3.7, 5.5)$$

and

$$y = (0, 1, 0, 0, 1, 1, 0, 0, 0, 1)$$

with fitted logistic regression model values given by

$$\begin{aligned}\widehat{\text{logit}} &= -7.3 + 1.3 \cdot x \\ &= (1.3, 0.6, \ -1.7, \ -5.4, 5.2, 1.6, \ -4.9, \ -4.3, \ -2.5, \ -0.1),\end{aligned}$$

Figure 6.1. Plots of logistic regression response values for observed independent covariate values x: dependent observed values y, estimated logit values, and estimated probability values. (**a**) y versus x, (**b**) logit versus x, (**c**) Probability versus x.

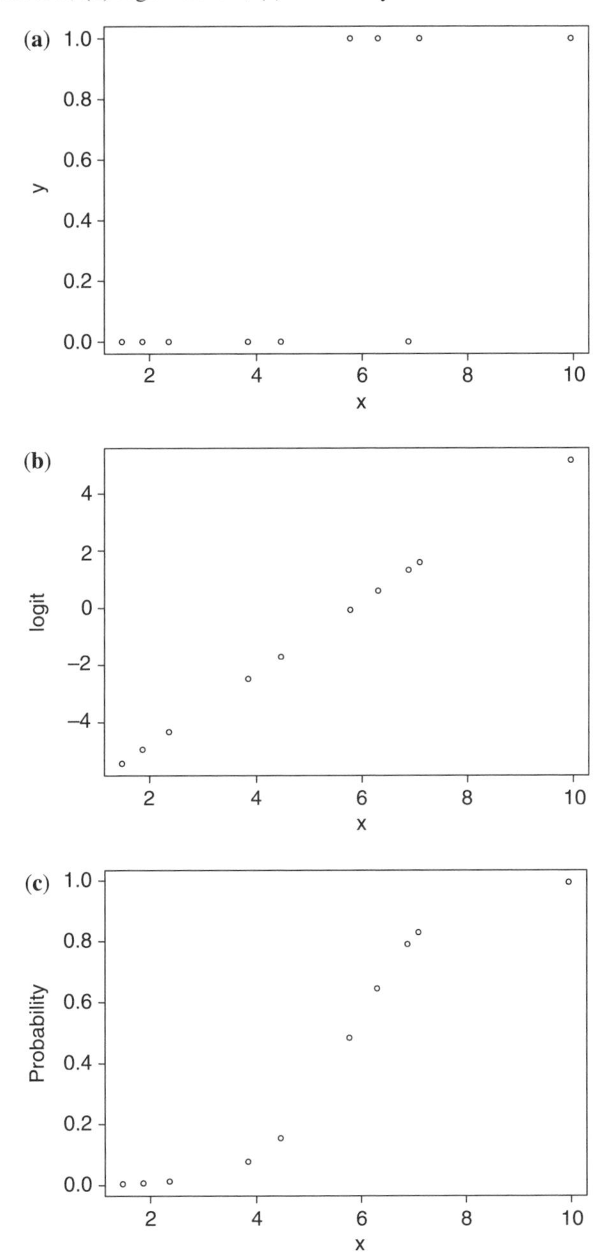

(i.e., $\hat{\beta}_1 = -7.3$ and $\hat{\beta}_2 = 1.3$) and predicted probability values

$$\text{Pr} = e^{\text{logit}}/(1 + e^{\text{logit}}) = (0.79, 0.65, 0.15, 0.01, 0.99, 0.83, 0.01, 0.01, 0.08, 0.48).$$

See Fig. 6.1 for plots of these values. It is apparent from these graphs that a cutoff point of, say, $p_c = 0.40 = 40\%$, will provide prediction values

$$y_p = (1, 1, 0, 0, 1, 1, 0, 0, 0, 1)$$

and an overall correct classification rate of $0.90 = 90.0\%$ accuracy, with 9 of 10 points correctly classified. The observation–prediction classification table is given by

Frequency Table

	Predicted = 0	Predicted = 1	
Observed = 0	5	1	$n_{y=0} = 6$
Observed = 1	0	4	$n_{y=1} = 4$
	$n_{y_p=0} = 5$	$n_{y_p=1} = 5$	$n = 10$

The probability cutoff point p_c can be chosen using various criteria. It can be chosen, for instance, to be the value that provides the best overall predictive accuracy, or a certain level of sensitivity or specificity (see text below), or a compromise between the two. A predictive logistic regression model can be recommended, therefore, from goodness-of-fit classification analysis not only with estimates for its parameters but also with a recommended cutoff point for its most effective application.

The overall **correct classification rate CC** and overall **error rate *E*** can be calculated from a sample dataset and specified cutoff point p_c by using the observation–prediction classification table as follows:

$$\text{CC} = (\text{frequency}_{0,0} + \text{frequency}_{1,1})/n,$$
$$E = (\text{frequency}_{0,1} + \text{frequency}_{1,0})/n.$$

In the example, $\text{CC} = \frac{9}{10} = 0.90 = 90\%$ and $E = \frac{1}{10} = 0.10 = 10\%$.

Specificity is the proportion of the responses with observed values equal to 0 that are predicted to be equal to 0. Recall that **type I error** is the probability of rejecting the null hypothesis in an experiment, if the null hypothesis is true. In this context, the null hypothesis is that the response is equal to 0. A type I error therefore is an observation equal to 0 that is predicted to be equal to 1. Hence, given the sample dataset and specified cutoff point p_c, the specificity and type I error α are given by

$$\text{Specificity} = \text{frequency}_{0,0}/n_{y=0},$$
$$\alpha = \text{frequency}_{0,1}/n_{y=0}.$$

In this example, specificity $= \frac{5}{6} = 0.833$ or 83.3% and type I error $\alpha = \frac{1}{6} = 0.167$ or 16.7%.

An **error of commission**, or **false positive**, on the other hand, is an observed value of 0 that is predicted to be equal to 1. The proportion of errors of commission therefore is given by

$$\text{Errors of commission} = \text{frequency}_{0,1/n_{y_p}=1}.$$

In the example, errors of commission $= \frac{1}{5} = 0.20$ or 20%.

Sensitivity is the proportion of the responses with observed values equal to 1 that are predicted to be equal to 1. Recall that **type II error** is the probability of not rejecting the null hypothesis, if the alternative hypothesis is true. The alternative hypothesis in this context is that the response is equal to 1. So the type II error is an observation equal to 1 that is predicted to be equal to 0. The sensitivity and type II error β for a given cutoff point p_c are given by

$$\text{Sensitivity} = \text{frequency}_{1,1}/n_{y=1},$$
$$\beta = \text{frequency}_{1,0}/n_{y=1}.$$

In the example, sensitivity $\frac{4}{4} = 1.0$ or 100% and type II error $\beta = \frac{0}{4} = 0.0$ or 0%.

An **error of omission**, or **false negative**, on the other hand, is an observed value of 1 that is predicted to be equal to 0. The proportion of errors of omission therefore is given by

$$\text{Errors of omission} = \text{frequency}_{1,0}/n_{y_p=0}.$$

In the example, errors of omission $= \frac{0}{5} = 0.0$ or 0%.

The **receiver operating characteristic (ROC) curve** is the graph of the function

$$\text{ROC} = \text{sensitivity}/(1 - \text{specificity})$$

obtained for varying cutoff points p_c. When the cutoff point $p_c = 0$, it is easy to see that sensitivity = 1.0 or 100% and specificity = 0.0 or 0%, so ROC = 1.0. As the cutoff point increases toward $p_c = 1$, sensitivity decreases toward 0.0 or 0% and specificity increases toward 1.0 or 100%, so (1 − specificity) also decreases toward

0.0 or 0%, and the ROC curve tends toward their limit. Best-fitting models tend to have higher sensitivity and specificity and hence high ROC curves with a large amount of area underneath them. The c statistic is a measurement of the area underneath the ROC curve and is sometimes used as an indicator of the fitness of the model. We have seen in Chapter 3 that sensitivity/(1 − specificity) describes **Bayes factors** that are useful for Bayesian statistical hypothesis testing. Therefore, the best-fitting models tend to be those with the highest c statistics from the ROC curves, specifically, those models with the largest Bayes factors.

Another statistic that is sometimes useful in evaluating goodness of fit of a logistic regression model is the concordance statistic (SAS Institute Inc. 1995). **Concordance** is the extent to which the order of the predicted response values y_{p} is in agreement with the order of the observed response values y for pairs of covariate values x. For example, suppose that a pair of covariate values x_1 and x_2 have observed response values $y_1 = 0 < y_2 = 1$. Then the predicted response values y_{1_p} and y_{2_p} are concordant with the observed response values if $y_{1_p} \leq y_{2_p}$. If $y_{1_p} > y_{2_p}$, then the predicted response values are discordant. The concordance statistic is the proportion of pairs of covariate values (x_i, x_j), $i \neq j$, whose observed and predicted responses are concordant. The concordance statistic ranges in value between 0.0 and 1.0. Concordance is calculated by considering all pairs of covariate values with different values of the response variable. If there are n_i 0s and n_j 1s in the sample dataset, there are a total of $n_i \cdot n_{\mathrm{j}} = n_{\mathrm{t}}$ pairs that need to be examined for concordance. Ties are not included in the calculation. If there are n_{t} total distinct pairs (x_i, x_j) with different observed response values (y_i, y_j), with n_{c} concordant pairs and n_{d} discordant pairs, then

$$\text{Concordance} = n_{\mathrm{c}}/n_{\mathrm{t}},$$
$$\text{Discordance} = n_{\mathrm{d}}/n_{\mathrm{t}}.$$

The example will serve to illustrate this concept. There are a total of six 0s and four 1s, so we need to examine $6 \cdot 4 = 24$ pairs. Of these 24 pairs, there are $5 \cdot 4 + 1 = 21$ concordant pairs (look at the five 0s on the left and four 1s on the right, and the lone 0 to the right with a singleton 1 to its right, in Fig. 6.1a). Alternatively, there are just three discordant pairs (look at the lone 0 on the right in the graph and the three 1s to its left). All other pairs are ties. Hence the concordance is $\frac{21}{24} = 0.875$ or 87.5%.

Several other statistics can be calculated from the concordance statistic n_{c}, the **Somer statistic** D, the **gamma statistic** γ, the **tau statistic** τ, and the c **statistic**, the area under a ROC curve given by the sensitivity/(1−specificity) for varying cutoff points

$$\text{Somers } D = (n_{\mathrm{c}} - n_{\mathrm{d}})/n_{\mathrm{t}},$$
$$\text{Gamma } \gamma = (n_{\mathrm{c}} - n_{\mathrm{d}})/(n_{\mathrm{c}} + n_{\mathrm{d}}),$$
$$\text{Tau } \tau = (n_{\mathrm{c}} - n_{\mathrm{d}})\Big/\left(\frac{1}{2} \cdot n \cdot (n-1)\right),$$
$$c = \left(n_{\mathrm{c}} + \frac{1}{2} \cdot (t - n_{\mathrm{c}} - n_{\mathrm{d}})\right)\Big/n_{\mathrm{t}}.$$

The gamma statistic γ is a particularly popular statistic for assessing the extent of agreement between observed and predicted values in the modeling of categorical data (Siegel and Castellan 1988).

A goodness-of-fit assessment would not be complete without a diagnostic analysis of the residuals. Further discussion of this issue can be found in Hosmer and Lemeshow (2000), SAS Institute, Inc. (1995), Dobson (1990), and McCullagh and Nelder (1996).

6.5 OTHER GENERALIZED LINEAR MODELS (GLMs)

Numerous other GLMs are available for analysis of natural resource datasets (McCullagh and Nelder 1996, Hardin and Hilbe 2001, S-Plus 2000). These GLMs satisfy a range of model assumptions for population sample datasets with varying error assumptions and link functions. All of these models satisfy the fundamental requirements for generalized linear models, with response variable values $y_1, y_2, \ldots, y_n$ assumed to have a distribution from an exponential family, a set of parameters $\boldsymbol{\beta}$ and explanatory variables $\mathbf{x}$, and a monotonic link function g such that $g(\mu_y) = \mathbf{x} \cdot \boldsymbol{\beta}$.

For binomial datasets, there are two other popular models besides the logit model of logistic regression: the probit model and the complementary log–log model. Probit models use an inverse cumulative normal distribution function as a link function

$$g(\mu_y) = \Phi^{-1}(\mu_y),$$

where $\mu_y = E[y]$ and its inverse is given by

$$f(x \cdot \beta) = \Phi(x \cdot \beta) = \frac{1}{\sigma \cdot \sqrt{2 \cdot \pi}} \int_{-\infty}^{x \cdot \beta} e^{[-(s-\mu)^2/(2 \cdot \sigma^2)]} ds.$$

The probit model has proved useful for certain kinds of biological and social science applications, such as the analysis of bioassay data. It provides a higher tail for the response curve than the logit model.

The complementary log–log model uses the link function

$$g(\mu_y) = \log(-\log(1 - \mu_y)).$$

It has also proved useful for certain kinds of biological applications, such as dose response models, and also differs from the logit and probit models in the tail of the response curve.

For count datasets of populations that are randomly dispersed, the Poisson regression GLM model has proved useful. Poisson regression uses a logarithmic link function and assumes that the means and variances are equal: $E[y] = \mu_y = \sigma_y^2 = E[(y - \mu_y)^2]$. If the count data are overdispersed, with $\mu > \sigma^2$, it is more appropriate to replace the Poisson regression model with the negative binomial GLM model, using also a logarithmic link function.

Finally, for categorical contingency data, with cell frequencies that are Poisson-distributed or multinomial-distributed with constraints, the multiplicative log-linear GLM model with a logarithmic link function has been shown to be useful. With this modeling approach, main effects and interaction effects can be examined, with testing procedures and with model comparison.

The interested reader should consult more specific references for a fuller picture of GLMs, such as Dobson (1990), McCullagh and Nelder (1996), Hardin and Hilbe (2001), and McCulloch and Searle (2001).

6.6 S-Plus OR R AND WinBUGS APPLICATIONS

Let's now illustrate the ideas of this chapter with an example. We first generate a standardized sample dataset *data1* from a habitat selection logistic regression "reality" with standardized habitat variables aspect, species, old.growth, rock, moss, temp, and moist, along with the logit function

$$\text{logit} = 5 \cdot \text{temp} + 10 \cdot \text{moist}$$

and binary response, as presented in Fig. 6.2.

6.6.1 Frequentist Logistic Regression Analysis in S-Plus and R

Next, let's use an a priori parsimonious model selection and inference strategy on dataset `data1`, analyzing the following collection of habitat selection models, with covariates as follows:

1. Model 1: {aspect, species, old.growth}
2. Model 2: {aspect, species, moss}
3. Model 3: {aspect, species, temp}
4. Model 4: {aspect, species, moist}
5. Model 5: {aspect, old.growth, temp}
6. Model 6: {species, old.growth, temp}
7. Model 7: {moist, temp} (correct model)
8. Model 8: {rock, moss, temp, moist}
9. Model 9: {} (null model)
10. Model 10: {aspect, species, old.growth, rock, moss, temp, moist} (full model).

The frequentist S-Plus and R logistic regression analysis of dataset `data1` is presented in Fig. 6.3.

The statistical results are presented in Fig. 6.4. Model 7, the correct model with the two biologically significant covariates, temp and moist, performed best with the lowest AIC = 20.08825. Model 9, with two specious covariates, rock and moss, as well as the two biologically significant covariates, temp and moist, performed second best with AIC = 23.09453. Model 10, the full model with all covariates, performed third best with AIC = 23.34713. The reader is encouraged to examine the covariates in each of the models for statistical significance at the 95% confidence level (i.e., statistically significant if $t_s \geq t_{df=n-k,\ \alpha=0.025} \cong 2$). The statistical results confirm that temp and moist are the two biologically significant covariates in the models.

Goodness-of-fit analysis statistics for the best-fitting correct model 7, including Nagelkerke's adjusted R^2 and concordance statistics, are presented in Fig. 6.5a. The partition and test results for the Hosmer–Lemeshow test are presented in Fig. 6.5b and 6.5c. The classification table for model 7 with results for varying

Figure 6.2. The S-Plus code used to generate the simulated habitat selection dataset `data1`. In R, use the same code (below) except for the replacement of sd for stdev.

```
> aspect <- sample(c(0,1),size=100,prob=c(.5,.5),replace=T)
> aspect
  [1] 1 1 0 1 0 0 1 1 1 0 ...
> table(aspect)
  0 1
 46 54
> species <-
sample(c("RW","Other"),size=100,prob=c(.2,.8),replace=T)
> species
  [1] "Other"  "RW"  "Other"  "Other"  "Other"  "RW"
 "Other"...
> for (i in 1:100) if (species[i]=="RW") species[i] <- 1
else species[i] <- 0
> species
  [1] 0 1 0 0 0 1 0 ...
> table(species)
  0  1
 81 19
> old.growth <- runif(100,0,1)
> old.growth )
  [1] 0.507387177 0.435674758 0.192782182 ...
> rock <- sample(c(0,1),size=100,prob=c(.8,.2),replace=T)
> rock
  [1] 0 0 0 0 0 0 0 0 1 0 0 0 0 0 0 0 1 0 ...
> table(rock)
  0 1
 87 13
> moss <- sample(c(0,1),size=100,prob=c(.6,.4),replace=T)
> moss
  [1] 1 0 1 0 0 1 1 0 0 1 ...
```

Figure 6.2. *Continued.*

```
> table(moss)
  0  1
 59 41
> temp <- runif(100,15,30)
> temp
  [1] 23.46148 24.44190 26.08371 20.44376 …
> moist <- runif(100,0,100)
> moist
  [1] 55.631869 55.883988 26.148716 …
> # Standardize and round off the habitat covariates
> aspect <- round((aspect-mean(aspect))/stdev(aspect),2)
> species <-round((species-mean(species))/stdev(species),2)
> old.growth <- round((old.growth-mean(old.growth))
/stdev(old.growth),2)
> rock <- round((rock-mean(rock))/stdev(rock),2)
> moss <- round((moss-mean(moss))/stdev(moss),2)
> temp <- round((temp-mean(temp))/stdev(temp),2)
> moist <- round((moist-mean(moist))/stdev(moist),2)
> # Generate the response
> logit <- 5*temp+10*moist
> prob <- exp(logit)/(1+exp(logit))
> response <- rbinom(100,1,prob)
> data1 <-
data.frame(aspect,species,old.growth,rock,moss,temp,
moist,logit,prob,response)
> data1
aspect species old.growth rock moss temp moist logit prob
     response
 1  0.92 -0.48  0.29 -0.38  1.19  0.24  0.27  3.90  0.98  1
 2  0.92 -0.48  0.04 -0.38 -0.83  0.47  0.28  5.15  0.99  1
 3 -1.08 -0.48 -0.81 -0.38  1.19  0.86 -0.87 -4.40  0.01  0
 4  0.92 -0.48 -0.35 -0.38 -0.83 -0.48  0.63  3.90  9.80  1
  ...
97 -1.08 -0.48  0.69 -0.38 -0.83 -1.07 -1.49 -20.25  0.01 0
98  0.92 -0.48  1.57 -0.38 -0.83 -1.17 -0.43 -10.15  0.01 0
99  0.92 -0.48  1.43 -0.38  1.19 -0.11  0.65   5.95  0.99 1
100 0.92 -0.48  1.04  2.57 -0.83  0.81  0.45   8.55  0.99 1
```

cutoff points are presented in Fig. 6.5d. The highest correct classification rate is 97.0%, which occurs at cutoff points $p_c = 0.250-0.400$. The sensitivity and specificity rates at those cutoff points are 98.3% and 95.2%, respectively.

6.6.2 Bayesian Analysis in WinBUGS

Let's next illustrate Bayesian statistical analysis of this same sample dataset *data1* using WinBUGS, with an a priori parsimonious model selection and inference strategy on the 10 candidate models described in Section 6.6.1. The WinBUGS code for the full model 10 is presented in Fig. 6.6a. The WinBUGS code for the other models can be obtained by removing the appropriate covariates from the program

Figure 6.3. The S-Plus and R frequentist logistic regression statistical analysis results of data `data1`, using predictive a priori parsimonious model selection and inference with AIC.

```
> model1 <-
glm(response~aspect+species+old.growth,data=data1,
   family=binomial(link=logit))
> summary(model1)
Coefficients:
                 Value    Std. Error   t value
(Intercept) 0.32406067  0.2029430 1.59680624
     aspect 0.06919870  0.2048259 0.33784163
    species 0.11264834  0.2093160 0.53817351
 old.growth 0.01032159  0.2042521 0.05053356
Null Deviance: 136.0584 on 99 degrees of freedom
Residual Deviance: 135.683 on 96 degrees of freedom
> model2 <-
glm(response~aspect+species+moss,data=data1,family=binomial
(link=logit))
> summary(model2)
Coefficients:
                 Value    Std. Error   t value
(Intercept) 0.32411220  0.2029473 1.59702638
     aspect 0.06746057  0.2061138 0.32729775
    species 0.11361225  0.2091867 0.54311404
       moss 0.01532510  0.2057874 0.07447057
Null Deviance: 136.0584 on 99 degrees of freedom
Residual Deviance: 135.6801 on 96 degrees of freedom
> model3 <-
glm(response~aspect+species+temp,data=data1,family=binomial
(link=logit))
> summary(model3)
Coefficients:
                Value   Std. Error   t value
(Intercept) 0.3827672  0.2229073 1.717158
     aspect 0.2153929  0.2278278 0.945420
    species 0.1581765  0.2322252 0.681134
       temp 0.9139369  0.2440666 3.744621
Null Deviance: 136.0584 on 99 degrees of freedom
Residual Deviance: 119.1157 on 96 degrees of freedom
> model4 <-
glm(response~aspect+species+moist,data=data1,family=binomial
(link=logit))
> summary(model4)
```

Figure 6.3. *Continued.*

```
Coefficients:
                  Value  Std. Error    t value
(Intercept)  0.5021480  0.3194692  1.5718196
     aspect -0.6271100  0.3558701 -1.7621879
    species  0.1943434  0.3257586  0.5965873
      moist  2.9624550  0.5771401  5.1329909
Null Deviance: 136.0584 on 99 degrees of freedom
Residual Deviance: 63.84085 on 96 degrees of freedom
> model5 <-
glm(response~aspect+old.growth+temp,data=data1,family=binomial
(link=logit))
> summary(model5)
Coefficients:
                 Value  Std. Error   t value
(Intercept) 0.3788146  0.2223539 1.7036562
     aspect 0.1956084  0.2251827 0.8686652
old.growth  0.1027766  0.2200893 0.4669770
       temp 0.9179548  0.2452318 3.7432122
Null Deviance: 136.0584 on 99 degrees of freedom
Residual Deviance: 119.3691 on 96 degrees of freedom
> model6 <-
glm(response~species+old.growth+temp,data=data1,family=
binomial(link=logit))
> summary(model6)
Coefficients:
                 Value   Std. Error   t value
(Intercept) 0.37978255  0.2219378 1.7112120
    species 0.12082129  0.2255746 0.5356155
old.growth  0.09316364  0.2202851 0.4229231
       temp 0.88917911  0.2415614 3.6809653
Null Deviance: 136.0584 on 99 degrees of freedom
Residual Deviance: 119.8398 on 96 degrees of freedom
> model7 <-
glm(response~moist+temp,data=data1,family=binomial(link=
logit))
> summary(model7)
Coefficients:
                Value Std. Error   t value
(Intercept)  1.192755  0.7549134 1.579990
      moist 15.402293  6.0247157 2.556518
       temp  9.900651  3.9654995 2.496697
Null Deviance: 136.0584 on 99 degrees of freedom
Residual Deviance: 14.08825 on 97 degrees of freedom
> model8 <-
glm(response~rock+moss+moist+temp,data=data1,family=binomial
(link=logit))
> summary(model8)
Coefficients:
```

Figure 6.3. *Continued.*

```
                  Value  Std. Error    t value
(Intercept)   1.2017172  0.8351867  1.4388606
       rock  -0.4960193  1.1227886 -0.4417745
       moss   0.6726261  0.8539822  0.7876348
      moist  15.8017869  6.1468826  2.5706993
       temp  10.4065574  4.1918030  2.4825969
Null Deviance: 136.0584 on 99 degrees of freedom
Residual Deviance: 13.09453 on 95 degrees of freedom
> model9 <-
glm(response~1,data=data1,family=binomial(link=logit))
> summary(model9)
Coefficients:
                Value  Std. Error   t value
(Intercept) 0.3227734  0.2026102 1.593076
Null Deviance: 136.0584 on 99 degrees of freedom
Residual Deviance: 136.0584 on 99 degrees of freedom
> model10 <-
glm(response~aspect+species+old.growth+rock+moss+temp+
moist,data=data1,family=binomial(link=logit))
> summary(model10)
Coefficients:
                  Value  Std. Error    t value
(Intercept)   1.2695298   1.977417  0.6420142
     aspect  -1.7935975   2.078134 -0.8630806
    species  -0.4856468   2.254010 -0.2154590
 old.growth   1.7032767   1.747489  0.9746996
       rock  -0.9681319   2.272320 -0.4260544
       moss   1.1692255   1.396520  0.8372423
       temp  12.6880509   5.549964  2.2861500
      moist  20.1631797   9.497663  2.1229622
Null Deviance: 136.0584 on 99 degrees of freedom
Residual Deviance: 7.347131 on 92 degrees of freedom
> aic1 <- model1$deviance+2*model1$rank
> aic2 <- model2$deviance+2*model2$rank
> aic3 <- model3$deviance+2*model3$rank
> aic4 <- model4$deviance+2*model4$rank
> aic5 <- model5$deviance+2*model5$rank
> aic6 <- model6$deviance+2*model6$rank
> aic7 <- model7$deviance+2*model7$rank
> aic8 <- model8$deviance+2*model8$rank
> aic9 <- model9$deviance+2*model9$rank
> aic10 <- model10$deviance+2*model10$rank
> c(aic1,aic2,aic3,aic4,aic5,aic6,aic7,aic8,aic9,aic10)
[1] 143.683 143.6801 127.1157 71.84085 127.3691 127.8398
20.08825 23.09453 138.0584 23.34713
```

Figure 6.4. Comparative frequentist statistics for the logistic regression habitat selection models.

Model	Covariates	k (= # of parameters)	Deviance	R^2	AIC	Likelihoods	Akaike Weights
1	aspect, species1, old.growth	4	135.6830	0.0050	143.6830	0.000000000000	0.000000000000
2	aspect, species1, moss	4	135.6800	0.0051	143.6800	0.000000000000	0.000000000000
3	aspect, species1, temp	4	119.1160	0.2096	127.1160	0.000000000000	0.000000000000
4	aspect, species1, moist	4	63.8410	0.6917	71.8410	0.000000000006	0.000000000004
5	aspect, old.growth, temp	4	119.3690	0.2067	127.3690	0.000000000000	0.000000000000
6	species1, old.growth, temp	4	119.8400	0.2014	127.8400	0.000000000000	0.000000000000
7	temp, moist	3	14.0880	0.9478	20.0880	1.000000000000	0.705030609734
8	rock, moss, temp, moist	5	13.0950	0.9517	23.0950	0.222350569667	0.156763957707
9	1 (null model)	1	136.0584	x	138.0584	0.000000000000	0.000000000000
10	aspect, speciel, old.growth, rock, moss, temp, moist (full model)	8	7.3470	0.9737	23.3470	0.196027563409	0.138205432555
							1.000000000000

Figure 6.5. Goodness-of-fit analysis results for the best-fitting correct model 7 with logistic regression analysis of dataset `data1` (CAS Institute Inc., 1995). (**a**) Goodness-of-fit analysis statistics for model 7. (**b**) Partition for Hosmer–Lemeshow test for model 7. (**c**) Hosmer–Lemeshow test results for model 7. (**d**) Classification results for model 7.

(**a**)
```
adjusted R^2 = 0.9478,
percent concordant = 99.5,
Somers' D = 0.991,
gamma = 0.992,
tau = 0.488,
number of pairs = 2436, and
c = 0.996
```

(**b**)

		response = 1		response = 0	
group	total	observed	expected	observed	expected
1	29	0	0.00	29	29.00
2	10	0	0.38	10	9.62
3	10	7	6.76	3	3.24
4	10	10	9.87	0	0.13
5	13	13	13.00	0	0.00
6	28	28	28.00	0	0.00

(**c**)

Chi-Square	DF	Pr > ChiSq
0.5624	4	0.9671

(**d**)

	Correct		Incorrect		Percentages				
Prob Level	Event	Non-Event	Event	Non-Event	Correct	Sensi-tivity	Speci-ficity	Fse POS	Fse NEG
0.050	58	36	6	0	94.0	100.0	85.7	9.4	0.0
0.100	58	37	5	0	95.0	100.0	88.1	7.9	0.0
0.150	58	38	4	0	96.0	100.0	90.5	6.5	0.0
0.200	58	38	4	0	96.0	100.0	90.5	6.5	0.0
0.250	57	40	2	1	97.0	98.3	95.2	3.4	2.4
0.300	57	40	2	1	97.0	98.3	95.2	3.4	2.4
0.350	57	40	2	1	97.0	98.3	95.2	3.4	2.4
0.400	57	40	2	1	97.0	98.3	95.2	3.4	2.4
0.450	56	40	2	2	96.0	96.6	95.2	3.4	4.8
0.500	56	40	2	2	96.0	96.6	95.2	3.4	4.8
0.550	56	40	2	2	96.0	96.6	95.2	3.4	4.8
0.600	56	40	2	2	96.0	96.6	95.2	3.4	4.8
0.650	55	40	2	3	95.0	94.8	95.2	3.5	7.0
0.700	53	40	2	5	93.0	91.4	95.2	3.6	11.1
0.750	53	40	2	5	93.0	91.4	95.2	3.6	11.1
0.800	53	40	2	5	93.0	91.4	95.2	3.6	11.1
0.850	53	40	2	5	93.0	91.4	95.2	3.6	11.1
0.900	52	40	2	6	92.0	89.7	95.2	3.7	13.0
0.950	50	41	1	8	91.0	86.2	97.6	2.0	16.3
1.000	0	42	0	58	42.0	0.0	100.0	.	58.0

code logit and prior statements, data code list, and initial values code list. The WinBUGS output for the leading model 7 is presented in Fig. 6.6b.

The WinBUGS comparative model output statistics are presented in Fig. 6.7. The DIC results are very similar to the AIC results for the comparative model frequentist analysis with the a priori parsimonious model selection and inference strategy for the collection of 10 candidate models. In both cases, we conclude that model 7, the "correct" model, is the best-fitting model. The inferences, of course, are different, with the Bayesian results providing posterior probability distributions for the parameters and permitting prior information to be included in the analysis as priors.

Figure 6.6. WinBUGS code and output for logistic regression models. (**a**) WinBUGS code for the full model 10. (**b**) WinBUGS output for the leading model 7.

(**a**)
```
1) Program code
   model
   {
   for(i in 1:n)
      {
      response[i] ~ dbin(p[i],1)
      logit(p[i]) <- beta0 + beta1*aspect[i]
          + beta2*species[i] + beta3*old.growth[i]
          + beta4*rock[i] + beta5*moss[i]
          + beta6*temp[i] + beta7*moist[i]}
      beta0 ~ dnorm(0,0.000000000001)
      beta1 ~ dnorm(0,0.000000000001)
      beta2 ~ dnorm(0,0.000000000001)
      beta3 ~ dnorm(0,0.000000000001)
      beta4 ~ dnorm(0,0.000000000001)
      beta5 ~ dnorm(0,0.000000000001)
      beta6 ~ dnorm(0,0.000000000001)
      beta7 ~ dnorm(0,0.000000000001)
      }

   2. Data
        list( aspect = c(0.92,0.92,-1.08,0.92,…,-1.08),
        species = c(-0.48,-0.48,-0.48,…,-0.48),
        old.growth = c(0.29,0.04,-0.81,…,1.04),
        rock = c(-0.38,-0.38,-0.38,…,2.57),
        moss = c(1.19,-0.83,1.19,…,-0.83),
        temp = c(0.24,0.47,0.86,…,0.81),
        moist = c(0.27,0.28,-0.87,…,0.45),
        response = c(1,1,0,…,1,0,0,1,1),
        n = 100)

   3. Initial values
        list(beta0 = 0,beta1 = 0,beta2 = 0,beta3 = 0,
        beta4 = 0,beta5 = 0,beta6 = 0,beta7 = 0)
```

Figure 6.6. *Continued*

(b)

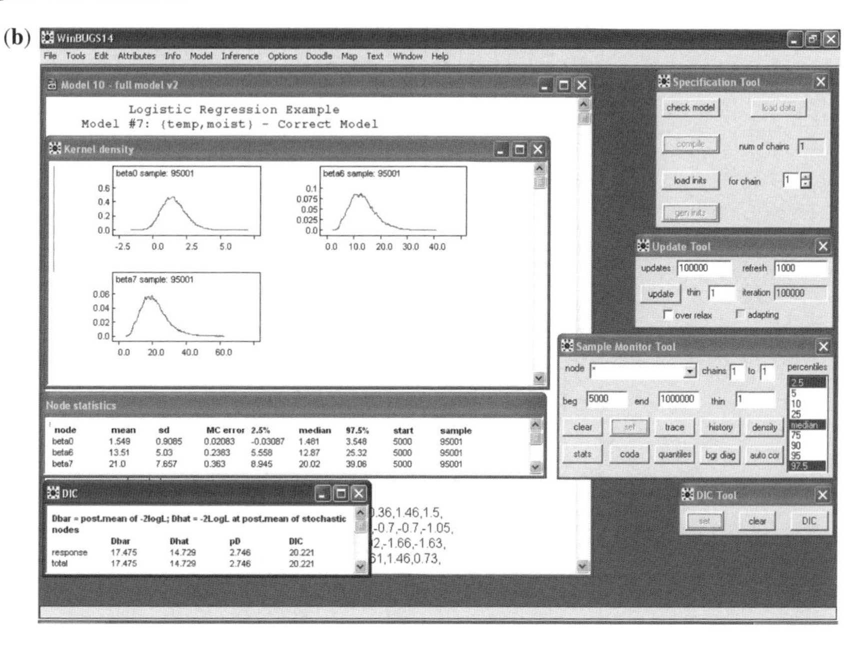

The frequentist results, on the other hand, provide probability assessments for the dataset, without the benefit of incorporating prior information formally into the analysis.

6.7 SUMMARY

We began this chapter with an introduction and overview of frequentist generalized linear modeling (GLMs). We discussed the importance of design for GLM modeling, emphasizing the contrast between design-based and model-based approaches. We presented GLM analysis for the exponential family of distributions, including the assumptions, ML fit, and statistics. We discussed the selection of models using varying selection criteria. We then devoted most of the rest of the chapter to a discussion of the most important GLM for the natural resource scientist, the logistic regression model, presenting its assumptions, ML fit, and statistics: parameter estimates and Wald tests, deviance, R^2, and AIC and AIC_c. We discussed model selection techniques using AIC and AIC weights, and goodness-of-fit methods such as the Hosmer–Lemeshow test, classification statistics, and residual analysis. We briefly mentioned some alternative GLMs such as the probit and complementary log–log models for binomial data and the Poisson and negative binomial models for count

Figure 6.7. Comparative Bayesian and frequentist statistics for the logistic regression habitat selection models.

Model	Covariates	*k* (= # of parameters)	p_D	DIC	DIC Weights	AIC	AIC Weights
1	aspect, species1, old.growth	4	4.092	143.8730	0.000000000000	143.6830	0.000000000000
2	aspect, species1, moss	4	4.060	143.8080	0.000000000000	143.6800	0.000000000000
3	aspect, species1, temp	4	4.064	127.2910	0.000000000000	127.1160	0.000000000000
4	aspect, species1, moist	4	3.931	71.9450	0.000000000004	71.8410	0.000000000004
5	aspect, old.growth, temp	4	4.063	127.5330	0.000000000000	127.3690	0.000000000000
6	species1, old.growth, temp	4	4.085	128.0530	0.000000000000	127.8400	0.000000000000
7	temp, moist	3	2.746	20.2210	0.671751567823	20.0880	0.705030609734
8	rock, moss, temp, moist	5	4.068	23.4200	0.135692131608	23.0950	0.156763957707
9	1(null model)	1	1.024	138.1070	0.000000000000	138.0584	0.000000000000
10	aspect, speciel, old.growth, rock, moss, temp, moist (full model)	8	3.043	22.7200	0.192556300565	23.3470	0.138205432555
					1.000000000000		1.000000000000

data. We concluded the chapter by illustrating the ideas of GLM using an example with logistic regression frequentist analysis in S-Plus and R and Bayesian statistical analysis in WinBUGS.

PROBLEMS

6.1 Conduct a frequentist statistical analysis of the sample dataset `data2` (Fig. 6.8) in S-Plus or R, using a predictive a priori parsimonious model selection and inference strategy with AIC_c. The dataset `data2` consists of habitat selection modeling data for salamander binary presence–absence response with covariates aspect, species, old.growth, rock, moss, temp, and moist. Use logistic regression GLM with the logit link function and binomial response for a comparative analysis of the following 12 models with linear covariates:

(a) {aspect, species, old.growth}
(b) {aspect, species, rock}
(c) {aspect, species, moss}
(d) {aspect, species, temp}
(e) {aspect, species, moist}
(f) {aspect, old.growth, temp}
(g) {rock, temp, moist}
(h) {moss, temp, moist}
(i) {old.growth, temp}
(j) {rock, moist}
(k) {rock, moss, temp,moist}
(l) {aspect, species, old.growth, rock, moss, temp, moist} (full model).

6.2 Conduct a frequentist statistical analysis of the sample dataset `data2` (Fig. 6.8) in S-Plus or R, using a descriptive a posteriori model selection–inference strategy. Use logistic regression for a comparative analysis of models, including best-subsets selection of the models with the 128 different combinations of linear covariates. Use AIC as the model selection criterion.

6.3 Conduct a Bayesian statistical analysis of the sample dataset `data2` (Fig. 6.8) in WinBUGS, using a predictive parsimonious a priori model selection–inference strategy with the deviance information criterion (DIC). Use logistic regression GLM with a logit link function and binomial response for a comparative analysis of the 12 models with the linear covariates prescribed in Problem 6.1 (above). Use normal priors on the beta parameters for the models, with a mean of 0 and precision of 10^{-6}, except for the poor-fitting overparameterized full model, where a somewhat informative prior with mean of 0 and precision of 0.10 may be required to ensure convergence. Compare the Bayesian results with the frequentist results in Problem 6.1.

Figure 6.8. Habitat selection dataset `data2` for Problems 6.1–6.6.

sample	aspect	species	old. growth	rock	moss	temp	moist	response
1	0.92	-0.48	0.29	-0.38	1.19	0.24	0.27	1
2	0.92	-0.48	0.04	-0.38	-0.83	0.47	0.28	0
3	-1.08	-0.48	-0.81	-0.38	1.19	0.86	-0.87	0
4	0.92	-0.48	-0.35	-0.38	-0.83	-0.48	0.63	0
5	-1.08	2.05	1.02	-0.38	-0.83	0.86	-0.40	1
6	-1.08	-0.48	0.69	-0.38	1.19	1.66	-0.37	1
7	0.92	-0.48	-1.34	-0.38	1.19	0.05	-1.55	0
8	0.92	-0.48	-0.25	-0.38	-0.83	0.36	0.05	1
9	0.92	2.05	-1.28	2.57	-0.83	1.46	0.95	0
10	-1.08	2.05	-0.51	-0.38	1.19	1.50	-1.30	1
11	-1.08	2.05	-0.37	-0.38	-0.83	-1.60	-0.95	0
12	0.92	-0.48	1.35	-0.38	1.19	-0.88	-1.40	1
13	-1.08	2.05	0.53	-0.38	-0.83	-1.65	-1.74	0
14	-1.08	-0.48	1.28	-0.38	-0.83	-1.30	0.82	1
15	-1.08	-0.48	-0.72	-0.38	-0.83	-0.62	-0.84	0
16	-1.08	-0.48	-0.32	-0.38	-0.83	-0.50	-1.73	0
17	-1.08	-0.48	1.03	2.57	-0.83	1.64	-1.80	1
18	0.92	-0.48	-0.37	-0.38	1.19	-0.30	-0.80	0
19	0.92	-0.48	0.27	-0.38	1.19	-0.37	1.77	1
20	0.92	-0.48	0.36	-0.38	1.19	-0.70	1.03	0
21	0.92	-0.48	0.15	2.57	-0.83	-0.70	1.34	0
22	-1.08	-0.48	0.14	-0.38	-0.83	-1.05	0.41	0
23	0.92	-0.48	1.79	-0.38	1.19	-0.78	0.79	1
24	0.92	-0.48	-0.28	-0.38	-0.83	-1.57	-1.84	0
25	-1.08	-0.48	1.67	-0.38	-0.83	0.41	-1.60	1
26	0.92	-0.48	1.97	-0.38	-0.83	-1.14	1.71	1
27	0.92	-0.48	-0.65	-0.38	-0.83	0.94	1.26	0
28	0.92	-0.48	1.91	-0.38	-0.83	-1.02	-0.93	1
29	-1.08	-0.48	1.73	-0.38	-0.83	0.91	1.55	1
30	0.92	-0.48	1.16	-0.38	1.19	-0.20	-0.36	1
31	-1.08	2.05	0.29	2.57	-0.83	0.02	0.65	0
32	0.92	-0.48	-0.26	-0.38	1.19	-1.66	0.71	0
33	0.92	-0.48	-0.42	-0.38	1.19	-1.63	0.39	0
34	-1.08	-0.48	0.00	-0.38	1.19	-0.67	0.52	0
35	0.92	-0.48	-0.07	-0.38	-0.83	-0.34	1.52	0
36	0.92	2.05	-1.05	-0.38	1.19	0.99	0.00	0
37	0.92	-0.48	-0.13	-0.38	-0.83	1.23	1.26	1
38	0.92	-0.48	0.26	-0.38	1.19	0.35	0.39	1
39	0.92	-0.48	-0.02	-0.38	-0.83	1.76	0.65	1
40	-1.08	-0.48	-0.56	2.57	1.19	-0.33	0.36	0
41	0.92	-0.48	0.89	2.57	-0.83	1.46	-1.16	1
42	-1.08	-0.48	-0.06	-0.38	1.19	-1.61	-0.02	0
43	0.92	-0.48	-1.23	-0.38	-0.83	1.46	0.12	0
44	-1.08	-0.48	0.43	-0.38	1.19	0.73	0.76	1
45	-1.08	-0.48	-0.24	-0.38	-0.83	1.12	-1.58	1
46	0.92	-0.48	-1.40	-0.38	-0.83	-0.66	0.48	0
47	-1.08	-0.48	-0.57	2.57	1.19	1.61	0.56	1
48	0.92	-0.48	0.64	-0.38	1.19	-1.70	1.09	0
49	0.92	-0.48	-1.15	-0.38	1.19	-0.92	-0.45	0
50	0.92	2.05	-0.53	-0.38	-0.83	-0.18	1.53	0

Figure 6.8. *Continued.*

51	0.92	-0.48	1.38	-0.38	1.19	0.89	0.63	1
52	0.92	-0.48	1.98	2.57	-0.83	0.15	-1.57	1
53	-1.08	-0.48	-1.35	-0.38	-0.83	0.32	0.90	0
54	0.92	2.05	0.85	-0.38	1.19	0.76	0.10	1
55	-1.08	-0.48	0.73	-0.38	-0.83	1.45	-0.42	1
56	-1.08	2.05	0.01	-0.38	1.19	-0.69	0.80	0
57	-1.08	-0.48	0.56	-0.38	-0.83	1.20	-0.83	1
58	-1.08	-0.48	-1.34	-0.38	-0.83	-1.59	-0.27	0
59	-1.08	-0.48	1.98	-0.38	1.19	0.34	-1.45	1
60	0.92	-0.48	-1.36	2.57	1.19	-1.48	1.42	0
61	-1.08	-0.48	0.73	-0.38	-0.83	0.05	0.25	1
62	0.92	-0.48	-0.80	-0.38	-0.83	0.40	-0.03	0
63	0.92	-0.48	-0.92	-0.38	1.19	0.66	1.38	0
64	0.92	-0.48	-1.02	-0.38	1.19	-1.40	-0.78	0
65	-1.08	-0.48	-1.35	-0.38	1.19	-0.11	0.12	0
66	-1.08	2.05	-0.03	-0.38	-0.83	0.16	-1.79	0
67	0.92	-0.48	0.20	-0.38	-0.83	-1.75	-0.79	0
68	0.92	-0.48	-0.16	-0.38	-0.83	1.04	0.94	0
69	-1.08	2.05	0.11	-0.38	-0.83	1.69	0.80	1
70	-1.08	-0.48	-1.10	2.57	-0.83	0.11	0.98	0
71	0.92	-0.48	0.00	-0.38	-0.83	-0.22	-1.59	1
72	-1.08	-0.48	-0.69	2.57	1.19	-0.46	-0.50	0
73	0.92	-0.48	1.53	-0.38	-0.83	0.71	1.22	1
74	0.92	-0.48	-1.27	-0.38	1.19	-1.17	0.57	0
75	0.92	-0.48	-0.17	-0.38	1.19	0.58	0.25	0
76	-1.08	-0.48	-0.65	2.57	-0.83	-0.94	0.43	0
77	-1.08	-0.48	-0.60	-0.38	1.19	-0.24	0.13	0
78	-1.08	-0.48	-0.67	-0.38	-0.83	1.10	0.18	0
79	-1.08	-0.48	-0.76	-0.38	-0.83	1.52	-0.98	1
80	-1.08	2.05	0.23	-0.38	-0.83	-0.43	0.96	1
81	0.92	-0.48	-1.05	-0.38	-0.83	-0.38	0.68	0
82	0.92	2.05	-0.95	-0.38	-0.83	0.17	0.28	0
83	0.92	-0.48	-1.12	-0.38	-0.83	0.25	-0.84	0
84	-1.08	-0.48	-1.16	-0.38	-0.83	1.24	1.29	1
85	0.92	2.05	-1.23	-0.38	1.19	0.35	1.53	0
86	-1.08	-0.48	-0.98	-0.38	-0.83	1.12	-0.83	0
87	-1.08	2.05	1.80	-0.38	-0.83	-1.59	-1.25	1
88	-1.08	-0.48	-1.32	-0.38	1.19	0.37	-1.43	0
89	-1.08	-0.48	1.33	-0.38	-0.83	-0.12	0.19	1
90	-1.08	2.05	1.13	-0.38	1.19	-0.27	0.77	1
91	0.92	-0.48	-1.18	-0.38	-0.83	1.03	-1.35	0
92	0.92	2.05	1.97	-0.38	-0.83	-1.40	0.24	1
93	-1.08	-0.48	-1.42	-0.38	1.19	0.91	-0.05	0
94	0.92	-0.48	-1.46	-0.38	1.19	-0.96	-1.62	0
95	0.92	2.05	-0.12	-0.38	1.19	-0.25	-0.31	0
96	-1.08	-0.48	0.08	-0.38	-0.83	0.88	0.32	1
97	-1.08	-0.48	0.69	-0.38	-0.83	-1.07	-1.49	0
98	0.92	-0.48	1.57	-0.38	-0.83	-1.17	-0.43	1
99	0.92	-0.48	1.43	-0.38	1.19	-0.11	0.65	1
100	0.92	-0.48	1.04	2.57	-0.83	0.81	0.45	1

6.4 Conduct model averaging on the results of Problems 6.1 and 6.3, using AIC_c and DIC weights. Calculate the unconditional estimates of the coefficients of old.growth and temp, with and without shrinkage, along with their unconditional estimates of standard error. Also calculate the importance of the covariates.

6.5 Conduct a goodness-of-fit classification analysis of the best-fitting model from the a priori model selection–inference results of Problems 6.1 and 6.3, compiling a classification table with sensitivity, specificity, correct classification, and $c = \text{sensitivity}/(1 - \text{specificity})$ proportions for probability cutoff points ranging from 0.05 to 0.95 in increments of 0.05. What is the optimal cutoff point, based on correct classification proportion?

6.6 Write a 3–5-page report, with attachments, summarizing the results of the analysis in Problems 6.1–6.5. Include the following sections in the report: abstract, introduction, statement of the problem, objectives, methods, results, discussion, conclusions, and references. Include figures and tables in an appendix. Specify the statistics used for the comparative analysis of the different logistic regression models, with tables. What can be inferred for the best-fitting model:

$$\texttt{logit} \sim \beta_0 + \beta_1 \cdot \texttt{aspect} + \beta_2 \cdot \texttt{species} + \beta_3 \cdot \texttt{old.growth} + \beta_4 \cdot \texttt{rock} + \beta_5 \cdot \texttt{moss} + \beta_6 \cdot \texttt{temp} + \beta_7 \cdot \texttt{moist}\ ?$$

Interpret the biological meaning of the statistical results from a management perspective.

7 Introduction to Mixed-Effects Modeling

For every complex problem, there is a solution that is simple, neat, and wrong.
—Henry L. Mencken

In this chapter we present an introduction to mixed-effects modeling. Mixed-effects modeling provides an efficient way of incorporating random effects, along with fixed effects, into the modeling of natural resource data. We begin this introduction by describing the types of dependent datasets suitable for mixed-effects modeling. We then describe mixed-effects generalizations of the mean model, with both fixed effects and random effects. We describe mixed-effects generalizations of the ANOVA and multiple linear regression models. We describe variance–covariance structures between groups and variance structures within groups with mixed-effects modeling. Finally, we describe covariance structures within groups with mixed-effects modeling, including serial and spatial correlation. We conclude with a brief introduction to nonlinear mixed-effects modeling. Throughout this chapter, the ideas will be illustrated with frequentist statistical analysis applications using the rich library of mixed-effects utilities available in S-Plus and R and with Bayesian statistical analysis applications in WinBUGS, employing the a posteriori exploratory model selection and inference strategy. Users of R will want to activate the linear and nonlinear mixed-effects modeling library by using the command `library (nlme)` to initiate the frequentist statistical analysis described in this chapter.

7.1 INTRODUCTION

Multiple linear regression and ANOVA are based on an assumption of independence for the sample datasets, requiring the residual errors to be independent. With many natural resource datasets, however, this assumption of independence is clearly violated. Mixed-effects modeling provides a tool for modeling such dependent datasets.

Contemporary Bayesian and Frequentist Statistical Research Methods for Natural Resource Scientists. By Howard B. Stauffer

It models fixed effects with estimates of parameters for treatment and covariates coefficient effects, along with random effects describing the variation between groups, or experimental units, of dependent observations in the dataset. There are many types of dependent datasets common in natural resource applications, such as pseudoreplicated data; temporal data such as repeated measures, time series, or longitudinal data; spatial data; hierarchical data; and metapopulation data, all with clustered dependences in the datasets. Random effects can be added to traditional fixed-effects models, such as ANOVA and multiple linear regression, to explain variation between the groups or clusters within the dependent datasets. The variance–covariance structure between the random effects that describe the groups can also be modeled, as can the variance–covariance structure within the groups.

Natural resource scientists have long been advised to avoid collecting dependent datasets because of the independence assumptions required by many classical statistical methods. Simple random sampling requires sampling units to be collected independently of each other. However, dependences are often an inherent characteristic of natural resource populations. Mixed-effects modeling provides natural resource scientists the tools to analyze these inherent dependences, sampling to incorporate them into their datasets, and analyzing them with mixed-effects modeling.

Scientists analyzing datasets with groups of dependent observations are confronted with two extreme alternatives. They can model (1) the entire population with a classical fixed-effects model, such as ANOVA or multiple linear regression or (2) each individual group of data with a fixed-effects model. Mixed-effects models provide an effective and efficient compromise between these two extremes, one that is more accurate than the fixed-effects modeling of the entire population because it is less biased for each group. But it is also more efficient than modeling of each individual group since there are fewer parameters to estimate and the errors in the parameter estimates will hence be smaller.

This chapter provides an introduction to the subject of mixed-effects modeling using examples and the powerful capabilities available in S-Plus or R for frequentist statistical analysis and in WinBUGS for Bayesian statistical analysis. There have been major theoretical advances in this field in recent years that have been implemented into S-Plus and R, and we will take advantage of this opportunity. The reader can investigate this subject in greater depth by referencing the more detailed account of mixed-effects modeling in S-Plus by Pinheiro and Bates (2000). This chapter illustrates the exploratory data analysis that is possible in S-Plus and R. It also illustrates how easily mixed-effects modeling can be incorporated into Bayesian statistical analysis in WinBUGS.

7.2 DEPENDENT DATASETS

We begin by illustrating the most common types of dependent datasets that are candidates for mixed-effects modeling: pseudoreplicated, temporally dependent, spatially dependent, hierarchical, and metapopulation datasets.

A first example of dependent data is **pseudoreplicated** data, consisting of measurements collected within clusters. It may indeed be convenient, or biologically interesting, to collect more than one measurement at a site or cluster center in a sample survey or experiment. However, clusters create dependences in datasets since measurements tend to be more similar within clusters than between clusters. The clusters create groups of dependent data within the dataset. A natural resource scientist may be interested in examining trends or relational patterns in the data that vary from group to group. Mixed-effects modeling will incorporate the group dependences into the analysis, using random effects to model the variation between groups.

A second example of dependent data is temporally dependent data that are collected at permanent plots or with subjects repeatedly measured in time. If the timepoints are few in number, such data are commonly called **repeated-measures data**. If there are larger numbers of timepoints, the data are called **time-series data**. Generally, repeated-measures and time-series data are called **longitudinal data**. Repeated-measures design is a commonly used experimental design for data that are collected repeatedly in time. With natural resource applications, data may be collected over time with respect to permanent sample plots or animal subjects. Scientists may be interested in examining trends over time that vary between subjects or plots. Mixed-effects modeling will incorporate the dependences in the data into the analysis, using random effects to model the variation between subjects or plots.

A third example of dependent data is **spatially dependent data**, measurements collected within close physical proximity that may be more dependent than those at greater distances. Such dependences have long been recognized in geologic data with geostatistical analysis. Natural resource scientists are increasingly recognizing such dependences in their data and the need to include assumptions of dependence in their analysis. Mixed-effects modeling will incorporate the dependences into the analysis, using random effects to model the variation between locations.

A fourth example of dependent data is **hierarchical data**. Many natural resource datasets are inherently hierarchical in structure. For example, collections of spores on leaves, on branches, on trees, positioned at microsites, in forest stands, in national forests can be conveniently sampled using a hierarchical multistage sampling design, such as a six-stage design for this example. With hierarchical structure, there are dependences in the data due to the grouping structure at the different stages. Measurements in closer proximity to each other at the different stages of the design may tend to be more similar, to be dependent. It will be more accurate to model these dependences in the analysis. Mixed-effects modeling can incorporate such dependences into the analysis.

A final example of dependent data is **metapopulation data**. Many natural resource populations consist of collections of subpopulations, distinguished by geographic or genetic differences. A metapopulation analysis provides a model for both the overall population and the individual subpopulations. Mixed-effects modeling provides models for both, with fixed effects describing the overall population, and the random effects describing the individual subpopulation differences.

In the next section, we will introduce linear mixed-effects modeling. We will generalize the classical fixed-effects ANOVA model to a mixed-effects model, incorporating random effects to model dependences in a dataset. Next, we will generalize the classical multiple linear regression model to include random effects. We will conclude the chapter with a brief introduction to nonlinear mixed-effects modeling. Throughout this chapter, we shall use S-Plus or R and WinBUGS examples to illustrate the ideas.

7.3 LINEAR MIXED-EFFECTS MODELING: FREQUENTIST STATISTICAL ANALYSIS IN S-Plus AND R

A good example is the best sermon.

—Benjamin Franklin

In this section we focus on linear mixed-effects modeling using S-Plus or R. We assume that the fixed effects and random effects can be modeled with linear forms. Later, we will generalize this modeling to include nonlinear mixed effects. We start with a mixed-effects generalization to the mean model for normal data. With mixed-effects modeling, we will assume that residual errors are independent and identically normally distributed.

7.3.1 Generalizations of Analysis of Variance (ANOVA)

For the first example, let's generalize the mean model to a mixed-effects model, using the dataset `Seedlings1` in S-Plus or R in Fig. 7.1 to illustrate. The dataset `Seedlings1` consists of Douglas fir seedling growth measurements (`growth`) taken from five trays (`Tray`) in a greenhouse experiment. `Seedlings1` is a grouped data frame, grouped by `Tray`, derived from a data frame `Seedlings`, in preparation for mixed-effects modeling in S-Plus or R.

Let's first ignore the obvious grouping created by the trays and analyze the dataset using the normal mean model. The normal mean model assumes that the experimental design for the `Seedlings1` dataset is completely randomized with Douglas fir seedling growth measurements randomly sampled from the greenhouse population (Fig. 7.2a). The design of this sample dataset is not completely randomized, obviously, because of the clusters of the seedlings in the trays (Fig. 7.2b). Ignoring this problem momentarily, the mean model is given by

$$y_i = \mu + e_i,$$

where μ is the mean growth and the residual errors are independent and identically normally distributed $e_i \sim \text{iid } N(0, \sigma)$ with mean 0 and standard deviation σ. For the analysis, we estimate the two parameters μ and σ using least-squares (LS) estimation in S-Plus or R (Fig. 7.2c). The estimated mean and standard deviation parameters are $\hat{\mu} = 6.01$ and $\hat{\sigma} = 1.93$, respectively, to two digits of accuracy.

Figure 7.1. The `Seedlings1` dataset, generated as grouped data from the data frame `Seedlings`, in S-Plus or R.

```
> Seedlings1 <- groupedData(growth~1|Tray,Seedlings)
> Seedlings1
Grouped Data:
growth ~ 1 | Tray
      growth Tray
  1 3.624421    1
  2 4.207856    1
  3 3.870086    1
  4 3.209784    1
  5 5.051546    2
  6 4.137267    2
  7 5.195171    2
  8 4.748900    2
  9 6.864213    3
 10 4.959441    3
 11 6.667230    3
 12 6.314804    3
 13 6.887878    4
 14 5.376680    4
 15 6.764992    4
 16 6.109793    4
 17 8.084306    5
 18 9.552097    5
 19 8.815998    5
 20 9.749536    5
```

The residual standard error is large and highly affected by the variation between the trays, as illustrated in Fig. 7.2d.

The problem with the mean model analysis is that it has ignored the obviously large amount of variation in the growth measurements between trays in the dataset. We might first attempt to model this variation by letting `Tray` be a "treatment" factor with five levels, 1–5. To do this properly, we would want to use a completely randomized experimental design, randomly assigning measurements of growth to the trays, viewed as fixed treatment levels. The one-factor, fixed-effects ANOVA model would then be

$$y_{ij} = \mu_j + e_{ij}, \quad i = \text{sample}, \quad j = \text{treatment (}\texttt{Tray}\text{)},$$

where $\mu_j = \mu + \tau_j$ is the jth `Tray` mean growth with treatment tray effect τ_j and the residual errors are $e_{ij} \sim$ iid $N(0, \sigma)$. For the analysis, let's estimate the six parameters μ_j, $j = 1$–5, and σ using LS estimation in S-Plus or R (Fig. 7.2e). The S-Plus or R default options for unordered and ordered treatment factors in the `lm` command are the Helmert and polynomial contrasts, respectively. However, the insertion of the character string `-1` in command formula `growth~Tray-1` (above) prescribes the cell means form option in S-Plus and R that estimates the treatment mean parameters μ_j. The estimated fixed-effects treatment mean parameters here are $\hat{\mu}_1 = 3.73$, $\hat{\mu}_2 = 4.78$, $\hat{\mu}_3 = 6.20$, $\hat{\mu}_4 = 6.28$, and $\hat{\mu}_5 = 9.05$, and the estimated standard deviation parameter is $\hat{\sigma} = 0.66$, respectively. This analysis has "captured" the variation between trays, reducing the residual error to one-third of its previous amount (Model 1. Seedlings1: $\hat{\sigma} = 1.93$; Model 2. Seedlings1: $\hat{\sigma} = 0.66$).

Figure 7.2. Graphs and analyses of the Seedlings1 dataset for the mean and fixed-effects models with S-Plus or R code. (**a**) Dotplot of the growth measurements (dotplot(Seedlings1$growth)) (use stripchart instead of dotplot in R). (**b**) Plot of the growth/Tray measurements (plot(Seedlings1)). (**c**) Analysis of the mean model for Seedlings1. (**d**) Boxplot of the mean model growth/Tray residual errors (plot(Seedlings$Tray,output1.Seedlings1$resid)). (**e**) Analysis of the fixed-effects model for the Seedlings1 dataset.

(**a**)

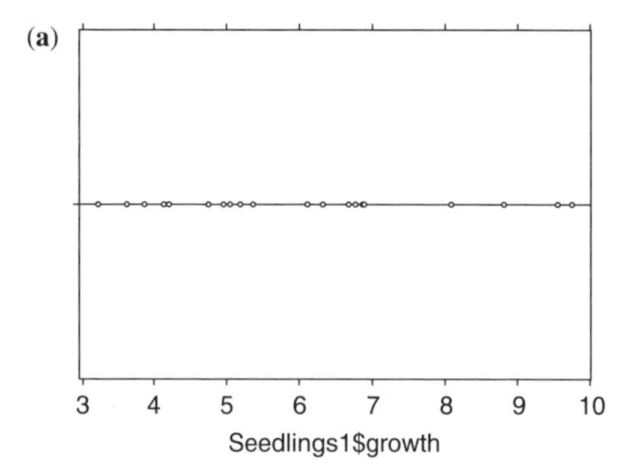

(**b**)

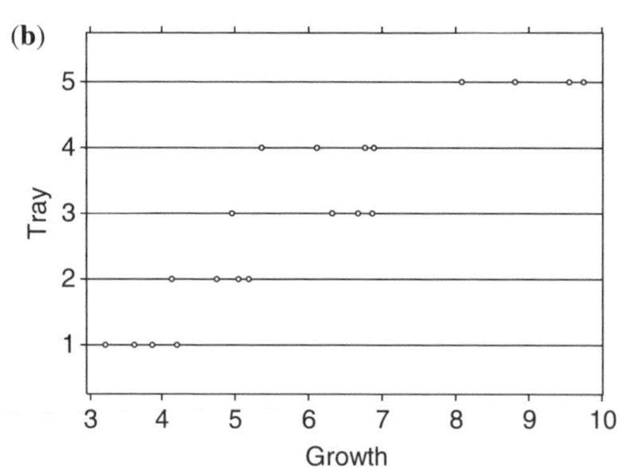

(**c**)

```
> # Model 1.Seedlings1: the mean model for Seedlings1.
> output1.Seedlings1 <- lm(growth~1,Seedlings1)
> summary(output1.Seedlings1)
Coefficients:
             Value  Std. Error  t value Pr(>|t|)
(Intercept)  6.0096  0.4318     13.9171  0.0000
Residual standard error: 1.93 on 19 degrees of freedom
```

Figure 7.2. *Continued.*

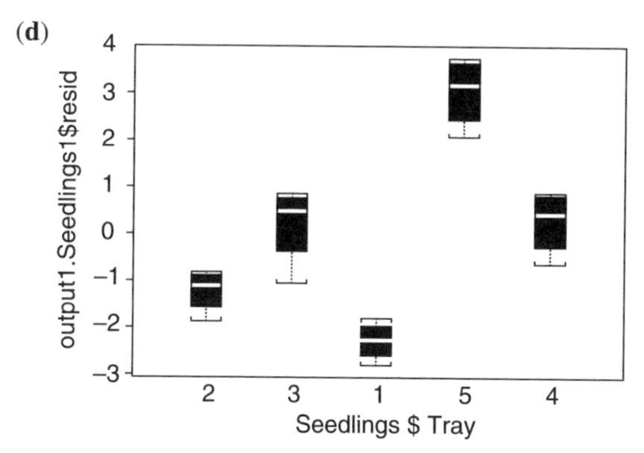

(e)
```
> # Model 2.Seedlings1:  a fixed-effects model
  (with cell means treatment effects) for Seedlings1.
> output2.Seedlings1 <- lm(growth~Tray-1, Seedlings1)
> summary(output2.Seedlings1)
Coefficients:
       Value  Std. Error  t value  Pr(>|t|)
Tray1  3.7280  0.3312     11.2576  0.0000
Tray2  4.7832  0.3312     14.4440  0.0000
Tray3  6.2014  0.3312     18.7266  0.0000
Tray4  6.2848  0.3312     18.9785  0.0000
Tray5  9.0505  0.3312     27.3299  0.0000
Residual standard error: 0.6623 on 15 degrees of freedom
Multiple R-Squared: 0.9917
F-statistic: 358.6 on 5 and 15 degrees of freedom, the
    p-value is 4.774e-015
```

However, we are not really interested in `Tray` as a treatment variable but rather as a "nuisance" variable. `Tray` is a variable that naturally and unavoidably occurs in the greenhouse population, causing variation in the response. It is not a treatment variable that can be manipulated by managers as a result of this experiment. We are not interested in determining which trays provide optimal growth response so that a manager could manipulate the environment and select only the optimally performing trays. We only need to acknowledge the natural occurrence of the trays in the population and the unavoidable effect of this occurrence on the variation in response. Furthermore, we cannot discern any particular patterns for its effect on the response. The trays appear to have a random effect on the response that isn't a function of their size or any other characteristic. If there were an effect on the response due to the size of the trays, we might then want to use the size of the tray as a variable in our modeling for a fixed effect. However, the variable `Tray` in our dataset has been measured only by its number as a label and, in that sense, it is a random effect. Additionally, the fixed-effects ANOVA analysis is inefficient in the sense that we have estimated five different parameters, one for each of the `Tray` mean effects on the response.

Again, we are not really interested in the effect of each tray, only in the variation between the trays. What we need is a way to estimate this variation only, with one parameter, say, σ_b (b for block), using a random-effects model.

So let's examine a mixed-effects model, to again analyze the `Seedlings1` dataset, this time incorporating a random effect to capture the variation of the response between trays. Here we would like to collect measurements that are randomized for each tray. The model will be

$$y_{ij} = \mu + b_j + e_{ij}, \quad i = \text{sample}, \ j = \text{group } (\texttt{Tray})$$

where μ is the mean growth, $b_j \sim$ iid $N(0, \sigma_b)$ are the j `Tray` growth random effects with the standard deviation parameter σ_b measuring the variation between the trays, and $e_{ij} \sim$ iid $N(0, \sigma)$ are the residual errors. We will estimate three parameters, μ, σ_b, and σ, using restricted ML (REML) estimation in S-Plus or R. Restricted ML (REML) integrates the likelihood function with respect to the fixed-effects parameters, and then estimates the random-effects parameters using maximal likelihood (Pinheiro and Bates 2000) with the linear mixed-effects command `lme` in S-Plus and R (see Fig. 7.3a). The estimated fixed-effects mean parameter is $\hat{\mu} = 6.01$, and the random-effects standard deviation parameters are $\hat{\sigma}_b = 1.98$ and $\hat{\sigma} = 0.66$. The standard deviation σ for the residual error is actually also a random-effects parameter. The random-effects model has captured most of the variation with the `Tray` grouping variable random effects. The addition of just one random-effects parameter to the mean model has reduced the noise to a third of its original amount in the mean model (`Model 1. Seedlings1:` $\hat{\sigma} =$ `1.93; Model 3. Seedlings1:` $\hat{\sigma} =$ `0.66`). Akaike's information criterion (AIC) along with the Bayesian information criterion (BIC) and the log likelihood are also included in the output, allowing us to compare models having the same fixed effects. None of the 95% confidence intervals for the estimates of the parameters contain 0, so the parameter estimates are all significant at a 95% confidence level. The predicted random effects b_j for each tray can be added to the estimated fixed effect $\hat{\mu}$ to obtain the mean growth estimates. For example, for `Tray` 1, $\hat{\mu} + b_1 = 6.01 - 2.22 = 3.79$ (Fig. 7.3a). The residual errors for this mixed-effects model are more reasonable, given the assumptions (Figs. 7.3b and 7.3c).

The sample dataset `Seedlings1` was simulated, using parameters $\mu = 6.0$, $\sigma_b = 1.5$ (with random effects $b_1 = -2.45772301$, $b_2 = -0.74215638$, $b_3 = 0.05353809$, $b_4 = 0.30999812$, and $b_5 = 2.83634318$), and $\sigma = 0.5$. The output statistics are compatible with these parameter values for the population.

Next, let's examine a second example, generalizing a one-factor fixed-effects ANOVA model to a mixed-effects model with one fixed effect and one random effect. We shall use the grouped data frame `Seedlings2` in S-Plus or R to illustrate the ideas (Fig. 7.4). The dataset `Seedlings2` consists of `growth` response measurements subjected to four levels, 1–4, of treatment variable `Treatment`. The dataset contains measurements from an experiment examining the effects of four nitrogen fertilizer treatment levels on Douglas fir seedling growth in five different trays in a greenhouse (Fig. 7.5a). We shall initially ignore the grouping variable

Figure 7.3. Analysis and graphs of the `Seedlings1` dataset, for the mixed-effects generalization of the mean model with S-Plus or R code. (**a**) Analysis of the mixed-effects mean model in S-Plus or R. (**b**) Boxplot of the mixed-effects mean model `growth/Tray` residual errors `(plot(Seedlings1$Tray,output3. Seedlings1$resid))`. (**c**) Scatterplot of the mixed-effects mean model growth standardized residual errors `(plot(output3. Seedlings1))`.

(**a**)

```
> # Model 3. Seedlings1: a random-effects model for
  Seedlings1.
> output3.Seedlings1 <- lme(growth~1,Seedlings1,
  random=~1|Tray)
> summary(output3.Seedlings1)
Linear mixed-effects model fit by REML

 Data: Seedlings1
       AIC      BIC    logLik
  61.66246 64.49578 -27.83123

Random effects:
 Formula:  ~1 | Tray
        (Intercept)  Residual
StdDev:    1.976765 0.6623128

Fixed effects: growth ~ 1
             Value Std.Error DF t-value   p-value
(Intercept) 6.0096 0.8963555 15 6.704482  <.0001
> intervals(output3.Seedlings1)
Approximate 95% confidence intervals

 Fixed effects:
               lower   est.    upper
(Intercept) 4.099063 6.0096 7.920137
 Random Effects:
  Level: Tray
                     lower     est.   upper
sd((Intercept)) 0.9693828 1.976765 4.03102

 Within-group standard error:
     lower      est.     upper
 0.4630699 0.6623128 0.9472829
> fixef(output3.Seedlings1)
  (Intercept)
      6.0096
> ranef(output3.Seedlings1)
  (Intercept)
1  -2.2192806
2  -1.1929007
3   0.1865856
4   0.2677224
5   2.9578734.
```

Figure 7.3. *Continued.*

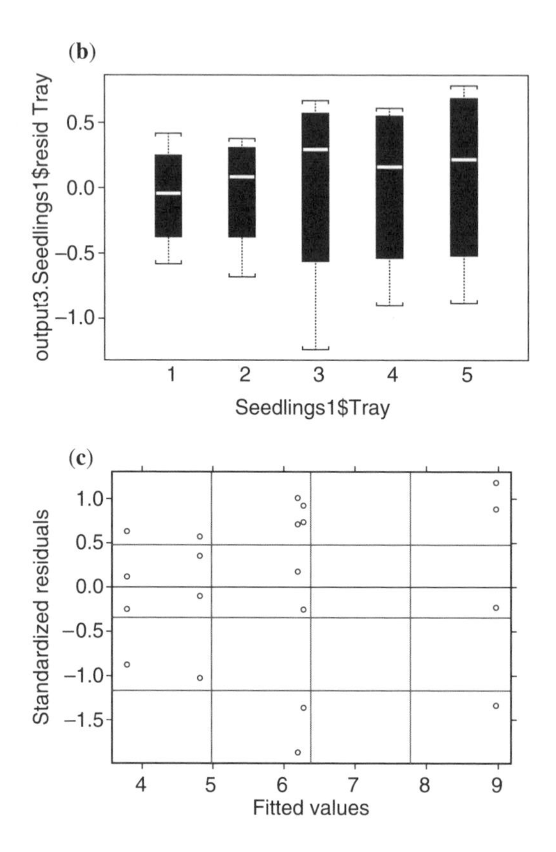

`Tray` that makes this dataset a candidate for mixed-effects modeling with a random effect for between-subject variation of the response, that of the groups created by the trays. Ignoring this `Tray` group effect for a moment, we first examine a fixed-effects one-factor ANOVA model.

We will assume a completely randomized experimental design for the fixed treatment levels. We model the dataset with a one-factor, fixed-effects ANOVA model:

$$y_{ij} = \mu_j + e_{ij}, \quad i = \text{sample}, \ j = \text{treatment } (\texttt{Treatment}),$$

where $\mu_j = \mu + \tau_j$ are the j treatment means and $e_{ij} \sim$ iid $N(0, \sigma)$. We estimate five parameters $\mu_j, j = 1$–4, for the fixed effects, and the residual error standard deviation σ, using LS estimation (see Fig. 7.5b). The estimated fixed-effects mean parameters for the treatments are $\hat{\mu}_1 = 7.93$, $\hat{\mu}_2 = 10.12$, $\hat{\mu}_3 = 12.29$, and $\hat{\mu}_4 = 14.24$, and the estimated residual standard deviation parameter is $\hat{\sigma} = 1.78$, respectively. We reject the null hypothesis that the treatment mean parameters are the same, based on the ANOVA F statistic. The estimated residual error standard deviation $\hat{\sigma} = 1.78$ is small, but we should compare that with a more accurate model that factors in the

Figure 7.4. The `Seedlings2` dataset, generated as grouped data from the data frame `Seedlings`, in S-Plus or R.

```
> Seedlings2 <- groupedData(growth~Treatment|Tray,Seedlings)
> Seedlings2
Grouped Data:
growth ~ Treatment | Tray
      growth Treatment Tray
 1  5.083296         1    1
 2  5.374564         1    1
 3  4.848325         1    1
 4  8.078080         2    1
 5  7.311337         2    1
 6  7.701099         2    1
 7  9.618617         3    1
 8 10.473698         3    1
 9 10.624043         3    1
10 12.192803         4    1
11 11.633321         4    1
12 10.493053         4    1
13  7.496540         1    2
14  7.290890         1    2
15  6.651376         1    2
16 10.412137         2    2
17  9.162686         2    2
18  9.981785         2    2
19 11.193206         3    2
20 11.745463         3    2
21 11.301645         3    2
22 13.091748         4    2
23 13.348608         4    2
24 13.463776         4    2
25  8.352266         1    3
26  8.957327         1    3
27  8.055527         1    3
28  9.396397         2    3
29 10.090525         2    3
30 10.297480         2    3
31 12.044342         3    3
32 12.263186         3    3
33 12.198116         3    3
34 14.835343         4    3
35 15.211385         4    3
36 14.692371         4    3
37  7.974991         1    4
38  8.435177         1    4
39  8.368959         1    4
40 10.969370         2    4
41 10.362955         2    4
42 10.241025         2    4
43 13.230619         3    4
```

Figure 7.4. *Continued.*

```
44 12.055935    3    4
45 12.761461    3    4
46 14.977464    4    4
47 14.727927    4    4
48 15.029507    4    4
49 10.698993    1    5
50 10.937124    1    5
51 10.468870    1    5
52 12.735983    2    5
53 12.512761    2    5
54 12.606732    2    5
55 14.779972    3    5
56 14.930798    3    5
57 15.131115    3    5
58 16.800580    4    5
59 16.572742    4    5
60 16.566120    4    5
```

response variation between the trays and better reflects the randomized block design of the experiment, with treatment measurements taken within the "subjects," the trays.

We need to account for the variation in response due to the trays. This "nuisance" variable creates groupings of dependent measurements in the dataset and can best be modeled as a random effect. Our model will be

$$y_{ij} = \mu_i + b_j + e_{ij}, \quad i = \text{treatment } (\texttt{Treatment}),\ j = \text{group } (\texttt{Tray}),$$

where μ_i are the treatment effects, $b_j \sim$ iid $N(0, \sigma_b)$ are the tray random effects with standard deviation parameter σ_b "measuring" the variation between the trays, and $e_{ij} \sim$ iid $N(0, \sigma)$. Here we will estimate six parameters, μ_i for $i = 1$–4, σ_b, and σ, using restricted ML estimation (REML) in S-Plus or R. The analysis is a two-way ANOVA with a fixed effect (`Treatment`) and a random effect (`Tray`), along with residual error that is also a random effect (see Fig. 7.5c). The estimated fixed-effects mean parameters are $\hat{\mu}_1 = 7.93$, $\hat{\mu}_2 = 10.12$, $\hat{\mu}_3 = 12.29$, and $\hat{\mu}_4 = 14.24$, and the random-effects standard deviation parameters estimates are $\hat{\sigma}_b = 1.85$ and $\hat{\sigma} = 0.45$. The random-effects model has captured most of the variation with the `Tray` grouping variable random effects. The addition of the one random-effects parameter to the fixed-effects one-factor ANOVA model has reduced the noise to less than one-third of its original amount in the mean model (`Model 1. Seedlings2:` $\hat{\sigma} = 1.78$; `Model 2. Seedlings2:` $\hat{\sigma} = 0.45$). The 95% confidence intervals tell us that the parameter estimates are all statistically significant at the 95% confidence level. The predicted random effects b_j for each tray can be added to the estimated fixed effects $\hat{\mu}_i$ to obtain the mean growth estimates for each treatment and tray. For example, for `Treatment` 1 and `Tray` 1, $\hat{\mu}_1 + b_1 = 7.93 + (-2.52) = 5.41$. The residual errors for this mixed-effects model are reasonable, given the assumptions (Fig. 7.5d).

The sample dataset `Seedlings2` was simulated, using parameters $\mu_1 = 8.0$, $\mu_2 = 10.0$, $\mu_3 = 12.0$, $\mu_4 = 14.0$, $\sigma_b = 1.5$ ($b_1 = -2.45772301$, $b_2 = -0.74215638$, $b_3 = 0.05353809$, $b_4 = 0.30999812$, and $b_5 = 2.83634318$), and $\sigma = 0.5$. The analysis results are compatible with these population parameters.

Figure 7.5. Graphs of the `Seedlings2` dataset for the mixed-effects generalization of the ANOVA model with S-Plus or R code. (**a**) Plot of the `growth/Tray` measurements for treatments 1–4 `(plot(Seedlings2))`. (**b**) Fixed-effects ANOVA analysis of the `Seedlings2` dataset in S-Plus or R. (**c**) Mixed-effects analysis of the `Seedlings2` dataset. (**d**) Scatterplot of the mixed-effects ANOVA growth standardized residual error `(plot(output2.Seedlings2))`.

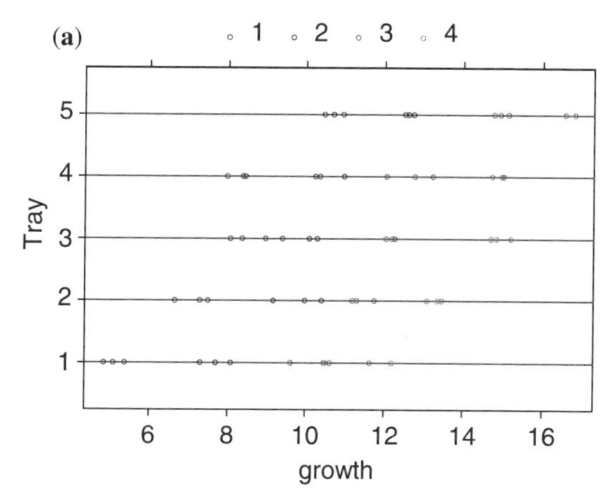

(**b**)
```
> # Model 1. Seedlings2: a fixed-effects ANOVA model
  (with cell means treatment effects) for Seedlings2.
> output1.Seedlings2 <- lm(growth~Treatment-1, Seedlings2)
> summary(output1.Seedlings2)

Coefficients:
                Value Std. Error t value Pr(>|t|)
Treatment1   7.9329   0.4584     17.3076   0.0000
Treatment2  10.1240   0.4584     22.0879   0.0000
Treatment3  12.2901   0.4584     26.8138   0.0000
Treatment4  14.2424   0.4584     31.0732   0.0000

Residual standard error: 1.775 on 56 degrees of freedom

Multiple R-Squared: 0.9778

F-statistic: 618 on 4 and 56 degrees of freedom, the p-
     value is 0
```

Figure 7.5. *Continued.*

(c) > # Model 2.Seedlings2: a mixed-effects model with fixed effects (cell means treatment effects) and with random effects.

```
> output2.Seedlings2 <- lme(growth~Treatment-1,
Seedlings2,random=~1|Tray)
> summary(output2.Seedlings2)
Linear mixed-effects model fit by REML
 Data: Seedlings2
       AIC      BIC    logLik
  114.2301 126.3823 -51.11507
Random effects:
 Formula:  ~ 1 | Tray
        (Intercept)  Residual
StdDev:    1.854025 0.4527051
Fixed effects: growth ~ Treatment - 1
              Value Std.Error DF  t-value p-value

Treatment1  7.93295 0.8373438 52  9.47394  <.0001
Treatment2 10.12402 0.8373438 52 12.09064  <.0001
Treatment3 12.29015 0.8373438 52 14.67754  <.0001
Treatment4 14.24245 0.8373438 52 17.00908  <.0001
> intervals(output2.Seedlings2)
Approximate 95% confidence intervals
 Fixed effects:
               lower      est.     upper
Treatment1  6.252695  7.932948  9.613202
Treatment2  8.443770 10.124024 11.804277
Treatment3 10.609894 12.290148 13.970401
Treatment4 12.562197 14.242450 15.922703
 Random Effects:
  Level: Tray
                  lower     est.    upper
sd((Intercept)) 0.9238433 1.854025 3.720771
 Within-group standard error:
     lower      est.     upper
 0.3735495 0.4527051 0.5486339
> fixef(output2.Seedlings2)
 Treatment1 Treatment2 Treatment3 Treatment4
   7.932948   10.12402   12.29015   14.24245
> ranef(output2.Seedlings2)
  (Intercept)
1  -2.5155410
2  -0.7155158
3   0.2177148
4   0.4450125
```

Figure 7.5. *Continued.*

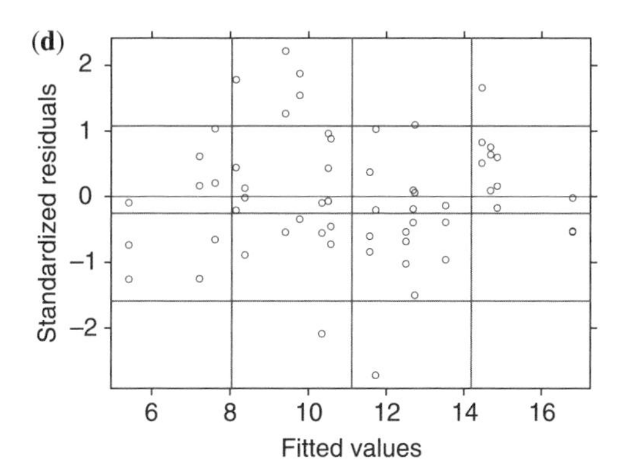

7.3.2 Generalizations of the Multiple Linear Regression Model

Let's turn next to a mixed-effects generalization of the multiple linear regression model. We shall illustrate with a linear regression example in S-Plus or R. We examine the grouped dataset `birds.density` that measures the density of bird counts (`density`) and amount of riparian habitat (`riparian`) at 100 sample sites, 25 each in four different watersheds (`Watershed`) (see Fig. 7.6). The scatterplot of the $(x, y) =$ (`riparian, density`) measurements (Fig. 7.7a) and a significant correlation estimate $\hat{\rho} = 0.83$ ($p < 0.01$) suggest that it is reasonable statistically to examine a linear relationship between the riparian covariate and the density response.

We start by examining a linear regression model to analyze the dataset `birds.density`. We will initially ignore the possible differences of the linear relationship between watersheds. Recall that the linear regression model assumes a linear relationship between an independent variable x (here, `riparian`) and a dependent variable y (`density`) satisfying

$$y_i = \beta_0 + \beta_1 \cdot x_i + e_i,$$

with the fixed-effects parameters β_0 and β_1 describing the y-intercept and slope of the linear relationship, respectively, and with the residual errors $e_i \sim$ iid $N(0, \sigma)$. We assume the following for the conditional population $y|x = \{y|(x, y), x \text{ is fixed}\}$:

1. $\mu_{y|x} = \beta_0 + \beta_1 \cdot x$ (linearity of the mean).
2. $\sigma_{y|x} = \sigma$ is constant (homoscedasticity).
3. $y|x$ are normally distributed $y|x \sim N(\mu_{y|x}, \sigma)$.
4. $y|x$ are randomly sampled (x can be either randomly sampled or fixed).
5. x are measured without error.

Figure 7.6. The grouped dataset `birds.density`, in S-Plus or R.

```
> birds.density
Grouped Data: density ~ riparian | Watershed

   riparian density Watershed
1     19.02   41.91         1
2     14.47   28.35         1
3     40.26  108.23         1
4      6.33   25.30         1
5     32.59   96.17         1
6     35.84   86.18         1
7      0.71    5.93         1
8     17.84   65.52         1
9     48.22  127.26         1
10    47.38   97.97         1
11     0.94   13.72         1
12    19.42   60.86         1
13    12.73   57.28         1
14    11.54   49.48         1
15    18.58   53.48         1
16    48.24  124.82         1
17     7.28   28.40         1
18     9.26   43.03         1
19    34.81   93.93         1
20    34.17   95.76         1
21     9.32   33.79         1
22    48.17  106.83         1
23    27.95   69.76         1
24     9.95   42.20         1
25    22.18   71.24         1
26     5.67   30.78         2
27    38.10   94.98         2
28    27.88   87.69         2
29    23.73   63.25         2
30    17.15   65.32         2
31     9.43   47.78         2
32    27.73   90.86         2
33    35.10  103.60         2
34    17.76   84.13         2
35    16.20   42.57         2
36    16.32   54.90         2
37    33.61   97.60         2
38    42.78  100.81         2
39    30.68   66.39         2
40    49.65  127.91         2
41    40.60   99.95         2
42     3.90   28.46         2
43    14.62   54.05         2
```

Figure 7.6. *Continued.*

```
44   33.47   89.46   2
45   13.89   44.03   2
46    0.52   31.67   2
47   17.79   73.66   2
48   18.58   47.24   2
49   46.26   96.27   2
50   23.14   66.59   2
51    2.90   97.13   3
52   16.40   76.50   3
53   46.62  115.22   3
54   32.60   88.83   3
55   45.79  101.77   3
56    0.77   86.96   3
57   35.49  104.45   3
58    7.10   74.43   3
59   16.06   80.79   3
60   20.45   98.57   3
61   40.20  112.34   3
62   24.73   71.06   3
63   11.87   83.61   3
64    1.20   59.09   3
65   29.37   83.62   3
66   14.92   81.89   3
67   46.51  103.35   3
68   10.88   81.13   3
69   48.45  114.51   3
70    6.11   78.33   3
71   45.24  110.05   3
72   24.87  104.97   3
73   19.50   83.66   3
74   26.05   98.45   3
75   15.45   85.75   3
76   36.27   74.21   4
77   18.46   50.41   4
78   22.58   68.70   4
79   36.15   99.59   4
80   43.30   99.54   4
81   36.41   95.53   4
82    0.12   53.18   4
83    9.52   55.06   4
84   44.70  120.84   4
85   33.34   99.05   4
86   14.41   62.28   4
87    6.91   51.33   4
88    0.88   53.91   4
89    0.56   48.70   4
90   12.51   54.80   4
91   20.37   71.53   4
92   41.98   97.21   4
```

Figure 7.6. *Continued.*

93	31.31	89.23	4
94	46.86	123.43	4
95	39.80	106.21	4
96	29.44	101.53	4
97	36.05	99.35	4
98	28.71	97.07	4
99	1.48	40.00	4
100	48.30	122.05	4

The S-Plus or R results are presented in Fig. 7.7b. The linear regression estimates $\hat{\beta}_0 = 39.95$ for the y intercept and $\hat{\beta}_1 = 1.55$ for the slope and the F statistic comparing this model with the null model are all statistically significant at the 95% confidence level. The R^2 statistic is 68.4%. The estimated residual standard deviation of

Figure 7.7. Graphs and analysis results of the `birds.density` dataset for the linear regression model with S-Plus or R code. **(a)** Scatterplot of the `(riparian,density)` measurements `(plot(birds.density$riparian,birds.density$density))`. **(b)** Linear regression analysis results. **(c)** Plot of the linear regression model `(abline(output1.birds.density$coef[1],output1.birds.density$coef[2]))`. **(d)** Scatterplot of the linear regression residual errors `(plot(output1.birds.density))`. **(e)** Plot of the linear regression model watershed residuals `(plot(as.numeric(birds.density$Watershed),output1.birds.density$resid))`.

(a)

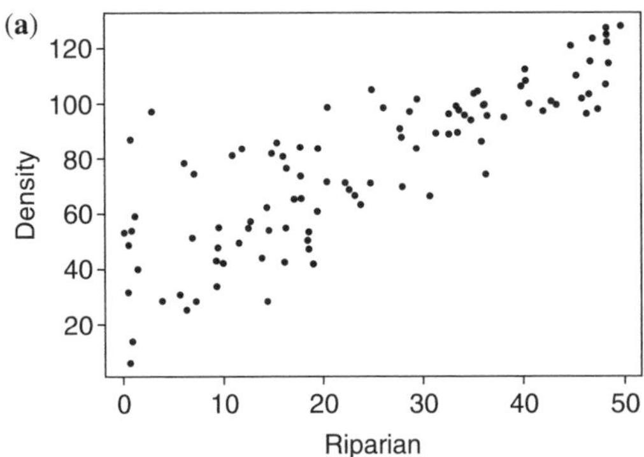

(b)

```
> output1.birds.density <-
  lm(density~riparian,birds.density)
> summary(output1.birds.density)
Coefficients:
                Value Std. Error t value Pr(>|t|)
(Intercept) 39.9512   3.0113    13.2669   0.0000
   riparian  1.5480   0.1063    14.5663   0.0000
Residual standard error: 15.81 on 98 degrees of freedom
Multiple R-Squared: 0.6841
F-statistic: 212.2 on 1 and 98 degrees of freedom, the p-value is
0
```

Figure 7.7. *Continued.*

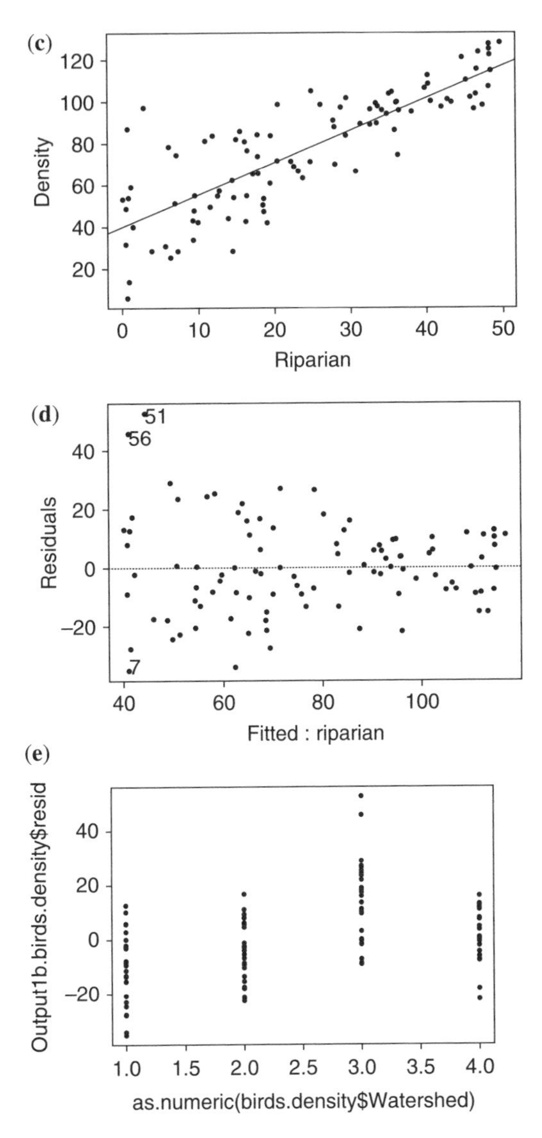

15.8 is somewhat large, and the residual errors include two outliers (Fig. 7.7d), but the linear regression model otherwise looks promising overall for the given dataset (Fig. 7.7c). However, we are ignoring the variation in the dataset due to the differences between the watersheds (Fig. 7.7e). Let's look at that next.

The linear regression model has ignored the variation between the watersheds. S-Plus and R provide scatterplots for each of the watersheds (Fig. 7.8a), where the sample data points are connected with a local regression smoother (loess).

Figure 7.8. Graphs of the `birds.density` dataset for the individual linear regression models with S-Plus or R code. (**a**) Scatterplot of the `(riparian, density)` measurements for each watershed `(plot(birds.density))`. (**b**) Linear regression analysis for each watershed. (**c**) Confidence intervals of the individual linear regression estimates of constant and slope `(plot(intervals(output2.birds.density)))`. (**d**) Plot of the individual linear regression models `(plot(augPred(output2.birds.density)))`. (**e**) Plot of the individual linear regression residuals `(plot(output2.birds.density))`.

(**a**)

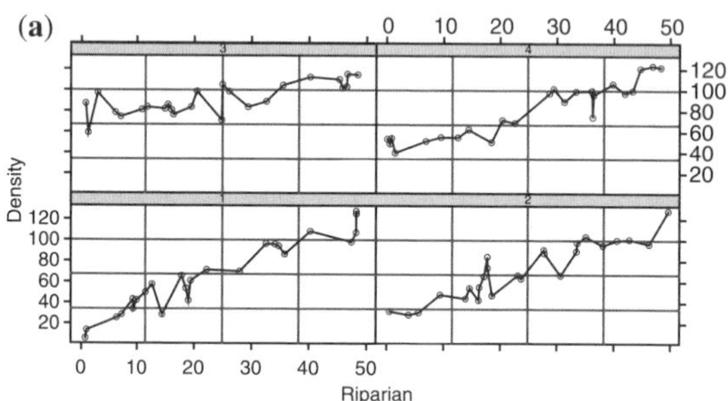

(**b**)

```
> output2.birds.density <-
  lmList(density~riparian,birds.density)
> summary(output2.birds.density)
Coefficients:
   (Intercept)
     Value Std. Error   t value       Pr(>|t|)
1 15.62620   3.626171   4.309284 4.099386e-005
2 26.85935   4.174599   6.433997 5.506136e-009
3 73.76478   3.681440  20.036937 0.000000e+000
4 42.45121   3.821747  11.107803 0.000000e+000
   riparian
      Value Std. Error   t value       Pr(>|t|)
1 2.1426628  0.1311543 16.336967 0.000000e+000
2 1.8500498  0.1516425 12.200072 0.000000e+000
3 0.7333648  0.1312088  5.589298 2.311067e-007
4 1.5200333  0.1272504 11.945215 0.000000e+000
Residual standard error: 9.974709 on 92 degrees
     of freedom
> intervals(output2.birds.density)
, , (Intercept)
     lower      est.    upper
1  8.424312 15.62620 22.82809
2 18.568242 26.85935 35.15047
3 66.453121 73.76478 81.07644
4 34.860892 42.45121 50.04153
, , riparian
     lower      est.     upper
1 1.8821791 2.1426628 2.4031465
2 1.5488747 1.8500498 2.1512250
3 0.4727729 0.7333648 0.9939567
4 1.2673031 1.5200333 1.7727636
```

Figure 7.8. *Continued.*

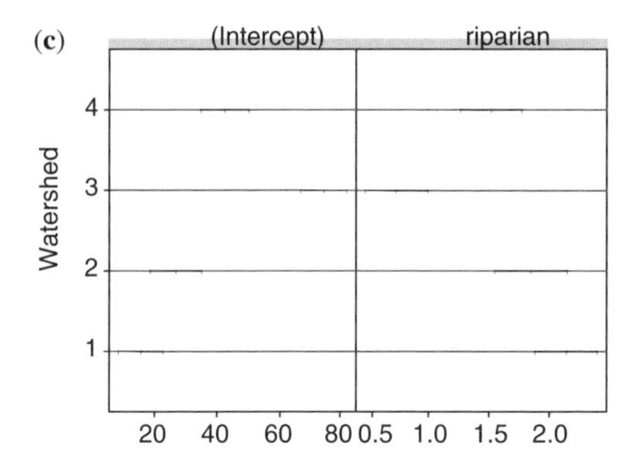

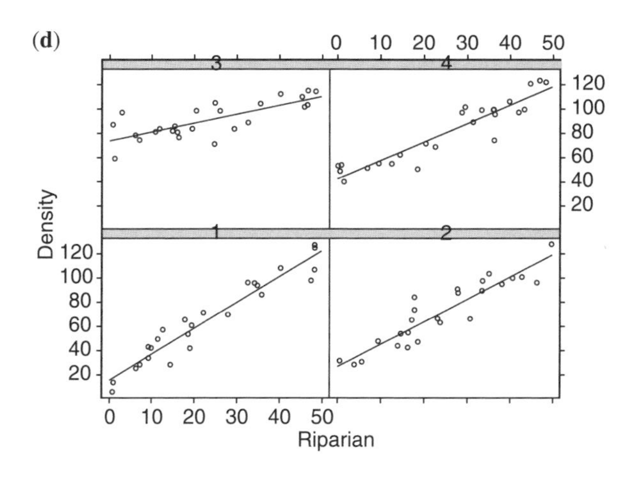

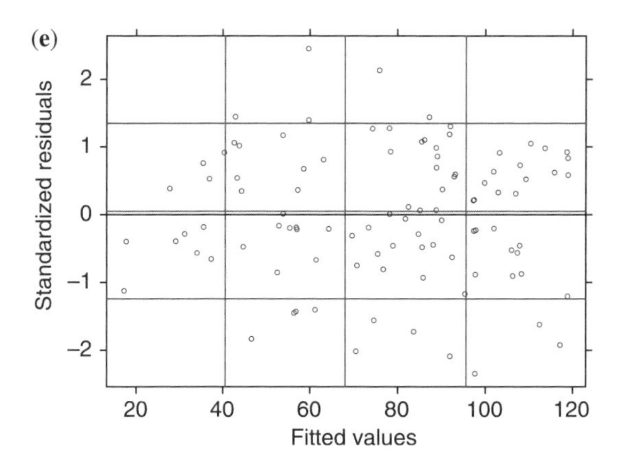

The linear relationships appear to be quite different between watersheds. The linear relationship in watershed 3 is flat, whereas that in watershed 1 is steep, with the others falling in between. Also, the watershed 1 relationship is generally lower and the watershed 3 relationship is higher, with the others in between. In fact, we should note that there is evidence of a possible negative correlation between the steepness and the height of the linear relationships for the watersheds; the steeper relationships have lower height, and vice versa. We shall examine more of that later. Let's model the individual linear regression relationships for each watershed and compare using the S-Plus or R `lmList` command (see Fig. 7.8b). The individual linear regression models for the watersheds all have statistically significant estimates for constant and slope. However, the estimates are quite distinct between watersheds, and it is apparent that there is a negative correlation between constants and slopes for the watersheds (Figs. 7.8c–7.8e).

We might be tempted to conclude our analysis at this point with separate linear regression models estimated for each of the four watersheds. However, this conclusion would be ignoring the fact that this pooled dataset was collected from watershed subpopulations of one larger population. We would prefer to have a holistic model that represents the entire population, but one that accounts for the distinctions between the watershed subpopulations. Furthermore, a statistical objection to our modeling of individual watersheds is that we have "over parameterized" our modeling effort. We have estimated too many parameters, nine in total, a constant and slope for each of four watersheds, plus an overall residual error standard deviation. Our individual estimates are relatively imprecise, since there are so many of them obtained from our dataset of fixed sample size. How might we more efficiently incorporate individual watershed differences into one model?

We will address this challenge by using the linear mixed-effects `lme` command in S-Plus or R to examine mixed-effects models incorporating random effects. We will start with the most general mixed-effects model. We assume a mixed-effects linear relationship between an independent variable x (`riparian`) and a dependent variable y (`density`) satisfying

$$y_i = (\beta_0 + b_0) + (\beta_1 + b_1) \cdot x_i + e_i,$$

where the fixed-effects parameters are β_0 for the constant and β_1 for the slope, the watershed random effects for the constant b_0 are iid $N(0, \sigma_{b0})$ and for the slope b_1 are iid $N(0, \sigma_{b1})$, and the residual errors e_i are iid $N(0, \sigma)$. We again assume the following for the conditional populations $y|x = \{y|(x, y), x \text{ is fixed}\}$ and their parameters:

1. $\mu_{y|x} = (\beta_0 + b_0) + (\beta_1 + b_1) \cdot x$ (linearity of the mean).
2. $\sigma_{y|x} = \sigma$ is constant (homoscedasticity).
3. $y|x$ are normally distributed: $y \sim N(\mu_{y|x}, \sigma)$.
4. $y|x$ are randomly sampled (x can be either randomly sampled or fixed).
5. x are measured without error.

Our S-Plus or R analysis results for this general mixed-effects linear regression model are given in Fig. 7.9a. The estimates for the parameters are $\hat{\beta}_0 = 39.61$ and $\hat{\beta}_1 = 1.56$

Figure 7.9. Analysis of the `birds.density` dataset for mixed-effects linear regression models with S-Plus and R code. (**a**) The general mixed-effects linear regression model analysis. (**b**) Generalized least squares linear regression analysis on the `birds.density` dataset. (**c**) Analysis for two reduced mixed-effects linear regression models. (**d**) Analysis of mixed-effects linear regression models with variance–covariance structures between-groups random effects. (**e**) Analysis of mixed-effects linear regression models with variance structures within-group random effects. (**f**) Approximate 95% confidence intervals for the added parameters in the `Power` and `Exp` mixed-effects linear regression models with variance–covariance structures between-groups random effects.

(**a**)
```
> output3a.birds.density <-
  lme(density~riparian,birds.density,random=
  ~riparian|Watershed)
> summary(output3a.birds.density)
Linear mixed-effects model fit by REML
 Data: birds.density
       AIC     BIC    logLik
  764.3522 779.862 -376.1761
Random effects:
 Formula:  ~ riparian | Watershed
 Structure: General positive-definite
                StdDev   Corr
(Intercept) 25.0600496 (Inter
   riparian  0.6033213 -1
   Residual  9.8268424
Fixed effects: density ~ riparian
               Value Std.Error DF  t-value p-value
(Intercept) 39.61387  12.67012 95 3.126558  0.0023
   riparian  1.56456   0.30887 95 5.065465  <.0001
> intervals(output3a.birds.density)
Approximate 95% confidence intervals
 Fixed effects:
                 lower      est.     upper
(Intercept) 14.4604973 39.613869 64.767241
   riparian  0.9513785  1.564558  2.177738
 Random Effects:
  Level: Watershed
                      lower        est.     upper
sd((Intercept))  11.0758796  25.0600496 56.700335
   sd(riparian)   0.2613419   0.6033213  1.392798
cor((Intercept),riparian)
                 -1.0000000  -0.9999996  1.000000
 Within-group standard error:
    lower     est.    upper
 8.524404 9.826842 11.32828
> fixef(output3a.birds.density)
 (Intercept) riparian
    39.61387 1.564558
> ranef(output3a.birds.density)
  (Intercept)     riparian
1  -23.542199  0.56677869
2  -12.931104  0.31131543
3   33.381047 -0.80364946
4    3.092257 -0.07444466
> output3a.birds.density$resid[,2]
        1         2        3         4        5
-14.6997 -18.56212 6.350703 -4.263033 10.63806
 ...
      96        97       98         99      100
 14.95493 2.925278 11.58271 -4.911494 7.371386
```

Figure 7.9. *Continued.*

(b)
```
> output1b.birds.density <-
  gls(density~riparian,birds.density)
> summary(output1b.birds.density)
Generalized least squares fit by REML
  Model: density ~ riparian
  Data: birds.density
       AIC     BIC  logLik
  839.7801 847.535 -416.89
Coefficients:
               Value Std.Error  t-value p-value
(Intercept) 39.95116  3.011336 13.26692  <.0001
   riparian  1.54804  0.106276 14.56631  <.0001
Residual standard error: 15.80779
Degrees of freedom: 100 total; 98 residual
> AIC(output1b.birds.density,output3a.birds.density)
                       df      AIC
output1b.birds.density  3 839.7801
output3a.birds.density  6 764.3522
> anova(output1b.birds.density,output3a.birds.density)
                       Model df      AIC     BIC    logLik
output1b.birds.density     1  3 839.7801 847.535 -416.8900
output3a.birds.density     2  6 764.3522 779.862 -376.1761
                         Test  L.Ratio p-value
output1b.birds.density
output3a.birds.density 1 vs 2 81.42794  <.0001
```

(c)
```
> output3b.birds.density) <-
lme(density~riparian,birds.density,random=~1|Watershed)
> summary(output3b.birds.density)
Linear mixed-effects model fit by REML
 Data: birds.density
       AIC      BIC    logLik
  808.3422 818.6821 -400.1711
Random effects:
 Formula:  ~ 1 | Watershed
        (Intercept) Residual
StdDev:     10.6895 12.74567
Fixed effects: density ~ riparian
               Value Std.Error DF  t-value p-value
(Intercept) 40.09028  5.871805 95  6.82759  <.0001
   riparian  1.54228  0.085854 95 17.96384  <.0001
> output3c.birds.density) <-
lme(density~riparian,birds.density,random=~riparian-1|Watershed)
> summary(output3c.birds.density)
Linear mixed-effects model fit by REML
 Data: birds.density
```

Figure 7.9. *Continued.*

```
       AIC      BIC    logLik
  840.9644 851.3043 -416.4822
Random effects:
 Formula:  ~ riparian - 1 | Watershed
        riparian Residual
StdDev: 0.1093365 15.57783
Fixed effects: density ~ riparian
              Value Std.Error DF  t-value p-value
(Intercept) 40.06737  2.970996 95 13.48618  <.0001
   riparian  1.54273  0.118340 95 13.03637  <.0001
> AIC(output3a.birds.density, output3b.birds.density,
 output3c.birds.density)
                       df      AIC
output3a.birds.density  6 764.3522
output3b.birds.density  4 808.3422
output3c.birds.density  4 840.9644
> anova(output3b.birds.density, output3a.birds.density)
                       Model df      AIC      BIC    logLik
Test  L.Ratio p-value
output3b.birds.density     1  4  808.3422 818.6821 -400.1711
output3a.birds.density     2  6  764.3522 779.8620 -376.1761 1 vs
2 47.99006  <.0001
> anova(output3c.birds.density, output3a.birds.density)
                       Model df      AIC      BIC    logLik
Test  L.Ratio p-value
output3c.birds.density     1  4 840.9644 851.3043 -416.4822
output3a.birds.density     2  6 764.3522 779.8620 -376.1761 1 vs
2 80.61224  <.0001
```

(d)
```
> output3d.birds.density <- lme(density~riparian,birds.
density, random=pdDiag(~riparian))
> summary(output3d.birds.density)
Linear mixed-effects model fit by REML
 Data: birds.density
       AIC      BIC    logLik
  777.6176 790.5425 -383.8088
Random effects:
 Formula:  ~ riparian | Watershed
 Structure: Diagonal
        (Intercept)  riparian Residual
StdDev:    24.27057 0.5761846 9.983725
Fixed effects: density ~ riparian
              Value Std.Error DF  t-value p-value
(Intercept) 39.71585  12.28571 95 3.232687  0.0017
   riparian  1.55969   0.29597 95 5.269745  <.0001
```

Figure 7.9. *Continued.*

```
> output3e.birds.density <-
lme(density~riparian,birds.density,random=pdIdent(~riparian))
> summary(output3e.birds.density)
Linear mixed-effects model fit by REML
 Data: birds.density
       AIC      BIC    logLik
  840.9341 851.274 -416.4671
Random effects:
 Formula:  ~ riparian | Watershed
 Structure: Multiple of an Identity
        (Intercept)  riparian Residual
StdDev:   0.1112594 0.1112594 15.57097
Fixed effects: density ~ riparian
   Value Std.Error DF  t-value p-value
(Intercept) 40.06944  2.970273 95 13.49015  <.0001
   riparian  1.54263  0.118750 95 12.99057  <.0001
> anova(output3e.birds.density,
output3d.birds.density,output3a.birds.density)
                       Model df       AIC      BIC
   logLik   Test         L.Ratio  p-value
output3e.birds.density     1  4  840.9341 851.2740 -416.4671
output3d.birds.density     2  5  777.6176 790.5425 -383.8088 1 vs
2        65.31649   <.0001
output3a.birds.density   3  6 764.3522 779.8620 -376.1761 2 vs
3        15.26547   0.0001
```

(e)

```
> output3f.birds.density <-
lme(density~riparian,birds.density,random=
~riparian|Watershed,weights=varFixed(~riparian))
> summary(output3f.birds.density)
Linear mixed-effects model fit by REML
 Data: birds.density
       AIC      BIC    logLik
  842.6914 858.2012 -415.3457
Random effects:
 Formula:  ~ riparian | Watershed
 Structure: General positive-definite
               StdDev   Corr
(Intercept) 29.3413291 (Inter
   riparian  0.7869081 -1
   Residual  3.5201119
Variance function:
 Structure: fixed weights
 Formula: ~ riparian
Fixed effects: density ~ riparian
               Value Std.Error DF  t-value p-value
(Intercept) 41.66034  14.70368 95 2.833327  0.0056
   riparian  1.48488   0.40205 95 3.693302  0.0004
```

Figure 7.9. *Continued.*

```
> output3g.birds.density <-
lme(density~riparian,birds.density,random=
~riparian|Watershed,weights=varPower(c(1), ~riparian))
> summary(output3g.birds.density)
Linear mixed-effects model fit by REML
 Data: birds.density
       AIC      BIC    logLik
  766.3125 784.4073 -376.1563
Random effects:
 Formula:  ~ riparian | Watershed
 Structure: General positive-definite
              StdDev   Corr
(Intercept) 24.9178019 (Inter
   riparian  0.5988831 -1
   Residual 10.2033478
Variance function:
 Structure: Power of variance covariate
 Formula:  ~ riparian
 Parameter estimates:
      power
 -0.01333092
Fixed effects: density ~ riparian
              Value Std.Error DF  t-value p-value
(Intercept) 39.54194  12.60424 95 3.137193  0.0023
   riparian  1.56672   0.30677 95 5.107091  <.0001
> output3h.birds.density <-
lme(density~riparian,birds.density,random=
~riparian|Watershed,weights=varConstPower(c(1),c(1),~riparian))
> summary(output3h.birds.density)
Linear mixed-effects model fit by REML
 Data: birds.density
       AIC      BIC    logLik
  768.3118 788.9915 -376.1559
Random effects:
 Formula:  ~ riparian | Watershed
 Structure: General positive-definite
              StdDev   Corr
(Intercept) 24.9187249 (Inter
   riparian  0.5989348 -1
   Residual  5.3503475
Variance function:
 Structure: Constant plus power of variance covariate
 Formula:  ~ riparian
 Parameter estimates:
     const       power
 0.9084218 -0.02648395
Fixed effects: density ~ riparian
```

Figure 7.9. *Continued.*

```
               Value Std.Error DF  t-value p-value
(Intercept) 39.53943  12.60478 95 3.136861  0.0023
   riparian  1.56680   0.30680 95 5.106870  <.0001
> output3i.birds.density <-
lme(density~riparian,birds.density,random=
~riparian|Watershed,weights=varExp(c(1),~riparian))
> summary(output3i.birds.density)
Linear mixed-effects model fit by REML
 Data: birds.density
       AIC      BIC    logLik
  766.3271 784.4218 -376.1635
Random effects:
 Formula:  ~ riparian | Watershed
 Structure: General positive-definite
               StdDev   Corr
(Intercept) 24.9751002 (Inter
   riparian  0.6000435 -1
   Residual 10.0271135
Variance function:
 Structure: Exponential of variance covariate
 Formula:  ~ riparian
 Parameter estimates:
        expon
 -0.0008244781
Fixed effects: density ~ riparian
               Value Std.Error DF  t-value p-value
(Intercept) 39.59932  12.63101 95 3.135086  0.0023
           riparian  1.56511   0.30725 95 5.093885  <.0001
```

(f)
```
> intervals(output3g.birds.density)
Approximate 95% confidence intervals
 Fixed effects:
                 lower      est.     upper
(Intercept) 14.5193518 39.541938 64.564524
   riparian  0.9576988  1.566722  2.175745
 Random Effects:
  Level: Watershed
                                lower        est.     upper
          sd((Intercept))  0.80584287  24.9178019 770.49370
             sd(riparian)  0.01741171   0.5988831  20.59883
cor((Intercept),riparian) -1.00000000  -0.9999997        NA
 Variance function:
            lower        est.     upper
power -0.2046213 -0.01333092 0.1779595
 Within-group standard error:
```

Figure 7.9. *Continued.*

```
     lower      est.     upper
  5.802689 10.20335 17.94139
 > intervals(output3i.birds.density)
 Approximate 95% confidence intervals
  Fixed effects:
                       lower       est.      upper
 (Intercept) 14.5235877 39.599321 64.675055
    riparian  0.9551386  1.565114  2.175089
  Random Effects:
   Level: Watershed
                                 lower         est.       upper
            sd((Intercept))  11.135798  24.9751002 56.013555
               sd(riparian)   0.262092   0.6000435  1.373763
 cor((Intercept),riparian)  -1.000000  -0.9999998  1.000000
  Variance function:
                 lower            est.         upper
 expon -0.01102098 -0.0008244781 0.009372027
  Within-group standard error:
     lower      est.     upper
  7.515348 10.02711 13.37836
```

for the constant and slope fixed effects, respectively, and $\hat{\sigma}_{b_0} = 25.06$, $\hat{\sigma}_{b_1} = 0.60$, and $\hat{\sigma} = 9.83$ for the constant, slope, and residual random effects. The 95% confidence intervals for these estimates are all significant. Note that the confidence interval for the estimated correlation between the constant and slope random effects contains 0, so the correlation is not statistically significant. The predicted watershed random effects for the constant are $b_0 = [-23.54, -12.93, 33.38, 3.09]$ and for the slope are $b_1 = [0.57, 0.31, -0.80, -0.07]$. The model estimates for data in watershed 1 are given by

$$
\begin{aligned}
y_i &= (\hat{\beta}_0 + b_0) + (\hat{\beta}_1 + b_1) \cdot x_i + e_i \\
&= (39.61 - 23.54) + (1.56 + 0.57) \cdot x_i + e_i \\
&= 16.07 + 2.13 \cdot x_i + e_i,
\end{aligned}
$$

so that the first data point in `birds.density` with $y_1 = 41.91$, $x_1 = 19.02$, and $e_1 = -14.70$ is described by the estimated model with

$$
41.91 = 16.07 + 2.13 \cdot (19.02) - 14.70,
$$

which is correct within roundoff error. The other data points in `birds.density` can be obtained similarly.

The estimated residual standard deviation $\hat{\sigma} = 9.83$ is considerably less than that of the fixed-effects linear regression model $\hat{\sigma} = 15.81$. We can use the S-Plus or R generalized LS command `gls` rather than the more restricted LS command `ls` to execute the original fixed-effects linear regression model analysis and also calculate

AIC so we can compare it with our mixed-effects model (Fig. 7.9b). We see that the AIC statistic is considerably lower (764.35 vs. 839.78) for the mixed-effects model. The `anova` command in S-Plus or R used in this context conducts a likelihood ratio test between nested models with equal fixed effects. Here our linear regression model is nested inside the more general mixed-effects model; both have the same fixed effects. We see that the likelihood ratio test rejected the null hypothesis that the two models are equal, providing yet more corroborative evidence along with the other statistics that the more general mixed-effects linear regression model better fits the data than does the simple fixed-effects linear regression model. Before finally, concluding that we have a best-fitting model, however, we should compare these results with two other reduced mixed-effects models and with mixed-effects models incorporating variance–covariance structures.

We have analyzed the most general mixed-effects model for the linear regression relationship, one with watershed random effects with respect to both constant and slope, `Model #3a.birds.density`. Now let's look at the two reduced mixed-effects models, one with watershed random effects for the constant only, `Model #3b.birds.density`, and the other with watershed random effects for the slope only, `Model #3c.birds.density` (Fig. 7.9c). Note that `Model #3b.birds.density` estimates only σ_{b_0}, assuming $\sigma_{b_1} = 0$, and `Model #3c.birds.density` estimates only σ_{b_1}, assuming $\sigma_{b_0} = 0$. The AIC and ANOVA likelihood ratio results indicate that the most general mixed-effects model, `Model #3a.birds.density`, is best-fitting. This is certainly what we expected from the original graphs and estimates of the linear regression models for the individual watersheds (Figs. 7.8c–7.8e).

7.3.3 Variance–Covariance Structure Between-Groups Random Effects

Both S-Plus and R provide the capabilities to model variance–covariance structure between the watershed group random effects. The graphs and estimates of the linear regression models for the individual watersheds have suggested a negative correlation between the constant random effects b_0 and the slope random effects b_1 in the general mixed-effects model. We will explore this possibility by examining model options for the variance–covariance structure between the random effects that are available in S-Plus and R.

Recall first the relationship between covariance σ_{ij} and correlation ρ_{ij} of two random variables indexed by i and j

$$\rho_{ij} = \sigma_{ij}/(\sigma_i \sigma_j),$$

where σ_i and σ_j are the standard deviations and σ_i^2 and σ_j^2 are the variances. The variance–covariance positive-definite matrix, illustrated in three dimensions, is of the positive-definite form

$$\begin{bmatrix} \sigma_1^2 & \sigma_{12} & \sigma_{13} \\ & \sigma_2^2 & \sigma_{23} \\ & & \sigma_3^2 \end{bmatrix}$$

The lower-left half of the matrix is omitted since the matrix must be symmetric: $\sigma_{ij} = \sigma_{ji}$. S-Plus and R have 5 pdMat structures (pdMat = positive − definite matrix) that can be modeled:

1. pdSymm: symmetric matrix structure (most general, the default)

$$\begin{bmatrix} \sigma_1^2 & \sigma_{12} & \sigma_{13} \\ & \sigma_2^2 & \sigma_{23} \\ & & \sigma_3^2 \end{bmatrix}$$

2. pdCompSymm: compound symmetry matrix structure, with equal variances σ_i^2 and equal covariances σ_{ij} (use for split-plot designs)

$$\begin{bmatrix} \sigma_i^2 & \sigma_{ij} & \sigma_{ij} \\ & \sigma_i^2 & \sigma_{ij} \\ & & \sigma_i^2 \end{bmatrix}$$

3. pdDiag: diagonal matrix structure, with covariances σ_{ij}=0

$$\begin{bmatrix} \sigma_1^2 & 0 & 0 \\ & \sigma_2^2 & 0 \\ & & \sigma_3^2 \end{bmatrix}$$

4. pdIdent: multiple of an identity matrix structure, with covariances σ_{ij}=0 and equal variances σ_i^2

$$\begin{bmatrix} \sigma_i^2 & 0 & 0 \\ & \sigma_i^2 & 0 \\ & & \sigma_i^2 \end{bmatrix}$$

5. pdBlocked: blocks of matrices down the diagonal (use for blocks of correlated random effects)

S-Plus and R model variance–covariance structures as standard deviation–correlation structures, using the lme command parameter syntax random=~pd`Structure` (~covariates) where `Structure` =*Symm*, *CompSymm*, `Diag`, `Ident`, or `Blocked` and the grouping structure is defaulted to that of the grouped database object (here, `Watershed` in `birds . density`).

In the `birds . density` example, we can model a two-dimensional positive-definite matrix for the two linear regression random effects

$$\begin{bmatrix} \sigma_1^2 & \sigma_{12} \\ & \sigma_2^2 \end{bmatrix}.$$

We illustrate these methods with a pdDiag structure, Model #3d.birds.density, and a pdIdent structure, Model #3e . birds . density in Fig. 7.9d. The Model

#3d . birds . density and #3e . birds . density AIC results are 777.62 and 840.93, respectively, compared to 764.35 for our most general Model #3a . birds . density with default summetric positive-definite structure. Also, the S-Plus and R ANOVA likelihood ratio tests indicate that Model #3a . birds . density is better-fitting than either of these two models with restricted variance–covariance structures. Note that we make this inference, based on AIC, despite the fact that the estimated correlation for the more general model wasn't statistically significant at the 95% confidence level.

We will not be able to illustrate the blocked `pdMat` variance–covariance structure between groups with this simple two-dimensional example, but refer the reader to Pinheiro and Bates (2000) for more complex examples.

7.3.4 Variance Structure Within-Group Random Effects

Next, we examine variance structures within the groups that we can model in S-Plus and R. If we examine the scatterplot of the linear regression model and residuals in Fig. 7.7 for the pooled dataset, we find that there is some evidence to suggest that the variance within the groups does not remain constant, a necessary assumption of linear regression modeling. Although two key outliers in particular contribute to this variation (see outlier points 61 and 66 in Fig. 7.7d), in general there does appear to be a decrease in the variance as the riparian covariate increases.

So let's explore the within-group variance structures that are available in S-Plus and R, to model such change in variance. There are 6 such varFunct classes available in S-Plus and R that model the within-group variance structure of the residual errors. The S-Plus and R syntax for implementing this structure is to add the following parameter to the lme command syntax

$$\texttt{weights} = \texttt{var}\mathtt{\textit{Structure}}([\mathtt{\textit{value}}],\mathtt{\textit{form}}),$$

where `value` is an optional initial value for the parameter in the model structure, `form` $=$ `~covariate` for an outer covariate that varies within the groups, or `form` $=$ `fitted(.)` for the model fitted values within the groups, and `Structure` $=$ `Fixed`, `Ident`, `Power`, `ConstPower`, `Exp`, or `Comb`. The default variance for the residual errors is assumed constant within the groups $\text{var}(e_{ij}) = \sigma^2$ or $\text{se}(e_{ij}) = \sigma$ for the standard deviation of the errors. S-Plus and R model a "variance" function g (covariates) for the variance as a function of the covariate that varies within the groups var $(e_{ij}) = g^2(\text{covariate}) \cdot \sigma^2$ or $\text{se}(e_{ij}) = g(\text{covariate}) \cdot \sigma$ so that the default variance function is $g(\text{covariate}) = 1$. The `varFunct` classes are

1. `varFixed`:
   a. Variance increases linearly with a covariate

$$\text{var}(e_{ij}) = \text{covariate}_{ij} \cdot \sigma^2,$$

so that $g(\text{covariate}) = \text{covariate}^{1/2}$.

b. Syntax is `weights = varFixed ([value,] ~covariate)`.
c. Note that there are no parameters for this structure;

2. `varIdent`:
   a. Variance is a function of the levels of a stratification variable s for the groups that is constant in value within the groups (i.e., an outer factor)

$$\mathrm{var}(e_{ij}) = \delta_{ij}^2 \cdot \sigma^2,$$

   so that $g\ (s_{ij}, \delta) = \delta_{ij}$.
   b. Syntax is `weights = varIdent ([value,] ~1|s)`.
   c. The parameters are the δ_{ij}.
3. `varPower`:
   a. Variance is a function of a power of the covariate

$$\mathrm{var}(e_{ij}) = |\mathrm{covariate}_{ij}|^{2\cdot\delta} \cdot \sigma^2,$$

   so that g (covariate, δ) = |covariate|.
   b. Syntax is `weights = varPower ([value,] ~covariate)`.
   c. The parameter is δ.
   d. This may be combined with stratification `weights = varPower ([value,] ~covariate|s)`.
4. `varConstPower`:
   a. Variance is a function of a constant plus a power of the covariate

$$\mathrm{var}(e_{ij}) = (\delta_1 + |\mathrm{covariate}_{ij}|^{\delta_2})^2 \cdot \sigma^2$$

   so that g (covariate, δ_1, δ_2) = $\delta_1 + |\mathrm{covariate}|^{\delta_2}$.
   b. Syntax is `weights = varConstPower ([value,] ~covariate)`.
   c. The parameters are δ_1 and δ_2.
   d. Note that this may be combined with stratification `weights = varConstPower ([value,] ~covariate|s)`.
5. `varExp`:
   a. Variance is a function of an exponential power of the covariate

$$\mathrm{var}(e_{ij}) = e^{(2\cdot\delta\cdot\mathrm{covariate}_{ij})} \cdot \sigma^2,$$

   so that g (covariate, δ) = $e_{ij}^{(\delta\cdot\mathrm{covariate})}$).
   b. Syntax is `weights = varExp ([value,] ~covariate)`.
   c. The parameter is δ.
   d. Note that this may be combined with stratification `weights = varExp ([value,] ~covariate|s)`.

6. `varComb`:
   a. Variance is a combination of `varFunct` structures (above)

$$\mathrm{var}(e_{ij}) = g_1^2 \cdot g_2^2 \cdot \sigma^2,$$

   so that g (covariate) $= g_1 \cdot g_2$.

   b. Syntax is `weights = varComb([value,] varStructure`$_1$`(), var Structure`$_2$`())`.
   c. Note that this may be combined with stratification.

We illustrate several of these varFunct structures in S-Plus and R with the `birds.density` dataset in Fig. 7.9e.

The AIC values are 842.69, 766.31, 768.31, and 766.33 for the `Fixed`, `Power`, `ConstPower`, and `Exp varFunct` models, respectively, compared to the general `Model #3a.birds.density` AIC of 764.35. Although the `Power` and `Exp` models are somewhat competitive with the general model, the AIC values nonetheless suggest that there is little advantage in adding a parameter to model the variance within groups. Furthermore, the 95% confidence intervals for the added parameters in the `Power` and `Exp` models indicate that these added parameters are statistically insignificant at the 95% confidence level (see Fig. 7.9f).

7.3.5 Covariance Structure Within-Group Random Effects: Time-Series and Spatially Dependent Models

Let's conclude the frequentist S-Plus and R analysis of the `birds.density` dataset by examining covariance structures within watershed groups. This structure models dependences of the response errors within groups. Our dataset suggests that there may some dependence, with the residual errors fluctuating up and down within the groups (see Figs. 7.8a and 7.8d); there may be an autocorrelation dependency of the response errors with respect to the riparian within-group covariate. So let's examine the statistics from analyses modeling covariance structures within the groups.

S-Plus and R have `corStruct` classes to model covariance structure within groups: (1) `Symm`, `CompSymm`, `AR1`, `CAR1`, and `ARMA(p,q)` for serial correlation and (2) `Exp`, `Gaus`, `Lin`, `Ratio`, and `Spher` for spatial correlation. The within-group covariance structure is modeled as

$$\mathrm{cor}(e_{ij},\, e_{ij'}) = h(d(p_{ij}, p_{ij'}), \rho),$$

where d is a distance function between position vectors p_{ij} and $p_{ij'}$, ρ is a vector of correlation parameters, and h is a correlation function. S-Plus and R model `corStruct` classes for within-group covariance structures using the parameter syntax

```
correlation = corStructure([ value,] form[,nugget=F])
```

where value = initial values for the parameters, `form = ~ covariate(s)[ |outer factor(s)]` or `~fitted(.)`, and `nugget` = F(default) or T for spatial structures.

Let's first focus on the serial correlation structures, sometimes called **temporal**, **time-series**, or **longitudinal structures**. S-Plus and R model serial correlation as the intraclass correlation coefficient

$$\mathrm{cor}(e_{ij}, e_{ij'}) = h(|p_{ij} - p_{ij'}|\rho) = \rho, \quad j \neq j'$$

using the empirical autocorrelation estimator of h at lag $k = |p_{ij} - p_{ij'}|$ with $j' = j + k$ and p_{ij} the position of residual e_{ij}

$$\hat{\rho}(k) = \frac{\sum_{i=1}^{n} \sum_{j=1}^{n_i - k} r_{ij} \cdot r_{i(j+k)}/N(k)}{\sum_{i=1}^{n} \sum_{j=1}^{n_i} r_{ij}^2/N(0)}$$

where $N(k)$ denotes the number of residual pairs and $r_{ij} = (y_{ij} - \hat{y}_{ij})/\hat{\sigma}_{ij}$ are the standardized residuals with $\sigma_{ij}^2 = \mathrm{var}\ (\varepsilon_{ij})$. S-Plus and R model serial covariance structures within groups with `corStruct` classes as follows:

1. `corSymm`: symmetric (default)
   a. $h(k, \rho) = \rho_k$.
   b. Syntax is
      `correlation = corSymm([value,] form = ~covariate).`
   c. Parameters are ρ_k.
   d. Note that this is overparameterized, to be used only as an exploratory tool to determine more parsimonious models.
2. `CompSymm`: compound symmetric
   a. $h(k, \rho) = \rho$.
   b. Syntax is
      `correlation = corCompSymm([value,] form = ~covariate).`
   c. Parameter is ρ.
   d. Note that this is usually unrealistic, since it is usually better to assume that the correlation decreases with the distance function, except for short time series or with split-plot experiments having observations collected at the same time within groups.
3. `AR1`: autoregressive, of order 1
   a. $h(k, \rho) = \rho^k$, $k = \text{lag}$.
   b. Syntax is
      `correlation = corAR1([value,] form = ~covariate).`
   c. Parameter is ρ, $-1 \leq \rho \leq +1$.
   d. Data e_t are collected at integer timepoints t, where the distance between observations e_t and e_s is the lag $k = |t - s|$.
   e. The correlation decreases in absolute value exponentially with lag k.
4. `CAR1`: continuous autoregressive, of order 1

a. $h(s, \rho) = \rho^s$, s continuous,
b. Syntax is
`correlation = corCAR1 ([ value,] form = ~covariate).`
c. Parameter is ρ, $-1 \leq \rho \leq +1$.
d. Data e_t are collected at continuous timepoints t, where the distance between observations e_t and e_s is the lag $k = |t-s|$.
e. The correlation decreases in absolute value exponentially with continuous time s.

5. `ARMA (p, q)`: autoregressive, moving-average model, of orders p and q, respectively, with
   a. Autoregressive model `AR(p)`: The current observation e_t is a linear function of the previous observations e_{t-i}, $i = 1, 2, \ldots, p$, plus a homoscedastic noise term a_t with $E[a_t] = 0$

$$e_t = \phi_1 \cdot e_{t-1} + \phi_2 \cdot e_{t-2} + \cdots + \phi_p \cdot e_{t-p} + a_t,$$

   where p is the order of the `AR(p)` model and the correlation parameters are $\phi = (\phi_1, \phi_2, \ldots, \phi_p)$,

   b. Moving-average model `MA(q)`: The current observation e_t is a linear function of iid noise terms

$$e_t = \theta_1 \cdot a_{t-1} + \theta_2 \cdot a_{t-2} + \cdots + \theta_q \cdot a_{t-q} + a_t,$$

   where q is the order of the `MA(q)` model and the correlation parameters are $\theta = (\theta_1, \theta_2, \ldots, \theta_q)$ with correlation function

$$h(k,\theta) = \frac{\theta_k + \theta_1 \cdot \theta_{k-1} + \cdots + \theta_{k-q} \cdot \theta_q}{1 + \theta_1^2 + \cdots + \theta_q^2}.$$

   c. `ARMA(p,q)` model = `AR(p)` + `MA(q)`

$$e_t = \sum_{i=1}^{p} \varphi_i \cdot e_{t-i} + \sum_{j=1}^{q} \theta_j \cdot a_{t-j} + a_t,$$

   with correlation parameters ρ given by the p autoregression parameters $\phi = (\phi_1, \phi_2, \ldots, \phi_p)$ and the q moving-average parameters $\theta = (\theta_1, \theta_2, \ldots, \theta_q)$.

   d. Syntax is `correlation = corARMA(value,form,p,q).`
   e. Examine the **partial autocorrelogram** to determine p for `AR(p)` and **autocorrelogram** to determine q for `MA(q)`.

Let's examine the ARMA correlation structure to model within-group correlation. We start by looking for significant correlation in the partial correlogram and correlogram in S-Plus and R to suggest order values for p and q for the `ARMA(p,q)` model autoregression and moving average (Figs. 7.10a and 7.10b). There is no particular

Figure 7.10. Correlograms, semivariogram, and mixed-effects linear regression analyses of the `birds.density` dataset for correlation within watershed group, with S-Plus and R code. (**a**) Partial correlogram of the riparian density measurements within watersheds `(acf(output3a.birds.density$resid [,2], type="partial",plot=T))`. (**b**) Correlogram of the riparian density measurements within watersheds `(acf(output3a.birds.density$resid [,2], type="correlation",plot=T))`. (**c**) Analysis of the `birds.density` dataset for mixed-effects linear regression models with S-Plus or R code to model ARMA correlation structure within-group `ARMA(1,0)`, `ARMA(0,1)`, and `ARMA(1,1)` models. (**d**) Semivariogram of the riparian density measurements within watersheds `(plot(Variogram(output3a.birds.density)))`. (**e**) Analysis of the `birds.density` dataset for mixed-effects linear regression models with S-Plus or R code to model exponential spatial correlation structure with a nugget for within-group error.

(**a**)

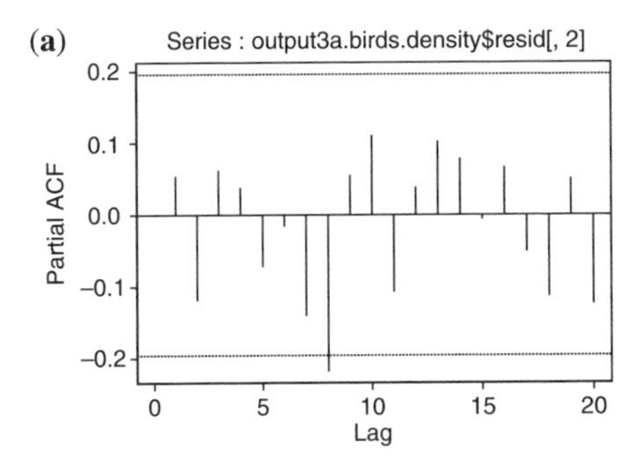

(**b**)

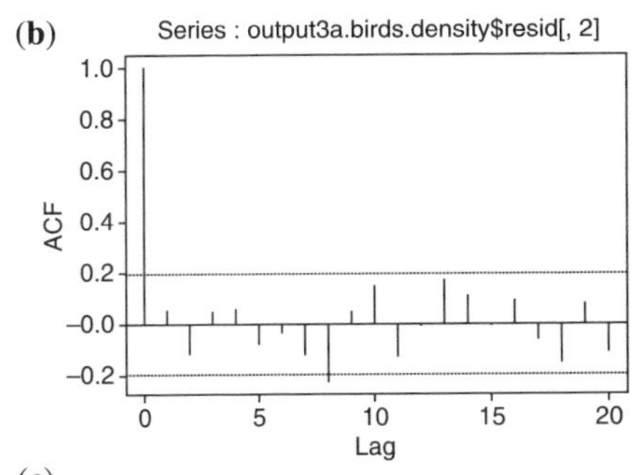

(**c**)

```
> output3j.birds.density <-
lme(density~riparian,birds.density,random=~riparian|
Watershed, correlation = corARMA(p=1,q=0))
> summary(output3j.birds.density)
Linear mixed-effects model fit by REML
 Data: birds.density
       AIC      BIC    logLik
  765.3195 783.4143 -375.6597
Random effects:
 Formula:  ~ riparian | Watershed
 Structure: General positive-definite
               StdDev   Corr
```

Figure 7.10. *Continued.*

```
(Intercept) 25.3959463 (Inter
   riparian  0.6154632 -1
   Residual  9.8587704
Correlation Structure: AR(1)
 Parameter estimate(s):
       Phi
 0.1124036
Fixed effects: density ~ riparian
               Value Std.Error DF  t-value p-value
(Intercept) 39.31033  12.84381 95 3.060644  0.0029
   riparian  1.57660   0.31467 95 5.010258  <.0001
> output3k.birds.density <-
lme(density~riparian,birds.density,random=~riparian|Watershed,cor
relation = corARMA(p=0,q=1))
> summary(output3k.birds.density)
Linear mixed-effects model fit by REML
 Data: birds.density
       AIC      BIC    logLik
  765.2296 783.3244 -375.6148
Random effects:
 Formula:  ~ riparian | Watershed
 Structure: General positive-definite
                StdDev   Corr
(Intercept) 25.3903745 (Inter
   riparian  0.6154962 -1
   Residual  9.8628136
Correlation Structure: ARMA(0,1)
 Parameter estimate(s):
    Theta1
 0.1218466
Fixed effects: density ~ riparian
               Value Std.Error DF  t-value p-value
(Intercept) 39.29546  12.84062 95 3.060246  0.0029
   riparian  1.57763   0.31468 95 5.013366  <.0001
> output3l.birds.density <-
lme(density~riparian,birds.density,random=~riparian|Watershed,cor
relation = corARMA(p=1,q=1))
> summary(output3l.birds.density)
Linear mixed-effects model fit by REML
 Data: birds.density
       AIC      BIC    logLik
  767.0067 787.6865 -375.5034
Random effects:
 Formula:  ~ riparian | Watershed
 Structure: General positive-definite
                StdDev   Corr
(Intercept) 25.2469789 (Inter
   riparian  0.6103889 -1
   Residual  9.8486897
```

Figure 7.10. *Continued.*

```
Correlation Structure: ARMA(1,1)
 Parameter estimate(s):
       Phi1    Theta1
 0.3423405 0.4603232
-
Fixed effects: density ~ riparian
               Value Std.Error DF  t-value p-value
(Intercept) 39.32461  12.76665 95 3.080261  0.0027
   riparian  1.57727   0.31216 95 5.052789  <.0001
> intervals(output31.birds.density)
Approximate 95% confidence intervals
 Fixed effects:
                 lower      est.     upper
(Intercept) 13.9796062 39.324610 64.669614
   riparian  0.9575551  1.577266  2.196977
 Random Effects:
  Level: Watershed
                                lower        est.     upper
          sd((Intercept))  11.0897558  25.2469789 57.477365
             sd(riparian)   0.2629409   0.6103889  1.416952
cor((Intercept),riparian)  -1.0000000  -0.9999990  1.000000
 Correlation structure:
            lower       est.     upper
  Phi1 -0.9734292 -0.3423405 0.8937733
Theta1 -0.8840817  0.4603232 0.9833345
 Within-group standard error:
    lower     est.    upper
 8.521228 9.84869 11.38295
```

(d)

Figure 7.10. *Continued.*

(e)
```
> output3m.birds.density <-
lme(density~riparian,birds.density,random=~riparian|Watershed,cor
relation=corExp(form=~riparian, nugget=T))
> summary(output3m.birds.density)
Linear mixed-effects model fit by REML
 Data: birds.density
       AIC      BIC    logLik
  766.1899 786.8696 -375.0949
Random effects:
 Formula:  ~ riparian | Watershed
 Structure: General positive-definite
              StdDev   Corr
(Intercept) 25.1410860 (Inter
   riparian  0.5994649 -1
   Residual  9.7604462
Correlation Structure: Exponential spatial
correlation
 Parameter estimate(s):
     range      nugget
 0.03811377 0.0000595548
Fixed effects: density ~ riparian
              Value Std.Error DF  t-value p-value
(Intercept) 39.45660  12.71146 95 3.104019  0.0025
   riparian  1.56396   0.30718 95 5.091287  <.0001
```

indication in the correlograms of significant autocorrelation falling outside the 95% confidence bands, but we will proceed anyway, examining the `ARMA (1,0)`, `ARMA (0,1)`, and `ARMA (1,1)` models, for the sake of illustration (see Fig. 7.10c). The AIC values for the three models are 765.32, 765.23, and 767.01, compared to the general `Model #3a.birds.density` AIC of 764.35, so the correlation modeling has not helped. Furthermore, the 95% confidence intervals for the correlation parameters both intersect 0, so these parameters are not statistically significant [we show the `ARMA (1,1)` results only]. From the graphs of the dataset, the correlograms, and these results, we can safely conclude that there is no further need to continue within-group correlation analysis, examining higher-order ARMA models.

S-Plus and R also include spatial correlation structures for within-group variation of the response residual errors with mixed-effects modeling that provide spatial statistics. S-Plus and R provide models of **isotropic** spatial correlation structures, which are continuous functions of distance between position vectors, in contrast to **anisotropic** spatial correlation structures, which are continuous functions of both distance and direction between position vectors. Response residual errors e_x are observed at position x and are assumed to be a function of the within-group covariate (`riparian`, in our example). Three **distance metrics** are available in S-Plus and R (where r is the dimension of the position vectors):

1. Euclidean or L_2: $d_{\mathrm{E}}(e_x, e_y) = \sqrt{\sum_{i=1}^{r}(x_i - y_i)^2}$

2. Manhattan or L_1: $d_{\text{Man}}(e_x, e_y) = \sum_{i=1}^{r} |x_i - y_i|$
3. Maximum distance: $d_{\max}(e_x, e_y) = \max_{i=1,2,\ldots,r} |x_i - y_i|$

Spatial correlation structures are represented by their **semivariogram** $\gamma(d(e_x, e_y), \rho) = 1/2 \cdot \text{var}(e_x - e_y)$ instead of their **correlation function** $h(d(e_x, e_y), \rho)$, where ρ is a parameter vector. They are complementary with respect to 1 if we assume the normally distributed residuals are standardized $e_x \sim \text{iid } N(0, 1)$, since

$$\begin{aligned}
\gamma(d(e_x, e_y), \rho) &= \tfrac{1}{2} \cdot \text{var}(e_x - e_y) = \tfrac{1}{2} \cdot E[(e_x - e_y)^2] \\
&= \tfrac{1}{2} \cdot (E[e_x^2] - 2 \cdot E[e_x \cdot e_y] + E[e_y^2]) \\
&= \tfrac{1}{2} \cdot (\text{var}(e_x^2) - 2 \cdot \text{cov}(e_x, e_y) + \text{var}(e_y^2)) \\
&= \tfrac{1}{2} \cdot (1 - 2 \cdot \text{cor}(e_x, e_y) + 1) = 1 - \text{cor}(e_x, e_y) \\
&= 1 - h(d(e_x, e_y), \rho).
\end{aligned}$$

So $\gamma(d, \rho) = 1 - h(d, \rho)$ and $\gamma(0, \rho) = 1 - h(0, \rho) = 0$. We note that semivariograms increase with increasing distance and that correlations decrease with increasing distance, both ranging between 0 and 1. An estimator of the semivariogram is given by

$$\hat{\gamma}(d) = \frac{1}{2 \cdot N(d)} \sum_{i=1^n} \sum_{\text{for all } j, j' \text{ such that } d(p_{ij}, p_{ij'})=d} (r_{ij} - r_{ij'})^2$$

where $N(d)$ denotes the number of residual pairs at a distance d from each other and $r_{ij} = (y_{ij} - \hat{y}_{ij})/\hat{\sigma}_{ij}$ are the standardized residuals with $\sigma_{ij}^2 = \text{var}(\varepsilon_{ij})$.

A nugget can be incorporated into the model, allowing a small discontinuity c_0, say, for measurement error, for γ at 0: $\gamma(d, \rho)$ -> c_0 as d -> 0 with $0 < c_0 < 1$. Thus $h(d, \rho)$ -> $1 - c_0$ as d -> 0 with $0 < c_0 < 1$.

There are five isotropic variogram models available in S-Plus and R (Pinheiro and Bates 2000, pp. 232–233):

1. Exponential: $\gamma(d, \rho) = 1 - \exp(-d/\rho)$, generalizing the CAR1 serial model
2. Gaussian: $\gamma(d, \rho) = 1 - \exp(-(d/\rho)^2)$
3. Linear: $\gamma(d, \rho) = 1 - (1 - d/\rho) \cdot I(d < \rho)$
4. Rational quadratic: $\gamma(d, \rho) = (d/\rho)^2/(1 + (d/\rho)^2)$
5. Spherical: $\gamma(d, \rho) = 1 - (1 - 1.5 \cdot (d/\rho) + 0.5 \cdot (d/\rho)^3) \cdot I(d < \rho)$

In the third and fifth models $I(d < \rho) = 1$ when $d < \rho$ and $= 0$ otherwise, and ρ is the **range parameter**. The S-Plus and R `corStruct` parameter syntax for spatial modeling is given by the command `correlation = corStruct(value, form, nugget, metric)`, where `value = c(range>0, 0≤nugget≤1)` and `value = numeric(0)` is the default, `form = ~position|group`, `nugget = T` (or `F`) (`nugget = F` is the default), and `metric = euclidean` (or `manhattan` or `maximum`). The five `corStruct` classes are `corExp`, corGaus, corLin, corRatio, and corSpher for the exponential, Gaussian, linear, rational quadratic, and spherical spatial structures, respectively.

Let's illustrate these ideas by examining the exponential spatial correlation structure for the watershed within-group residual error in S-Plus and R. An assessment of model form based on the semivariogram (Fig. 7.10d) is inconclusive, but let's try an exponential model with a nugget to fit this semivariogram (see Fig. 7.10e). The AIC doesn't compare favorably with the general mixed-effects model, increasing slightly to 766.19 compared to 764.35 for the general Model #3a.birds.density. Removing the nugget would likely decrease the AIC, possibly by up to two units. The estimates for the range and nugget are very small, close to insignificant, and the semivariogram suggested no obvious form, so we risk overfitting models to our sample data. The reader is encouraged to experiment with other spatial structures, with and without nuggets. We conclude at this point that the general model, Model #3a.birds.density, has provided the best fit to our dataset. To complete the analysis process, we should examine goodness of fit of this model and competing models.

To further pursue the subject of spatial statistics, the interested reader is encouraged to explore the capabilities of GS+ software (GS+ : Geostatistics for the Enviromental Sciences 1998). With this software, spatial data can be more fully analyzed, modeling dependences associated with each spatial location. The analyst can use both isotropic and anisotropic semivariograms and kriging to model spatially dependent data and display the results on maps. **Kriging** provides interpolated values for points not physically sampled, based on the semivariograms. Spherical, exponential, linear, and Gaussian forms are available in GS+ for the model forms. GS+ provides a friendly and powerful platform for spatial statistical analysis.

7.4 NONLINEAR MIXED-EFFECTS MODELING: FREQUENTIST STATISTICAL ANALYSIS IN S-Plus AND R

Let's conclude the discussion of frequentist mixed-effects modeling by briefly examining nonlinear mixed-effects modeling in S-Plus and R. We illustrate this application by adding a trigonometric fixed-effects component to the linear model, one that incorporates a periodic effect with amplitude, frequency, and initial value parameters. Then we will add random effects to the amplitude, frequency, and initial value, as well as the parameters of the linear component. As usual, the random effects are due to the watershed groups for the mixed-effects nonlinear model. First we need to define a nonlinear function in S-Plus or R, namely, trig, which is a function of the covariate riparian and the parameters alpha, beta, amplitude, frequency, and initial (Fig. 7.11a). We begin the nonlinear analysis with a fixed-effects model using the nonlinear least-squares command nls in S-Plus or R (Fig. 7.11b). Note that we needed to declare "reasonable" initial values for the parameters of the nonlinear nonlinear model that will be approximated with computer iteration. Plots of the residuals of our results are not satisfactory because of the watershed differences (Fig. 7.11c), so let's next examine fixed-effects models for each watershed group (Fig. 7.11d). We examine and compare confidence intervals for the estimates of the watershed parameters, and plot the results (Fig. 7.11e).

Figure 7.11. Analysis and plots of the `birds.density` dataset for nonlinear mixed-effects modeling, with S-Plus and R code. (**a**) Creation of the function `trig`. (**b**) Nonlinear analysis with a fixed-effects model using the nonlinear least-squares command `nls`. (**c**) Plots of the residuals of the nonlinear analysis with a fixed-effects model `(plot(output3r.birds.density, Watershed~resid(.), abline = 0))`. (**d**) Nonlinear analysis with fixed-effects models for each of the `watershed` groups. (**e**) Plot of the confidence intervals for the estimates of the watershed parameters for the nonlinear analysis with a fixed-effects model for each of the watershed groups `(plot(intervals(output3s.birds.density)))`. (**f**) Nonlinear analysis of a general mixed-effects model. (**g**) Nonlinear analysis of the mixed-effects model without the frequency and initial random effects. (**h**) Plot of the nonlinear analysis of the mixed-effects model without the frequency and initial random effects `(plot(output3u.birds.density,Watershed~resid(.),abline = 0))`. (**i**) Plot of the nonlinear analysis of the mixed-effects model without the frequency and initial random effects `(plot(output3u.birds.density))`.

(**a**)
```
> trig <-
function(riparian,beta0,beta1,amplitude,frequency,initial)
{beta0+beta1*riparian+amplitude*sin(frequency*riparian+initial)}
```

(**b**)
```
> output3r.birds.density <-
nls(density~trig(riparian,beta0,beta1,amplitude,
frequency,initial),data=birds.density,start=
c(beta0=0,beta1=1,amplitude=1,frequency=1,initial=0))
> summary(output3r.birds.density)
Parameters:
                Value Std. Error    t value
    beta0  39.941100   3.197260  12.492300
    beta1   1.538780   0.112157  13.719900
amplitude   1.928800   2.469680   0.780994
frequency   0.970548   0.073624  13.182500
  initial  -0.323010   2.095020  -0.154180
Residual standard error: 16.0041 on 95 degrees of
    Freedom
> AIC(output3r.birds.density)
[1] 845.2271
```

(**c**)

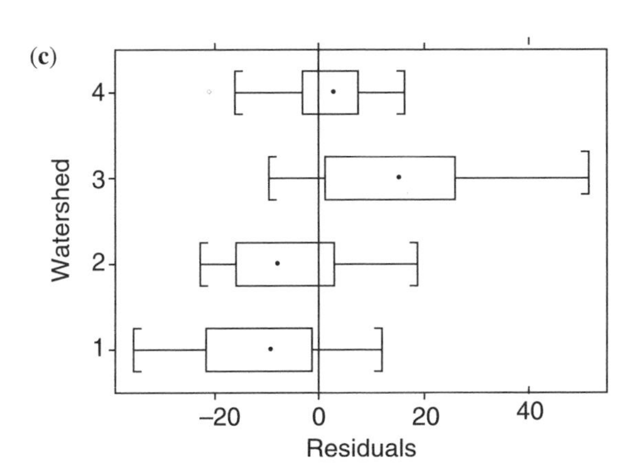

Figure 7.11. *Continued.*

(d)

```
> output3s.birds.density <-
nlsList(density~trig(riparian,beta0,beta1,
amplitude,frequency,initial),data=birds.density,start=
c(beta0=1,beta1=1,amplitude=1,frequency=1,initial=0))
> summary(output3s.birds.density)
Coefficients:
   beta0

     Value Std. Error   t value
1 18.83749   3.947709  4.771751
2 26.47180   4.098069  6.459579
3 74.70761   3.713535 20.117654
4 38.58659   4.206675  9.172706
   beta1
      Value Std. Error   t value
1 2.0427932  0.1560863 13.087589
2 1.8322263  0.1482873 12.355922
3 0.6921013  0.1460254  4.739595
4 1.6375958  0.1371046 11.944135
   amplitude
      Value Std. Error   t value
1 -7.611789   3.024820 -2.516444
2  8.393187   2.557652  3.281599
3  3.385827   2.886739  1.172890
4  7.826901   3.250627  2.407812
   frequency
      Value Std. Error  t value
1 1.3723188 0.02584900 53.08982
2 0.7690112 0.02490963 30.87205
3 1.1426520 0.05871857 19.45981
4 0.8147036 0.02292680 35.53499
   initial
      Value Std. Error   t value
1 -6.289314  0.7050452 -8.920440
2  6.330052  0.6821391  9.279709
3 -2.333616  1.7265739 -1.351588
4  1.584669  0.7254161  2.184496
Residual standard error: 9.334566 on 80 degrees
of freedom
```

On the basis of these results, we examine a general nonlinear mixed-effects model (Fig. 7.11f). Note the `fixed` = ... and `random` = ... parameters required for the `nlme` command. We have chosen the pdDiag variance–covariance structure for the parameters between the watershed groups to avoid overparameterizing our model. We chose our starting values using the parameter estimates from the fixed-effects model. On the basis of AIC, our nonlinear mixed-effects model is certainly an improvement over the nonlinear fixed-effects model (AIC = 788.11 vs. 845.23).

Figure 7.11. *Continued.*

(e)

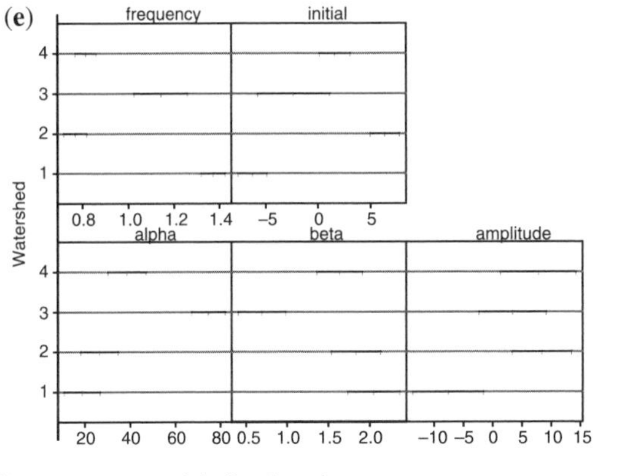

(f)

```
> output3t.birds.density <-
nlme(density~trig(riparian,beta0,beta1,amplitude,
frequency,initial),data=birds.density,fixed=
beta0+beta1+amplitude+frequency+initial~1,random=
pdDiag(beta0+beta1+amplitude+frequency+initial~1),
start=c(beta0=40,beta1=1.5,amplitude=2,frequency=1, initial=-.5))
> summary(output3t.birds.density)
Nonlinear mixed-effects model fit by maximum likelihood
  Model: density ~ trig(riparian, beta0, beta1, amplitude,frequency,
initial)
 Data: birds.density
       AIC      BIC   logLik
  788.1079 816.7648 -383.054
Random effects:
 Formula: list(beta0 ~ 1,beta1 ~ 1,amplitude ~ 1,
  frequency ~ 1,initial ~ 1)
 Level: Watershed
 Structure: Diagonal
            beta0     beta1 amplitude   frequency
StdDev: 21.27617 0.5084634   2.342042 2.4525e-006
               initial Residual
StdDev: 0.00008268379 9.465879
Fixed effects: beta0 + beta1 + amplitude + frequency + initial ~ 1
             Value   Std.Error DF  t-value  p-value
    beta0 38.72984  11.08887 92   3.49268  0.0007
    beta1  1.59636   0.26966 92   5.91992  <.0001
amplitude  3.87286   1.91800 92   2.01922  0.0464
frequency  1.12129   0.02304 92  48.67029  <.0001
  initial -4.64529   0.65231 92  -7.12130  <.0001
 Correlation:
            beta0  beta1 ampltd frqncy
    beta1 -0.038
amplitude -0.018  0.034
frequency -0.030  0.033 -0.081
  initial  0.043 -0.040  0.119 -0.844
> intervals(output3t.birds.density)
Approximate 95% confidence intervals
```

Figure 7.11. *Continued.*

```
 Fixed effects:
                  lower       est.      upper
     beta0  17.2640381  38.729842  60.195646
     beta1   1.0743519   1.596356   2.118359
 amplitude   0.1600013   3.872857   7.585712
 frequency   1.0766935   1.121291   1.165889
   initial  -5.9080309  -4.645292  -3.382553
 Random Effects:
  Level: Watershed
       lower          est.         upper
sd(beta0)      1.042019e+001 2.127617e+001 4.344217e+001
sd(beta1)      2.429429e-001 5.084634e-001 1.064180e+000
sd(amplitude) 3.899996e-001 2.342042e+000 1.406452e+001
sd(frequency) 7.600425e-154 2.452588e-006 7.914276e+141
sd(initial)   5.187060e-101 8.268379e-005 1.318012e+092
 Within-group standard error:
    lower     est.    upper
 8.155306 9.465879 10.98706
```

(g)
```
> output3u.birds.density <-
nlme(density~trig(riparian,alpha,beta,amplitude,
frequency,initial),data=birds.density,fixed=
alpha+beta+amplitude+frequency+initial~1,random=
pdDiag(alpha+beta+amplitude~1),start=c(alpha=
40,beta=1.5,amplitude=2,frequency=1,initial=0))
> summary(output3u.birds.density)
 Data: birds.density
       AIC      BIC   logLik
  784.1079 807.5545 -383.054
Random effects:
 Formula: list(alpha ~ 1, beta ~ 1, amplitude ~ 1)
 Level: Watershed
 Structure: Diagonal
           alpha      beta amplitude Residual
StdDev: 21.27607 0.5084604  2.342273 9.465866
Fixed effects: alpha + beta + amplitude + frequency + initial ~ 1
              Value Std.Error DF   t-value p-value
    alpha  38.72990  11.08882 92   3.49270  0.0007
     beta   1.59635   0.26966 92   5.91993  <.0001
amplitude   3.87311   1.91806 92   2.01929  0.0464
frequency   1.12127   0.02304 92  48.66934  <.0001
  initial  -4.64502   0.65228 92  -7.12119  <.0001
> intervals(output3u.birds.density)
Approximate 95% confidence intervals
 Fixed effects:
              lower       est.      upper
    alpha  17.264192  38.729902  60.195612
     beta   1.074350   1.596351   2.118352
amplitude   0.160139   3.873106   7.586073
frequency   1.076672   1.121270   1.165868
  initial  -5.907698  -4.645015  -3.382332
```

Figure 7.11 *Continued.*

```
 Random Effects:
  Level: Watershed
                   lower        est.      upper
    sd(alpha) 10.4204190 21.2760703 43.440784
     sd(beta)  0.2429626  0.5084604  1.064081
sd(amplitude)  0.3874602  2.3422729 14.159499
 Within-group standard error:
    lower     est.    upper
 8.155139 9.465866 10.98726
```

(h)

(i)

The insignificant estimates of the frequency and initial random effects suggest that we delete them from the model (Fig. 7.11g). Our model results are an improvement, with the lower AIC = 784.11 and the statistically significant parameter estimates as indicated by the 95% confidence intervals. Plots of the results are also indicative of this improvement (Fig. 7.11h and 7.11i).

Unfortunately, we cannot compare AICs between the linear and nonlinear mixed-effects models since the fixed effects are different. However, looking at the estimates of the fixed-effects parameters for the nonlinear function, we conclude that there is some evidence that a nonlinear mixed-effects model is an improvement over the linear mixed-effects models that we have examined, at least for this dataset. We are in danger, however, of overfitting this dataset, with a nonlinear model having 11 parameters for a dataset with $n = 100$ observations. We would want to explore this issue further before drawing a firm conclusion, analyzing goodness of fit and examining further sample datasets, if possible, to see whether there really is the "periodic" effect described by the trigonometric function in this model for the population.

We could continue with this exploratory analysis, but we will stop and leave it to the curious reader to further investigate this fascinating and powerful world of linear and nonlinear mixed-effects modeling. We refer you to Pinheiro and Bates (2000) and Wolfinger (2000) for further details and examples.

7.5 CONCLUSIONS: FREQUENTIST STATISTICAL ANALYSIS IN S-Plus AND R

7.5.1 Conclusions: The Analysis

We have conducted a partial exploratory a posteriori model selection analysis of the `birds.density` dataset, using mixed-effects modeling in S-Plus or R. Our modeling efforts provided linear and nonlinear regression models based on the fundamental form

`density ~ riparian`,

generalized to include random effects due to the watershed groups. Our linear results are summarized in Fig. 7.12. The general mixed-effects model, `Model #3a.birds.density`, with constant and slope random effects and general defaulted variance–covariance structures, was the best-fitting model, with an Akaike weight of 26.254%. Two other models with temporal covariance structure within groups also performed competitively, the `ARMA(1,0) Model #3j.birds.density` and `ARMA(0,1) Model #3k.birds.density`, having Akaike weights of approximately 16% each. However, given the limited sample size of the dataset, and the descriptive a posteriori model selection and inference strategy we used, we may have experienced some overfitting, particularly with the ARMA models.

7.5.2 Conclusions: The Reality of the Dataset

The dataset `birds.density` was simulated using the fixed-effects linear relationship $y = 10 + 1.5^{*}x + N(0, 30)$, with watershed random effects for the constant of $N(0, 20)$ ($b_0 = [-9.51,\ -1.78,\ 39.76,\ 18.75]$) and for the slope of $N(0, 0.50)$ ($b_1 = [0.38,\ 0.26,\ -0.64,\ -0.17]$). Hence $\beta_0 = 10.0$, $\beta_1 = 1.5$, $\sigma_0 = 20.0$, $\sigma_1 = 0.50$, and $\sigma = 30.0$. It is interesting to compare this "reality" with the mixed-effects modeling results. Of the best-fitting models, we see that the estimated general mixed-effects model is compatible with the reality, whereas the models with covariance structures within the watershed groups are not but are examples of model overfitting. There is a specious correlation in the dataset induced because the watersheds with the flatter trends were higher; that is, there is a negative correlation imposed between the constant and slope random effects. So the general mixed-effects model was better-fitting than the `pdDiag` mixed-effects model used for the simulated data.

Figure 7.12. Frequentist mixed-effects modeling results for the `birds.density` dataset, with the number of parameters (k), AIC, and Akaike weights.

Model Label	Description	k	AIC	Akaike Weights
1b	Linear regression model	3	839.78	< 0.0005%
3a	General mixed-effects model, with constant and slope random effects	6	764.35	26.254%
3b	Mixed-effects model, with constant random effects only	4	808.34	< 0.0005%
3c	Mixed-effects model, with slope random effects only	4	840.96	< 0.0005%
3d	General mixed-effects model, with `pdDiag` between-group covariance structure	5	777.62	0.034%
3e	General mixed-effects model, with `pdIdent` between-group covariance structure	4	840.93	< 0.0005%
3f	General mixed-effects model, with `varFixed` within-group variance structure	6	842.93	< 0.005%
3g	General mixed-effects model, with `varPower` within-group variance structure	7	766.31	9.853%
3h	General mixed-effects model, with `varConstPower` within-group variance structure	8	768.31	3.625%
3i	General mixed-effects model, with `varExp` within-group variance structure	7	766.33	9.755%
3j	General mixed-effects model, with `corARMA(1,0)` within-group covariance serial structure	7	765.32	16.164%
3k	General mixed-effects model, with `corARMA(0,1)` within-group covariance serial structure	7	765.23	16.908%
3l	General mixed-effects model, with `corARMA(1,1)` within-group covariance serial structure	8	767.01	6.943%
3m	General mixed-effects model, with `corExp` within-group covariance spatial structure	7	766.19	10.463%
				100.000%

7.6 MIXED-EFFECTS MODELING: BAYESIAN STATISTICAL ANALYSIS IN WinBUGS

Let's illustrate Bayesian mixed-effects modeling in WinBUGS by returning to the mixed-effects multiple linear regression example for the `birds.density` dataset in Section 7.3.2. The WinBUGS code for the full mixed-effects model with random effects for both the constant and slope in the linear regression model (cf. `output3a.birds.density`) is presented in Fig. 7.13a. The Bayesian WinBUGS output results with the noninformative priors (Fig. 7.13b) are similar to the frequentist results, although the posterior samples of the random-effects parameters are skewed and have means and medians that differ somewhat from those of the frequentist mode estimates (e.g., frequentist $\hat{\sigma}_{ML} = 9.83$ versus Bayesian mean of $\sigma = 10.08$ and median of $\sigma = 10.03$, frequentist $\hat{\sigma}_{0ML} = 25.06$ versus Bayesian mean of $\sigma_0 = 32.03$ and median of $\sigma_0 = 26.97$, and frequentist

Figure 7.13. Bayesian statistical analysis of the `birds.density` dataset for linear mixed-effects modeling, with WinBUGS. (**a**) WinBUGS code for the full mixed-effects model, with

(**a**)
```
1) Program code
model
{for(i in 1:n)
   {density[i] ~ dnorm(mu[i],tau)
   mu[i] <- (beta0+eta0[watershed[i]])
        + (beta1+eta1[watershed[i]])*riparian[i]}
for (j in 1:4)
   {eta0[j] ~ dnorm(0,tau0)
   eta1[j] ~ dnorm(0,tau1)}
beta0~ dnorm(0,0.0000000000000001)
beta1 ~ dnorm(0,0.0000000000000001)
tau ~ dgamma(0.001,0.001)
tau0 ~ dgamma(0.001,0.001)
tau1 ~ dgamma(0.001,0.001)
sigma <- 1/sqrt(tau)
sigma0<-1/sqrt(tau0)
sigma1<-1/sqrt(tau1)}
2. Data
list(
riparian = c(19.02,14.47,40.26,…,1.48,48.3),
density = c(41.91,28.35,108.23,…,40.0,122.05),
watershed =c(1,1,1,…,2,2,2,…,3,3,3,…,4,4,4,…),
n = 100)
3. Initial values
list(beta0=1,beta1=1,tau=1,tau0=1,tau1=1)
```

(**b**)

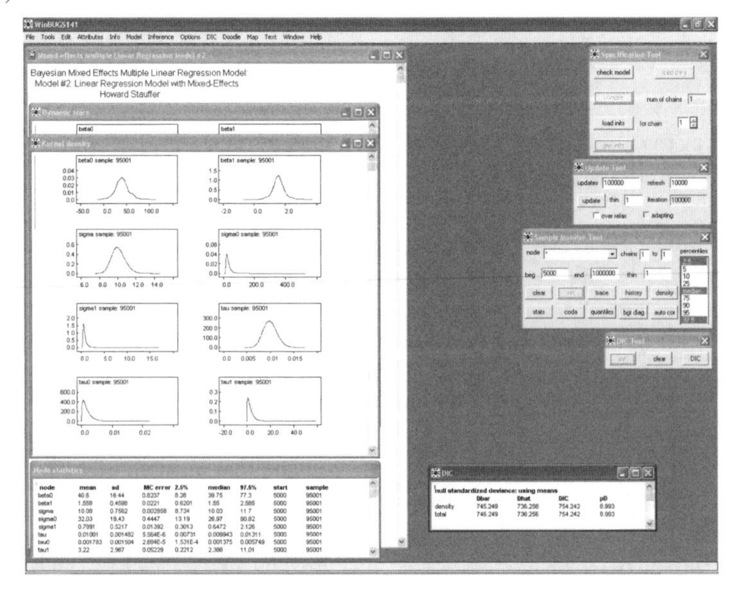

Figure 7.14. Frequentist and Bayesian statistics for the four linear regression mixed-effects models, with AIC and Akaike weights, and DIC and DIC weights.

Model	Fixed Effects	Random Effects	k	p_D	AIC	Akaike Weights	DIC	DIC Weights
1b	β_0, β_1	σ	3	3.032	839.7801	0.0000%	839.950	0.0000%
3a	β_0, β_1	$\sigma, \sigma_0, \sigma_1$	6	8.993	764.3522	100.0000%	754.242	100.0000%
3b	β_0, β_1	σ, σ_0	4	5.991	808.3422	0.0000%	799.999	0.0000%
3c	β_0, β_1	σ, σ_1	4	4.929	840.9644	0.0000%	839.321	0.0000%
					Total:	100.0000%	Total:	100.0000%

Note: k = number of parameters
p_D = Bayesian number of parameters

$\hat{\sigma}_{1\text{ML}} = 0.6033$ versus Bayesian mean of $\sigma_1 = 0.7891$ and median of $\sigma_1 = 0.6472$. The DICs of this and the other Bayesian WinBUGS mixed-effects model analyses comparable to the frequentist `output3b.birds.density` and `output3c.birds.density` with constant random effects and with slope random effects only indicate a ranking very similar to the frequentist AIC ranking of the models 3a, 3b, and 3c (see Fig. 7.14). The DIC criterion tends to slightly overfit sometimes, as is illustrated with the DIC ranking of the mixed-effects model 3c with slope random effect slightly ahead of the fixed-effects linear regression model 1b, contrary to the AIC ranking. As the reader can see from this example, mixed effects are remarkably easy to incorporate into Bayesian modeling with WinBUGS.

7.7 SUMMARY

Mixed-effects modeling is an important, innovative statistical tool that generalizes fixed-effects methods such as ANOVA and multiple regression. It efficiently and parsimoniously incorporates random effects into models, along with fixed effects, to describe dependences in datasets. Mixed-effects modeling is appropriate for dependent datasets such as pseudoreplicated, temporally dependent, spatially dependent, hierarchical, and metapopulation datasets. In this chapter we described mixed-effects generalizations of the mean, ANOVA, and linear regression models. We described the modeling of variance–covariance structures between groups with mixed-effects models. We also described the modeling of variance and covariance structures within groups with mixed-effects models, including serial and spatial correlation. Finally, we briefly introduced the topic of nonlinear mixed-effects modeling. Mixed-effects modeling with frequentist statistical analysis was illustrated in S-Plus and R and, with Bayesian statistical analysis, was illustrated in WinBUGS. Mixed-effects modeling can be generally incorporated into other modeling methods, such as generalized linear models (GLMs) and marked data analysis.

PROBLEMS

7.1 Conduct a frequentist statistical analysis of the S-Plus dataset `Orthodont` in S-Plus or R (see Fig. 7.15). The `Orthodont` dataset available in S-Plus

Figure 7.15. The `Orthodont` dataset in S-Plus (Pinheiro and Bates 2000, Potthoff and Roy 1964).

Sample	Distance	Age	Subject	Sex
1	26	8	M01	Male
2	25	10	M01	Male
3	29	12	M01	Male
4	31	14	M01	Male
5	21.5	8	M02	Male
6	22.5	10	M02	Male
7	23	12	M02	Male
8	26.5	14	M02	Male
9	23	8	M03	Male
10	22.5	10	M03	Male
11	24	12	M03	Male
12	27.5	14	M03	Male
13	25.5	8	M04	Male
14	27.5	10	M04	Male
15	26.5	12	M04	Male
16	27	14	M04	Male
17	20	8	M05	Male
18	23.5	10	M05	Male
19	22.5	12	M05	Male
20	26	14	M05	Male
21	24.5	8	M06	Male
22	25.5	10	M06	Male
23	27	12	M06	Male
24	28.5	14	M06	Male
25	22	8	M07	Male
26	22	10	M07	Male
27	24.5	12	M07	Male
28	26.5	14	M07	Male
29	24	8	M08	Male
30	21.5	10	M08	Male
31	24.5	12	M08	Male
32	25.5	14	M08	Male
33	23	8	M09	Male
34	20.5	10	M09	Male
35	31	12	M09	Male
36	26	14	M09	Male
37	27.5	8	M10	Male
38	28	10	M10	Male
39	31	12	M10	Male
40	31.5	14	M10	Male
41	23	8	M11	Male
42	23	10	M11	Male
43	23.5	12	M11	Male
44	25	14	M11	Male
45	21.5	8	M12	Male
46	23.5	10	M12	Male
47	24	12	M12	Male
48	28	14	M12	Male
49	17	8	M13	Male

Figure 7.15. *Continued.*

50	24.5	10	M13	Male
51	26	12	M13	Male
52	29.5	14	M13	Male
53	22.5	8	M14	Male
54	25.5	10	M14	Male
55	25.5	12	M14	Male
56	26	14	M14	Male
57	23	8	M15	Male
58	24.5	10	M15	Male
59	26	12	M15	Male
60	30	14	M15	Male
61	22	8	M16	Male
62	21.5	10	M16	Male
63	23.5	12	M16	Male
64	25	14	M16	Male
65	21	8	F01	Female
66	20	10	F01	Female
67	21.5	12	F01	Female
68	23	14	F01	Female
69	21	8	F02	Female
70	21.5	10	F02	Female
71	24	12	F02	Female
72	25.5	14	F02	Female
73	20.5	8	F03	Female
74	24	10	F03	Female
75	24.5	12	F03	Female
76	26	14	F03	Female
77	23.5	8	F04	Female
78	24.5	10	F04	Female
79	25	12	F04	Female
80	26.5	14	F04	Female
81	21.5	8	F05	Female
82	23	10	F05	Female
83	22.5	12	F05	Female
84	23.5	14	F05	Female
85	20	8	F06	Female
86	21	10	F06	Female
87	21	12	F06	Female
88	22.5	14	F06	Female
89	21.5	8	F07	Female
90	22.5	10	F07	Female
91	23	12	F07	Female
92	25	14	F07	Female
93	23	8	F08	Female
94	23	10	F08	Female
95	23.5	12	F08	Female
96	24	14	F08	Female

Figure 7.15. *Continued.*

```
 97      20     8    F09   Female
 98      21    10    F09   Female
 99      22    12    F09   Female
100    21.5    14    F09   Female
101    16.5     8    F10   Female
102      19    10    F10   Female
103      19    12    F10   Female
104    19.5    14    F10   Female
105    24.5     8    F11   Female
106      25    10    F11   Female
107      28    12    F11   Female
108      28    14    F11   Female
```

(Pinheiro and Bates 2000, Potthoff and Roy 1964) consists of the response variable `distance`, covariate variable `age`, and clustering categorical variables `Sex` and `Subject` nested in `Sex`. The variable `distance` is a measurement of the length between the pituitary gland and the pterygomaxillary fissure, two easily measured points on x-rays of skulls of 27 children, 16 males, and 11 females, taken from ages 8 to 14. Use an a priori parsimonious model selection and inference strategy. First analyze and compare fixed-effects constant, linear, and quadratic models of `distance` with `age`:

$$\texttt{distance} = \beta_0 + \mathrm{N}(0, \sigma),$$

$$\texttt{distance} = \beta_0 + \beta_1 \cdot \mathit{age} + \mathrm{N}(0, \sigma),$$

$$\texttt{distance} = \beta_0 + \beta_1 \cdot \mathit{age} + \beta_2 \cdot \mathit{age}^2 + \mathrm{N}(0, \sigma).$$

Use the generalized least-squares `gls` command for the linear regression analysis. Use AIC to compare the models, constructing a table with AIC and Akaike weights to compare the fixed-effects models.

7.2 Next, on the best-fitting fixed-effects model(s) from Problem 7.1 (above), analyze and compare mixed-effects models, adding random effects due to `Sex`, `Subject`, and `Subject` nested in `Sex` (use `|Sex/Subject` in S-Plus or R). Use an a priori model selection–inference strategy on the mixed-effects models, the different random effects models with the same fixed-effects. Construct a table with AIC and Akaike weights to compare the models, using AIC.

7.3 Examine the clustered datasets in `Orthodont`, using the S-Plus or R commands `plot(Orthodont)` and `plot(augPred(output))`. Do the model analysis results substantiate the visual differences observed between the clustered datasets in `Orthodont`?

7.4 Conduct a Bayesian statistical analysis of the S-Plus dataset `Orthodont` in WinBUGS. Use an a priori parsimonious model selection–inference strategy. First analyze and compare fixed effects constant, linear, and quadratic models of `distance` with `age`:

$$\mathtt{Distance} = \beta_0 + \mathrm{N}(0, \sigma),$$
$$\mathtt{Distance} = \beta_0 + \beta_1 \cdot \mathit{age} + N(0, \sigma),$$
$$\mathtt{Distance} = \beta_0 + \beta_1 \cdot \mathit{age} + \beta_2 \cdot \mathit{age}^2 + N(0, \sigma).$$

Use DIC to compare the models, constructing a table with DIC and DIC weights to compare the various fixed-effects models.

7.5 Next, on the best-fitting fixed-effects model(s) from Problem 7.4, analyze and compare mixed-effects models, adding random effects due to `Sex`, `Subject`, and `Subject` nested in `Sex`. Use an a priori model selection–inference strategy on the mixed-effects models, the different random-effects models with the same fixed-effects. Use DIC to compare the models, expanding the table of Problem 7.4 with DIC and DIC weights to compare the models.

7.6 Do the model analysis results substantiate the visual differences observed between the clustered datasets in `Orthodont`?

8 Summary and Conclusions

In this final chapter, we will present a summary of the contemporary Bayesian and frequentist statistical research methods for natural resource scientists that have been introduced in this book, and draw some final conclusions. We introduced this subject in Chapter 1 with three important case studies in natural resource science and offered some partial solutions to these problems, using traditional frequentist methods of statistical analysis: parameter estimation, hypothesis testing, and linear regression analysis. More complete solutions to these case studies were presented throughout the remainder of this book.

Besides reviewing these methods, we cautioned the reader in Chapter 1 to provide adequate attention to the planning and conclusion phases of a natural resource project that surround the data collection phase. It is particularly important to consider the statistical design in the planning phase and the statistical analysis and inference methods in the conclusion phase of a project. Chapters 2–4 provided an introduction to an alternative approach to the traditional frequentist approach to statistical analysis and inference: Bayesian statistical analysis and inference. Chapter 5 presented two contrasting strategies for model selection and inference and illustrated them using multiple linear regression analysis models. Chapters 6 and 7 introduced two important contemporary methods of statistical analysis: generalized linear modeling and mixed-effects modeling, illustrated with both frequentist and Bayesian approaches to statistical analysis and inference. These contemporary approaches and methods, Bayesian statistical analysis and inference, alternative strategies for model selection and inference, generalized linear modeling, and mixed-effects modeling, provide the natural resource scientist with an enriched toolbox of contemporary statistical research methods for the analysis of their datasets.

8.1 SUMMARY OF SOLUTIONS TO CHAPTER 1 CASE STUDIES

Chapter 1 presented three case studies that served to illustrate some of the most important statistical problems currently of interest to natural resource scientists.

Contemporary Bayesian and Frequentist Statistical Research Methods for Natural Resource Scientists. By Howard B. Stauffer

These problems required solutions beyond the scope of the more common traditional statistical methods available to the natural resource scientist. It has been the mission of this book to present more advanced contemporary statistical research methods that address these problems and enhance the toolbox of statistical methods available to the natural resource scientist. We will summarize these methods in this chapter and draw some final conclusions.

8.1.1 Case Study 1: Maintenance of a Population Parameter Above a Critical Threshold Level

Case study 1 in Chapter 1 addressed the problem of assessing the maintenance of a population parameter above a critical threshold level. We illustrated this problem with a proportion parameter p and critical threshold p_c. We looked at the example of a timber company required to maintain the proportion of its timberlands occupied by nesting Northern Spotted Owl pairs above a specified threshold level p_c determined by biologists to ensure the viability of the local population of owls. Failure to maintain the population parameter p above the critical threshold level p_c should result in a decision of "corrective action," the reduction of timber harvesting. Alternatively, success at maintaining the parameter p above the threshold level should result in a decision of "no action," of maintaining the timber harvesting at current levels.

We looked at several common traditional statistical methods for analyzing datasets obtained from monitoring the population. The first traditional method consisted of a sampling survey and parameter estimation. The parameter estimate $\hat{p}$ based on the sample could be compared with the critical threshold level p_c in order to draw a conclusion. Or, the confidence interval $[\hat{p} - E, \hat{p} + E]$, where E is the sampling error, based on a specified level of confidence, could be compared with the critical threshold level p_c in order to draw a conclusion. Either of these methods poses some interesting and difficult challenges in application and interpretation. Conclusions could be ambiguous, with estimates and confidence intervals close to or overlapping the critical threshold level. Frequentist interpretations are difficult to comprehend for many natural resource scientists in the context of repeated sampling, since these methods provide interpretations based on probabilities for the sample dataset, rather than probabilities for the parameter. They also provide no estimates of risk.

The second common traditional statistical method for analyzing datasets obtained from monitoring the population consisted of an experiment and hypothesis testing. The null hypothesis could be formulated that the proportion parameter was equal to the critical threshold level with an alternative hypothesis that the proportion parameter was below the critical threshold level. Again, there are problems with this method used as a solution to the case study. Frequentist conclusions can be ambiguous, with the burden of proof placed on "corrective action." Additionally, a p value can be close to the level of significance α. Frequentist interpretations pose a challenge of interpretation to many natural resource scientists in the context of repeated experimentation, since this method also provides an interpretation based on probabilities for the sample dataset, rather than probabilities for the parameter. As with parameter estimation, this method provides no estimates of risk.

Throughout this book, we have introduced some contemporary statistical methods that address these concerns and provide more satisfactory solutions to case study 1. Bayesian statistical analysis, presented in Chapters 2–4, provides probability distribution assessments for the parameter. We saw that a choice of conjugate beta priors for the proportion parameter p provide beta posterior solutions for binomially distributed binary datasets. The Bayesian statistical solution provides inferences with probability statements for parameters in the critical threshold problem, and it provides estimates of risk. Hence, Bayesian statistical analysis provides a contemporary alternative solution for case study 1.

8.1.2 Case Study 2: Estimation of the Abundance of a Discrete Population

Case study 2 presented the problem of estimating the abundance of a discrete population. The abundance is measured with count data, nonnegative integer measurements that are not normally distributed. However, the traditional method of estimating mean abundance and confidence intervals is based on normal distributional assumptions for the data, or, more specifically, for the mean estimates. In the course of this book, we have provided more contemporary methods for solution of this problem. We presented the Poisson model, which provides a solution if the population is randomly distributed both spatially and temporally. This model can be readily analyzed with Bayesian statistical analysis, using either a conjugate gamma prior and posterior solution or a Markov Chain Monte Carlo (MCMC) solution in WinBUGS for the Poisson distributed data. If the distribution of the population is aggregated spatially and temporally, the negative binomial model may be used to model the data, using an MCMC solution in WinBUGS, providing a solution to the problem in case study 2. These ideas were presented in Chapters 2 and 4.

8.1.3 Case Study 3: Habitat Selection Modeling of a Wildlife Population

Case study 3 presented the problem of the habitat selection modeling of a wildlife population. The objective of habitat selection modeling is to model the response of a wildlife population as a function of habitat attributes such as vegetation, geologic, and climatic variables. In Chapter 5 we examined a traditional approach to habitat selection modeling using multiple linear regression. This method is appropriate for response data that are normally distributed, such as biomass or abundance measurements. For binary response data such as presence–absence observations that are binomially distributed, the contemporary method of generalized linear modeling that was presented in Chapter 6 is more appropriate. The method of generalized linear modeling can be applied to response data that are constrained and not necessarily normally distributed. It provides linear models for constrained response measurements by using link functions. Multiple linear regression provides a traditional method, and the generalized linear model of logistic regression provides a more contemporary method for the habitat selection modeling of wildlife populations. With both of these methods, either frequentist or Bayesian approaches to the statistical modeling may be used. Either of the two strategies for model selection and inference that are described in Chapter 5 may also be used with either of these methods: a priori parsimonius model selection and inference using AIC and a posteriori model selection and inference. The former is generally more appropriate for predictive modeling, whereas the latter is more appropriate for descriptive modeling.

8.2 APPROPRIATE APPLICATION OF STATISTICS IN THE NATURAL RESOURCE SCIENCES

In this section we will summarize some recommendations for the appropriate application of statistics in the most common areas of neglect and misuse of statistics in the natural resource sciences. These common areas of misapplication include

1. Insufficient planning and development of sufficiently rigorous statistical design for data collection
2. The overuse of hypothesis testing
3. The misuse of observational data
4. Inadequate attention given to analysis, interpretation, and conclusions
5. The overuse of a posteriori "data dredging" methods for comparison of statistical models
6. The misuse of traditional statistical methods on dependent data

These problem areas have been identified and discussed in previous chapters.

A leading area of neglect in the use of statistics in the natural resource sciences has been the lack of sufficient planning and failure to develop a sufficiently rigorous statistical design for data collection. A significant amount of attention should be dedicated to the initial planning phase of a data collection project, particularly to the development of a rigorous statistical design for the data collection. With a sample survey, the sampling design should be clearly identified before the data are collected. Consideration should be given to incorporating a design that provides unbiased estimators that are efficient, with acceptable bounds on error. If the estimators are biased, that bias should be estimable. The principles of randomization should be applied in the sampling design as appropriate. Many effective sampling designs, such as simple random sampling, stratified sampling, cluster sampling, systematic sampling, multistage sampling, multiphase sampling, variable-probability sampling, and adaptive sampling, are available for use with data collection. Sample size should be estimated to provide an adequate bound on sampling error to meet the objectives of the study. Further details on survey sampling design and analysis are given in Cochran (1977), Scheaffer et al. (1996), Thompson (1992), and Sarndal et al. (1992).

Similarly, with experiments, adequate attention should be devoted to the specification of rigorous experimental design before data collection. Again, randomization and replication are of paramount concern. Effective experiments should be conducted under controlled conditions constraining the values of attributes affecting the response that are not of interest in the study. Such controlled conditions are often difficult to attain in natural resource settings. The number of replicates for the experiment should be determined prior to data collection and should be sufficient to ensure a specified level of power for a given confidence level to statistically distinguish biologically important differences in the response between treatment levels. The distinction between biologically and statistically significant differences should be minimized. Further details on experimental design and analysis are given in Hicks (1993) and Kuehl (1994).

Hypothesis testing, which has been overused with natural resource science data, should be applied to scientific hypotheses, not statistical hypotheses, in natural resource experiments. The mean growth of seedlings due to different fertilizer treatments can be described with clearly defined scientific hypotheses, whereas the differences in the abundance of wildlife between two specific forest stands can be described only by a statistical hypothesis. "Silly" null hypotheses that are patently false should be avoided at all cost. There will clearly be some amount of difference in the abundance of wildlife between two specific forest stands, no matter how small, and a sufficiently large amount of data replication will be able to statistically distinguish between these differences. In this case where the alternative hypothesis is clearly true, hypothesis testing will provide statistical results that depend on the magnitude of biological difference and the amount of data collected rather than whether there is a biologically significant difference. This misuse of hypothesis testing can be avoided by placing emphasis on estimation of the difference with a sufficient large sample size to provide the precision required to detect biologically significant differences.

The data collected in many natural resource studies is observational, rather than experimental. Observational data, unlike experimental data, are collected without rigorous adherence to randomization. With rigorous experimental design, experimental units must be both randomly selected from a population and randomly assigned to treatment levels. These two requirements are often violated in natural resource studies, particularly the first requirement, where the small amount of population remaining from an original population does not provide adequately representative units for an "experiment." This situation is illustrated with studies comparing old-growth with young-growth habitat in forest stands along the coast of northern California. The few stands of old-growth habitat that remain extant clearly do not provide a representative sample from the original population that existed several centuries ago before human intervention. So inferences obtained from analysis of observational data obtained in studies comparing conditions in these remaining old-growth stands with conditions in other stands must be regarded with due caution. See Johnson (1999) for further discussion of this issue.

Be sure to devote adequate attention to analysis, interpretation, and the drawing of conclusions at the end of natural resource data collection projects. The methods of analysis should be clearly identified prior to data collection so that the statistical design for the data collection can be specified for the analysis to address the objectives of the study. The analysis should receive an adequate amount of attention in order to sufficiently extract the biological information from the sample dataset. The analysis, interpretation, and conclusions drawn from the dataset are a key component of the study and should be treated as such.

"Data dredging" methods of analysis have been overused in the natural resource sciences. This strategy for identifying models for comparison after the collection of data can often be helpful in detecting attributes and models of importance for biological populations that are not well understood. However, data dredging tends to overfit sample data and compound error in the model selection process. Results from this strategy should be viewed tentatively, as descriptive rather than predictive. To avoid overfitting of data and compounding of error, use the a prior parsimonious model selection–inference strategy with information-theoretic criteria such as Akaike's information criterion (AIC) and the deviance information criterion (DIC). This strategy can be particularly effective if the population of interest is somewhat understood biologically so that a collection of reasonably plausible models can be hypothesized prior to data collection.

Avoid using traditional methods of statistical analysis for dependent data. Traditional methods often require assumptions of independence that are violated with dependent sample datasets. Historically, natural resource scientists have been cautioned to avoid collecting dependent datasets for that reason. Now however, with the development of methods such as mixed-effects modeling, natural resource scientists can be encouraged to collect sample datasets that are representative of the biological processes of interest even if those systems contain dependences. As we have seen, mixed-effects modeling can then be used to model these dependences by incorporating random effects that explain the variation between the clustered groups in the dependent data.

The objective of this book has been to present an introduction to contemporary methods of statistical analysis that address many of these problem areas of neglect and misuse of statistics in the natural resource sciences.

8.3 STATISTICAL GUIDELINES FOR DESIGN OF SAMPLE SURVEYS AND EXPERIMENTS

Here are some important statistical guidelines for the natural resource scientist for the design of sample surveys and experiments, in the planning phase of the project:

1. Projects proposing to estimate parameters, such as abundance, presence or absence, productivity, survival, recruitment, movement, or fitness, should indicate the following prior to data collection:
   a. The amount of **precision** at a specified confidence level required to meet the objectives of the project
   b. The **sample size** required to realize that level of precision

2. Projects proposing to conduct an experiment with a comparison of treatments using hypothesis testing should include a power analysis, prior to data collection, indicating
   a. The **effect size**, the biologically significant difference of interest in the alternative hypothesis, required to meet the objectives of the project
   b. The **confidence level**
   c. The **number of replicates** required to attain a specified level of **power** in the experiment
3. Any project proposing to detect and estimate a trend in a parameter, such as abundance, presence or absence, productivity, survival, recruitment, movement, or fitness, should indicate the following prior to data collection:
   a. The **effect size**, the amount of decrease or increase in the trend that is of biological interest
   b. The **number of replicates** required at a specified **confidence level** and **power** to detect this trend, based on a power analysis.

8.4 TWO STRATEGIES FOR MODEL SELECTION AND INFERENCE

In Chapter 5, we contrasted two alternative strategies for model selection and inference. Traditional model selection and inference have been based on a descriptive a posteriori model selection–inference strategy. In this strategy, models are selected, analyzed, and compared after data collection. Burnham and Anderson's (1998, 2002) important book has identified an important alternative strategy for model selection and inference based on predictive a priori parsimonious model selection and inference using information-theoretic criteria. In their strategy, a parsimonious collection of models is prescribed for analysis prior to data collection. We suggest that each strategy has its place and is effective in appropriate contexts.

The a posteriori strategy is useful for the descriptive modeling of sample datasets. A multitude of a posteriori methods may be utilized with this strategy. Scatterplots and other graphs may be examined, and correlations may be assessed for the sample dataset, along with other traditional statistics. Stepwise and best-subsets selection methods may be employed. Powerful algorithms perform these procedures with fast and efficient software programs that implement these methods. These methods can be temptingly easy to use with current software, and therein lies a danger, for these methods are prone to compounded error and model overfitting. Hence, results based on this strategy of model selection should be interpreted with caution and viewed as descriptive and preliminary, tentative, and subject to further validation. Results stemming from this strategy can be used to formulate hypotheses for model selection with additional sample datasets. Sometimes pejoratively called "data dredging" by some statisticians because of its susceptibility to overuse, this a posteriori strategy is nonetheless highly useful, particularly for the exploratory data analysis of populations that are not well understood.

The second alternative strategy for model selection and inference suggested by Burnham and Anderson is appropriate for populations that are better understood. Burnham and Anderson recommend identifying a collection of candidate models a priori to data collection. The collection of candidate models should be parsimonious: relatively small in number, with each model having a relatively small number of covariates. Burnham and Anderson also recommend the use of the information-theoretic criterion, Akaike's information criterion (AIC) or the corrected AIC_c (Akaike 1973) for evaluating the frequentist relative fit of statistical models to sample datasets. This relative measure of the Kullback–Liebler distance between the model and reality provides a way, along with the restriction of the analysis to a parsimonious collection of candidate models, of avoiding the problems associated with the compounded error and model overfitting of the traditional a posteriori strategy of model selection and inference. The Akaike information criterion effectively separates signal from noise in a sample dataset. The model with the lowest AIC value is the best-fitting model among the collection of candidate models to the population reality represented by the sample dataset. The statistical inferences from this strategy are hence dependent on both the collection of candidate models and the sample dataset. Therefore, this strategy is most suitable for populations that are better understood biologically, to ensure that viable leading models are well represented among the collection of candidate models. Such a strategy has been successfully applied to endangered species populations such as the Northern Spotted Owl and the Marbled Murrelet, which have more recently been the source of numerous studies by natural resource scientists. If populations are not sufficiently understood, a collection of candidate models may possibly not contain any well-fitting models. This strategy will then break down and accomplish little beyond ranking a collection of poorly fitting models. It is also possible, if a variety of contrasting models are not well chosen, that many or most of the models will be reasonably well fitting, and the strategy will then merely rank a collection of well-fitting models, many of which might be suitably useful for natural resource managers. Hence, good judgment suggests that this strategy should be applied to populations for which there is at least some reasonable level of biological understanding.

Of course, regardless of strategy and competitiveness of ranking, goodness-of-fit methods should always be included in the analysis to ensure that the best-fitting models are indeed also well-fitting.

8.5 CONTEMPORARY METHODS IN STATISTICAL ANALYSIS I: GENERALIZED LINEAR MODELING AND MIXED-EFFECTS MODELING

Two important contemporary methods in statistical analysis developed over the last several decades enable natural resource scientists to more accurately and effectively analyze their datasets. These methods model errors in data that are not necessarily normally or independently distributed. Furthermore, they can be implemented using either a frequentist or a Bayesian approach to analysis and interpretation.

Generalized linear modeling (GLM) models error structures that may not be normally distributed, such as with binomially distributed binary datasets, and Poisson or negatively binomially distributed count datasets. The GLM models use link functions to connect linear models to response data that may be restricted in value. With frequentist analysis, maximum-likelihood (ML) estimation is used to estimate model parameters for GLMs, in contrast to least-squares (LS) estimation, which is used to estimate model parameters for multiple linear regression. Frequentist software such as S-Plus or R can be used to analyze GLMs.

Bayesian statistical analysis can also be used to analyze linear regression models using posterior distributions for the model parameters that are based on prior distributions for the parameters and likelihood functions for the model data. Bayesian statistical software such as WinBUGS can also be used to analyze GLMs.

Dependent data, sampled in groups, can be analyzed using random effects generated parsimoniously with a minimum number of parameters in mixed-effects modeling, with either frequentist software S-Plus or R or Bayesian statistical software WinBUGS. Random effects can be estimated in mixed-effects modeling with variance and covariance components accounting for the variation between and within the groups in the data.

These breakthroughs in technology have provided the natural resource scientist with the analysis tools needed to model nonnormally distributed dependent natural resource data. Prior to these developments, natural resource scientists have had to either avoid the collection of such datasets or ignore the problems associated with its analysis. Now natural resource scientists can be encouraged to collect such datasets, with designs that reflect these natural structures in biological populations, since appropriate techniques are available for its analysis.

8.6 CONTEMPORARY METHODS IN STATISTICAL ANALYSIS II: BAYESIAN STATISTICAL ANALYSIS USING MCMC METHODS WITH WinBUGS SOFTWARE

In 1763 the Reverend Thomas Bayes first identified the basic concept of conditional probability that is now summarized in a theorem named in his honor. Yet, more than two centuries transpired before methods were fully developed to utilize the power of his theorem in providing statistical analysis solutions for sample datasets. The general application of Bayes' theorem has been an exceedingly challenging problem, with mathematical solutions for general statistical problems formulated using the Bayesian method of parameter estimation proving to be limited. In the mid nineteenth century, however, advances in computer technology created conditions conducive to the development of algorithms that provided general Bayesian solutions using MCMC simulation methods. Metropolis et al. (1953), Hastings (1970), and others provided a collection of MCMC techniques for producing representative samples from the posterior distribution of parameters for statistical models. More recently, the development of WinBUGS software has provided a practical tool for the natural resource scientist to access this methodology.

Bayesian methodology turns frequentist statistical inference on its head. Instead of providing frequentist probabilistic statements about data, conditional on assumed parameter values, Bayesian statistical inference provides probability statements for parameters, conditional on sample datasets. Bayesian statistical analysis incorporates prior information into the process. It combines prior information about parameters along with sample data and statistical models for the data to provide a posterior distribution for the parameters. Parameter estimation, hypothesis testing, and model selection and inference can all be achieved using Bayesian statistical analysis and inference. Bayesian statistical analysis provides a method of inference similar to that of the scientific method; human inquiry can advance incrementally and cumulatively as new data are collected revising existing knowledge represented by prior distributions with adjustments represented by posterior distributions. Bayesian statistical analysis has come of age, and we can look forward to exciting new possibilities with its application to problems in the natural resource sciences.

8.7 CONCLUDING REMARKS: EFFECTIVE USE OF STATISTICAL ANALYSIS AND INFERENCE

This book has been written to provide contemporary Bayesian and frequentist statistical research methods for the appropriate use of statistics in natural resource science. The author has appealed to scientists to bring increased attentiveness and rigor to statistical design in the planning phase and to statistical analysis and interpretations in the conclusions phase of natural resource data collection projects. Hypothesis testing techniques using frequentist statistical inference should be used only in proper context, for randomized, replicated, controlled experiments, challenging scientific, non-"silly," null hypotheses. The results of frequentist parameter estimation and hypothesis testing should be properly interpreted, as probability statements for datasets, not as probability statements for parameters or hypotheses.

Probability statements for parameters, however, may be obtained by using Bayesian statistical analysis and inference. Bayesian hypothesis testing uses Bayes factors for analysis and may be integrated into a decision-theoretic context. Two alternative strategies for model selection, an a posteriori descriptive strategy and an a priori parsimonious predictive strategy that avoids compounded error and overfitting, can be utilized to select best-fitting models. Contemporary methods such as generalized linear modeling and mixed-effects modeling provide effective tools for modeling dependence and nonnormality in datasets. The mission of this book has been to communicate these new ideas to you so that they can provide more appropriate and effective statistical solutions to many important contemporary natural resource problems.

8.8 SUMMARY

In this final chapter, we have summarized the topics presented in this book, beginning with the original three case studies in Chapter 1. We reviewed these problems and summarized the solutions presented throughout the book. We next summarized the appropriate use of statistics in six important areas in natural resource science with: (1) sufficient planning and specification of statistical design prior to data collection; (2) the proper use of hypothesis testing; (3) the appropriate analysis of observational data; (4) adequate attention given to analysis, interpretation, and conclusions after data collection; (5) the proper use of a posteriori model selection and inference "data dredging" methods for comparison of statistical models; and (6) the appropriate analysis of dependent and nonnormally distributed sample datasets. We presented three important guidelines for sample surveys and experiments, addressing issues of sample size and replication prior to data collection. We summarized the relative merits and appropriate use of two strategies for model selection and inference, a posteriori description and a priori predictive model selection and inference using AIC and DIC. We summarized the key features of two important contemporary statistical methods, generalized linear modeling and mixed-effects modeling, which are useful for analyzing data with errors that are not normally or independently distributed. We highlighted the alternative approach to statistical analysis using Bayesian statistical analysis and inference. We concluded with final remarks about the most effective use of contemporary Bayesian and frequentist statistical research methods for the natural resource scientist at the beginning of this twenty-first century.

APPENDIX A
Review of Linear Regression and Multiple Linear Regression Analysis

In this appendix we review the concepts of linear regression and multiple linear regression analysis.

A.1 INTRODUCTION

Let's begin this review by discussing the **linear regression model**, based on a single covariate, an independent "predictor" variable x, and a response, a dependent variable y, with a linear relationship given by

$$y_i = \beta_0 + \beta_1 \cdot x_i + e_i,$$

where $e_i \sim$ iid $N(0, \sigma)$ are independent and identically distributed error (stochastic) elements or "residuals," normally distributed with mean 0 and standard deviation σ. Assume the x to be fixed and measured without error. There are two parameters, β_0, β_1, along with σ, that can be estimated with estimates $b_0 = \hat{\beta}_0$ and $b_1 = \hat{\beta}_1$ along with $s = \hat{\sigma}$ in this linear regression model. The assumptions for the linear regression model are as follows for the conditional population $y|x = \{y|(x, y) \text{ is in the population}\}$:

1. $\mu_{y|x} = \beta_0 + \beta_1 \cdot x$ (linearity of the mean).
2. $\sigma_{y|x} = \sigma$ is constant (homoscedasticity).
3. $y|_x$ is normally distributed: $y|_x \sim N(\mu_{y|x}, \sigma)$ (normality).
4. $y|_x$ are randomly sampled and independent, with x either randomly sampled or fixed, and measured without error.

The data consist of ordered pairs of x and y sample measurements (x_i, y_i), $i = 1,2, \ldots, n$, from the larger population.

Contemporary Bayesian and Frequentist Statistical Research Methods for Natural Resource Scientists. By Howard B. Stauffer

We can begin the analysis by examining the sample dataset graphically, looking at an (x, y) scatterplot for the linearity of the data. The **Pearson sample correlation statistic**

$$r = \hat{\rho} = \frac{\sum_{i=1}^{n} (x_i - \bar{x}) \cdot (y_i - \bar{y})}{\sqrt{\sum_{i=1}^{n} (x_i - \bar{x})^2} \cdot \sqrt{\sum_{i=1}^{n} (y_i - \bar{y})^2}},$$

which estimates the **Pearson correlation parameter** in the case of a population of finite size N

$$\rho = \frac{\sum_{i=1}^{N} (x_i - \mu_x) \cdot (y_i - \mu_y)}{\sqrt{\sum_{i=1}^{N} (x_i - \mu_x)^2} \cdot \sqrt{\sum_{i=1}^{N} (y_i - \mu_y)^2}},$$

should first be examined for statistical significance before a linear regression modeling analysis is conducted. Correlation, varying between -1 and $+1$, "measures" whether there is a significant linear relationship between the variables x and y. Values near $+1$ or -1 suggest a linear relationship with a positive or negative slope, respectively, whereas values near 0 are indicative of no linear relationship. The analyst should generally proceed with a linear regression modeling analysis only if there is significant nonzero correlation. The correlation can be tested for significance

$$H_0: \rho = 0$$

by using the test statistic

$$t_s = r \cdot \sqrt{\frac{n-2}{1-r^2}}.$$

This test statistic is t_{n-2}-distributed, if the null hypothesis is true.

The linear regression model with a single covariate x can be generalized to the **multiple linear regression model** with $p \geq 1$ "independent" "predictor" covariates $\{x_1, x_2, \ldots, x_p\}$ and dependent response y

$$y_i = \beta_0 + \beta_1 \cdot x_{i1} + \beta_2 \cdot x_{i2} + \cdots + \beta_p \cdot x_{ip} + e_i,$$

where $e_i \sim \text{iid } N(0, \sigma)$ are independent and identically distributed error (stochastic) elements or "residuals," normally distributed with mean 0 and standard deviation σ.

There are $k = p+1$ parameters, $\beta_0, \beta_1, \ldots,$ and β_p, along with σ, which can be estimated with estimates $b_0 = \hat{\beta}_0$, $b_1 = \hat{\beta}_1, \ldots,$ and $\hat{\beta}_p$, along with the residual standard error $s_{y|x} = \hat{\sigma}$, in this multiple linear regression model. All the linear regression results described in this appendix will also generalize to the multiple linear regression model (Seber 1977, Draper and Smith 1981, Manly 1994, Hocking 1996, Ryan 1997, Cook and Weisberg 1999).

A.2 LEAST-SQUARES FIT: THE LINEAR REGRESSION MODEL

For linear regression analysis, a least-squares (LS) method is used to fit the linear regression model to the sample data, minimizing the **goodness-of-fit profile** and providing unbiased, minimum variance estimators for the parameters β_0 and β_1. In Section 2.2, we examined other ways of fitting models to data: maximum-likelihood (ML) estimators that maximize the likelihood profile and Bayesian methods. In this chapter, with linear regression modeling, we focus on LS fit.

The **least-squares (LS) fit** of the linear regression model estimates the parameters β_0 and β_1 by minimizing the sum of squared residuals, the goodness-of-fit profile

$$\text{GOF} = \sum_{i=1}^{n} (y_i - (\beta_0 + \beta_1 \cdot x_i))^2 = \sum_{i=1}^{n} e_i^2$$

(Hilborn and Mangel 1997). Taking partial derivatives of the GOF profile with respect to the unknown parameters β_0 and β_1 and setting these partial derivatives equal to 0, the solution for the **parameter estimators** is obtained with the normal equations

$$\hat{\beta}_1 = \frac{\sum_{i=1}^{n} (x_i - \bar{x}) \cdot (y_i - \bar{y})}{\sum_{i=1}^{n} (x_i - \bar{x}) \cdot (x_i - \bar{x})} = \frac{\text{SP}_{xy}}{\text{SS}_x}$$

$$\hat{\beta}_0 = \bar{y} - \hat{\beta}_1 \cdot \bar{x}.$$

The second equation can also be obtained from the fact that the mean average or centroid $(\bar{x}, \bar{y})$ of the x and y sample values falls on the regression line. Estimators for the **standard errors** of $\hat{\beta}_1$ and $\hat{\beta}_0$ are given by

$$\text{se}_{\hat{\beta}_1} = \hat{\sigma} \cdot \sqrt{\frac{1}{(n-1) \cdot s_x^2}},$$

$$\text{se}_{\hat{\beta}_0} = \hat{\sigma} \cdot \sqrt{\frac{1}{n} + \frac{\bar{x}^2}{(n-1) \cdot s_x^2}},$$

where the estimator for the standard deviation of the residual error, the **residual standard error**, $\hat{\sigma} = s_{y|x}$ is

$$\hat{\sigma} = \sqrt{\frac{\text{GOF}}{n-2}} = \sqrt{\frac{\sum_{i=1}^{n} (y_i - (\hat{\beta}_0 + \hat{\beta}_1 \cdot x_i))^2}{n-2}} = \sqrt{\frac{\sum_{i=1}^{n} e_i^2}{n-2}} = s_{y|x}.$$

The estimators for β_0 and β_1 are **BLUE (the best linear unbiased estimators)**; that is, they are unbiased estimators of minimum variance.

Similarly, for multiple linear regression analysis, a LS method is used to fit the multiple linear regression model to sample data, minimizing the **goodness-of-fit profile** and providing unbiased, minimum variance estimators for the parameters, $\beta_0, \beta_1, \ldots, \beta_p$.

A.3 LINEAR REGRESSION AND MULTIPLE LINEAR REGRESSION STATISTICS

In this section we discuss eight leading statistics that form the basis for evaluating linear and multiple linear regression models:

1. The coefficient estimates and their significance using either confidence intervals or t tests
2. The coefficient of determination R^2
3. The residual standard error $s_{y|x}$
4. The ANOVA F test
5. The adjusted R^2
6. The Mallows C_p
7. The Akaike information criterion (AIC) and the corrected Akaike information criterion (AIC_c)
8. The Bayesian information criterion (BIC)

A.3.1 Estimates of Coefficients and Their Significance: Confidence Intervals and t Tests

The estimates for slope ($\hat{\beta}_1$) and y intercept ($\hat{\beta}_0$) for the linear regression model can be tested for statistical significance at the $P = 1-\alpha$ level of confidence with the null hypotheses respectively

$$H_0: \beta_0 = 0$$
$$H_0: \beta_1 = 0$$

in two equivalent ways (where the degrees of freedom df are given by df $= n - 2$):

1. Determine whether 0 is outside the confidence intervals:

$$\hat{\beta}_0 \pm t_{\mathrm{df},1-\alpha/2} \cdot \mathrm{se}_{\hat{\beta}_0}$$
$$\hat{\beta}_1 \pm t_{\mathrm{df},1-\alpha/2} \cdot \mathrm{se}_{\hat{\beta}_1}$$

2. Examine the significance of the t statistics:

$$t_s = \frac{\hat{\beta}_0}{\mathrm{se}_{\hat{\beta}_0}}$$
$$= \frac{\hat{\beta}_1}{\mathrm{se}_{\hat{\beta}_1}}$$

Reject H_0 if and only if the confidence interval does not contain 0. This occurs if and only if the test statistic falls within the rejection region, or equivalently if the p value is less than or equal to the type I error α. Otherwise, do not reject H_0.

It is important to determine whether the estimates for β_0 and β_1 are statistical significant. If the estimate for β_1 is not statistically significant, then β_1 is statistically equivalent to 0 and the linear term for x should be eliminated, reducing the model to the **null model**:

$$y_i = \beta_0 + e_i.$$

If the estimate for β_0 is not statistically significant, then β_0 is statistically equivalent to 0 and the constant term should be eliminated from the model and the **ratio model** used instead if β_1 is statistically significant:

$$y_i = \beta_1 \cdot x_i + e_i.$$

A.3.2 The Coefficient of Determination R^2

The **coefficient of determination** R^2 is another important statistic used to evaluate the linear model with the LS fit. It varies between 0 and 1, with a higher R^2 value indicating a linear relationship. For the linear regression with one independent variable, $R = r$, the Pearson correlation coefficient, whereas with multiple linear regression with more than one independent covariate, R generalizes the Pearson correlation coefficient to more than one dimension. The R^2 index "measures" the proportion of the variation of y "captured" by the regression model

$$R^2 = \frac{\mathrm{SS}_{\mathrm{regression}}}{\mathrm{SS}_{\mathrm{total}}} = \frac{\sum_i (\hat{y}_i - \bar{y})^2}{\sum_i (y_i - \bar{y})^2}$$
$$= 1 - \frac{\mathrm{SS}_{\mathrm{error}}}{\mathrm{SS}_{\mathrm{total}}} = 1 - \frac{\sum_i (y_i - \hat{y}_i)^2}{\sum_i (y_i - \bar{y})^2}$$

where the three components of variation are given by

$$\mathrm{SS}_{\mathrm{regression}} = \sum_i (\hat{y}_i - \bar{y})^2,$$
$$\mathrm{SS}_{\mathrm{error}} = \sum_i (y_i - \hat{y}_i)^2,$$
$$\mathrm{SS}_{\mathrm{total}} = SS_y = \sum_i (y_i - \bar{y})^2,$$

with

$$\mathrm{SS}_{\mathrm{total}} = \mathrm{SS}_y = \mathrm{SS}_{\mathrm{regression}} + \mathrm{SS}_{\mathrm{error}}$$

and $\hat{y}_i = \hat{\beta}_0 + \hat{\beta}_1 \cdot x_i$.

Figure A.1 displays scatterplots of (x,y) samples from populations satisfying the assumptions of linear models, with four prototype cases of R^2 values exhibited with extremes of slope and error. The highest R^2 values occur with datasets having maximal slope and minimal error. The lowest R^2 values occur with datasets having minimal slope and maximal error. Moderate R^2 values occur with datasets having moderate slope and moderate error. The R^2 index effectively "measures" both steepness and slope.

However, R^2 alone does not provide a definitive indicator of fit, despite common misperception. Datasets may have a high R^2, with a large amount of steepness and small amount of error, yet be curvilinear, rather than linear, in shape. Additionally, most natural resource datasets will contain an appreciable amount of stochastic error that cannot readily be explained by identifiable and measurable independent covariates, so there is a limit to the amount of error that a model should minimize. Yet a linear regression model may still prove useful and provide a reasonable fit, with a significant amount of error estimated by $s_{y|x}$. A natural resource relationship may be gradual, with moderate slope, and with a moderate amount of R^2 value, say, between 20% and 60%, yet still be statistically and biologically significant.

Models with high R^2 values may overfit sample datasets and not serve well as predictive models (Burnham and Anderson 1998). As the number of independent covariates in a model increases, R^2 values also increase, since the model will have

Figure A.1. Scatterplot of (x, y) sample datasets from populations satisfying the assumptions of linear regression, with extremes of slope and error, and their effects on the estimated coefficients of determination R^2. (**a**) High slope, low error: `x <- runif(80,0,100), y1 <- 30+5.0*x+rnorm(80,0,20)`, $R^2 = 0.992$. (**b**) Low slope, low error: `x <- runif(80,0,100), y2 <- 30+1.2*x+rnorm(80,0,20)`, $R^2 = 0.984$. (**c**) high slope, high error: `x <- runif(80,0,100), y3 <- 30+5.0*x+rnorm(80,0,40)`, $R^2 = 0.812$. (**d**) Low slope, high error: `x <- runif(80,0,100), y4 <- 30+1.2*x+rnorm(80,0,40)`, $R^2 = 0.429$. (**e**) Comparison of all prototype cases (a)–(d) above.

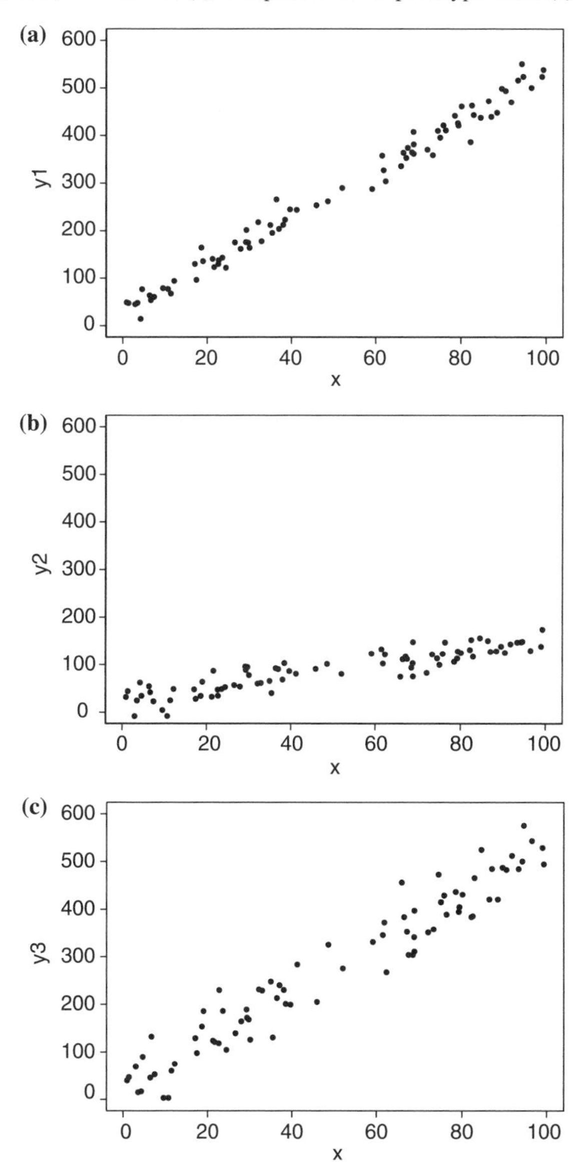

Figure A.1. *Continued.*

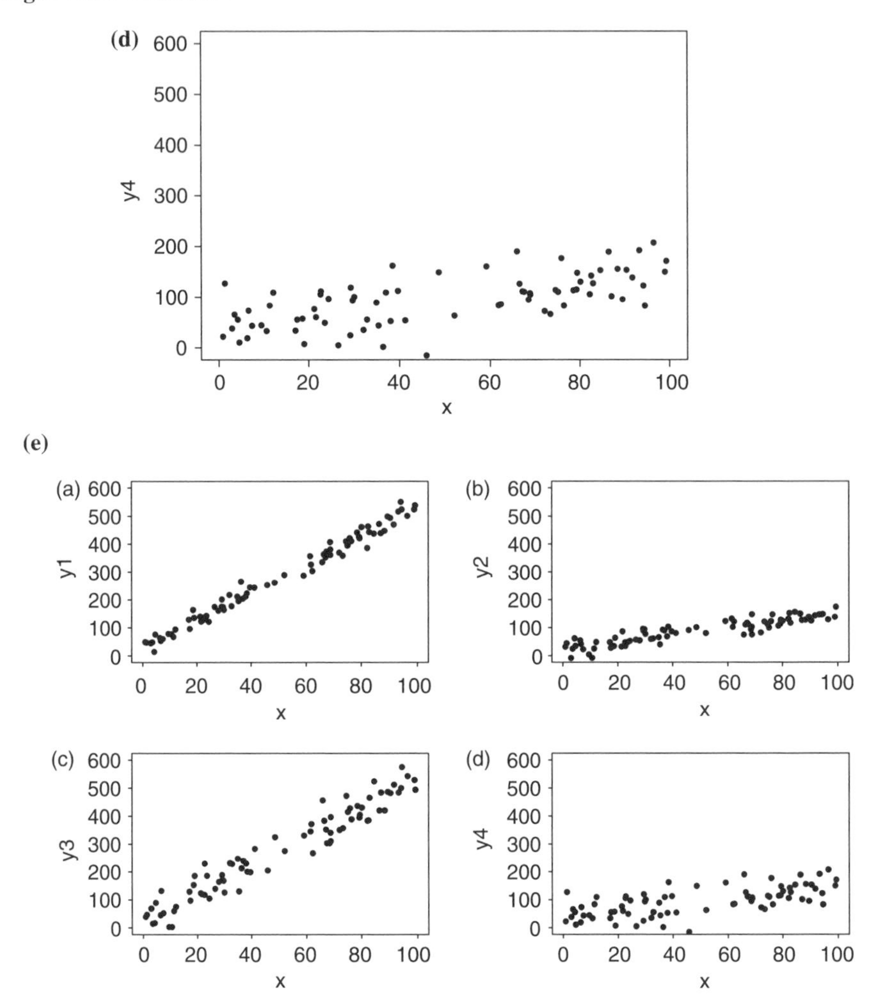

reduced error in fitting sample datasets. However, a model that overfits a sample dataset may not necessarily fit the population data very well. A model with $n-1$ covariates can in some cases exactly fit a dataset of sample size n. If the model more closely fits a sample dataset with additional independent covariates, the error, or bias, will be less and the R^2 value may be higher. However, the precision of the estimates of the parameters in the model may be reduced because of more parameter coefficients to estimate with the same size sample dataset. The precision of the estimates may also be reduced if the covariates are correlated with each other, thereby reducing the "stability" and increasing the error of the estimates. So, although a higher R^2 value may in general be a worthwhile indicator of a better-fitting model, there is a danger of overfitting models using R^2 as a sole criterion for model fitting.

In conclusion, the analyst should examine the R^2 statistic but use it judiciously along with other statistics to evaluate the fit of models. Biological models with a significant amount of stochasticity may have low R^2 values, yet be well- or best-fitting, and, conversely, models with high R^2 statistics may overfit sample datasets and be overparameterized. Models with best-fitting relationships to sample datasets are not exclusively characterized by high R^2 values.

A.3.3 The Residual Standard Error $s_{y|x}$

Another important statistic useful for evaluating the fit of models with the linear regression and multiple linear regression models is the **residual standard error** $s_{y|x} = \hat{\sigma}$ is the approximately unbiased estimator of the standard deviation parameter σ for the error given by

$$s_{y|x} = \sqrt{\frac{\sum_{i=1}^{n}(y_i - \hat{y}_i)^2}{(n-p-1)}} = \sqrt{\frac{\sum_{i=1}^{n}(y_i - (\hat{\beta}_0 + \sum_{j=1}^{p}\hat{\beta}_j \cdot x_j))^2}{(n-p-1)}}$$

$$= \sqrt{\frac{\sum_{i=1}^{n} e_i^2}{(n-p-1)}},$$

where n is the sample size and p is the number of covariates in the model. Its square $s_{y|x}^2$ is the unbiased estimator of the variance σ^2. In the case of linear regression, $p = 1$ and the parameters are β_0, β_1, and σ. Since $s_{y|x}$ estimates the standard deviation of the error for the model, better-fitting models will tend to have lower residual standard errors. However, analogous to increases in R^2, this is true only up to a point. As with increases in R^2, $s_{y|x}$ tends to decrease with increasing numbers of covariates that reduce the amount of error in the fit of the model with the sample dataset. However, again, there is a danger of overfitting that must be considered in evaluating the fit of the model. So look for models that decrease $s_{y|x}$ but also examine other statistics such as the adjusted R^2, the Mallows C_p, and AIC, which penalize models with too many "independent" covariates for overfitting sample datasets.

A.3.4 The F Test

As with experimental data and analysis of variance (ANOVA), an F **test** can be conducted for linear regression and multiple linear regression analysis, calculating the components of variance for the regression model, the regression (regr), error (resid), and total components, in the following table

Components of Variance	df	SS	MS	F	p Value
regr	df_{regr}	SS_{regr}	MS_{regr}	F_{s}	—
resid	df_{resid}	SS_{resid}	MS_{resid}		
Total	df_{total}	SS_{total}			

The degrees of freedom (df) formulas for the linear regression model are given by

$$\begin{aligned} \text{df}_{\text{regr}} &= 1, \\ \text{df}_{\text{resid}} &= n - 2, \\ \text{df}_{\text{total}} &= n - 1, \end{aligned}$$

where n is the sample size. Note that

$$\text{df}_{\text{total}} = \text{df}_{\text{regr}} + \text{df}_{\text{resid}}.$$

For multiple regression with p independent variables, the degrees of freedom formulas are given by

$$\begin{aligned} \text{df}_{\text{regr}} &= p, \\ \text{df}_{\text{resid}} &= n - p - 1, \\ \text{df}_{\text{total}} &= n - 1. \end{aligned}$$

The sums of squares (SS) formulas are given by

$$\text{SS}_{\text{regr}} = \sum_{i=1}^{n} (\hat{y}_i - \bar{y})^2,$$

$$\text{SS}_{\text{resid}} = \sum_{i=1}^{n} (y_i - \hat{y}_i)^2,$$

$$\text{SS}_{\text{total}} = \sum_{i=1}^{n} (y_i - \bar{y})^2.$$

As usual

$$\text{SS}_{\text{total}} = \text{SS}_{\text{regr}} + \text{SS}_{\text{resid}}.$$

The mean-squares (MS) formulas are

$$\begin{aligned} \text{MS}_{\text{regr}} &= \text{SS}_{\text{regr}}/\text{df}_{\text{regr}}, \\ \text{MS}_{\text{resid}} &= \text{SS}_{\text{resid}}/\text{df}_{\text{resid}}, \end{aligned}$$

and the F_s-test statistic is

$$F_s = MS_{regr}/MS_{resid}.$$

The null and alternative hypotheses are given by

$$H_0\colon y_i = \beta_0 + e_i \qquad \text{(null model); i.e., } \beta_1 = 0,$$
$$H_A\colon y_i = \beta_0 + \beta_1 \cdot x_i + e_i \quad \text{(linear regression model)}$$

(i.e., $\beta_1 \neq 0$). The F test may be significant for covariates even without good model fit since it assesses whether a model with covariates fits the data better than does the null model without covariates, so other statistics should also be examined. If the F-test results are insignificant; however, that is a good indication that the model with the covariate is not helpful.

A.3.5 Adjusted R^2

If R^2 is not always a reliable statistic for evaluating the competitive fit of models because of possible overfitting, what statistics will serve in its place? The **adjusted R^2 statistic** compensates for the problem of overfitting by extracting a "penalty" for the use of too many covariates. The adjusted R^2 statistic is given by the formula

$$\text{Adjusted}\, R^2 = R^2_{adj} = 1 - \frac{n-i}{n-k} \cdot (1 - R^2),$$

where n = sample size, $i = 1$ if there is intercept and 0 otherwise, and k = the number of parameters. Models with the highest adjusted R^2 are the best-fitting models. It is clear from the formula that increasing numbers of covariates negatively affect the magnitude of R^2_{adj}. You can avoid the problem of models overfitting sample datasets by examining the adjusted R^2 statistic rather than the R^2 statistic. However, the adjusted R^2 statistic still may tend to overfit models to sample datasets, favoring models with too many variables. In Sections A.3.6–A.3.8, we will examine other statistics, Mallows' C_p, AIC, AIC_c, and BIC that most effectively address the problem of overfitting.

A.3.6 Mallows' C_p

Another statistic designed to assess LS fit for models with normal residuals having constant variance is the **Mallows C_p statistic**

$$C_p = p + (n-p) \cdot \frac{(\hat{\sigma}^2 - \hat{\sigma}^2_{full})}{\hat{\sigma}^2},$$

where p is the number of covariates in the model, the full model is the model with all the explanatory covariates under consideration, and

$$\hat{\sigma}^2 = \frac{\sum_{i=1}^{n} (y_i - \hat{y}_i)^2}{n} = \frac{n-p-1}{n} \cdot s^2_{y|x}.$$

Models with a lower Mallows C_p value are better-fitting, or more accurately, models with $C_p \cong p$ are the best-fitting models. The Mallows C_p provides approximately the same ranking for models as AIC and AIC_c (see Section A3.7). If the full model is not overfitting, then presumably models with C_p approximately equal to p will also not be overfitting. Models with C_p below p will have underestimated the error and be overfitting. Because of these rather questionable assumptions, we recommend the use of AIC and AIC_c (below) as the most rigorous and theoretically justified approach to model fitting of natural resource datasets.

A.3.7 Akaike's Information Criterion: AIC and AIC_c

We now describe the important information-theoretic statistic that is most effective for evaluating the competitiveness of models at fitting natural resource data. Akaike's information criterion was developed in the early 1970s by the Japanese mathematician Hirotugu Akaike (1973). It is an information-theoretic measurement of the **Kullback–Liebler distance** between a model and reality. **Akaike's information criterion (AIC)** and the **corrected Akaike information criterion (AIC_c)** for multiple linear regression are given by the formulas

$$\begin{aligned}
\text{AIC} &= n \cdot \log\left(s_{y|x}^2 \cdot \frac{n-p-1}{n}\right) + 2 \cdot k \\
&= n \cdot \log\left(s_{y|x}^2 \cdot \frac{n-k+1}{n}\right) + 2 \cdot k, \\
\text{AIC}_\text{c} &= n \cdot \log\left(s_{y|x}^2 \cdot \frac{n-p-1}{n}\right) + 2 \cdot k + 2 \cdot \frac{k \cdot (k+1)}{(n-k-1)} \\
&= n \cdot \log\left(s_{y|x}^2 \cdot \frac{n-k+1}{n}\right) + 2 \cdot k + 2 \cdot \frac{k \cdot (k+1)}{(n-k-1)},
\end{aligned}$$

where p is the number of covariates and $k = p + 2$ is the number of parameters (including the intercept and σ). For the linear regression model with the parameters β_0, β_1, and σ, $p = 1$ and $k = 3$. AIC is the linear Taylor series approximation of the Kullback–Leibler distance, whereas AIC_c is a second-order Taylor series approximation. Since AIC_c is more precise, it should always be used in preference to AIC, particularly for datasets with small sample size. The best-fitting model has the lowest AIC_c value.

The AIC_c criterion can be used to determine the most parsimonious model, the one with the most optimal combination of minimal bias and maximal precision. It penalizes a model with too many covariates from overfitting sample data. As we have emphasized, other traditionally popular regression statistics, such as R^2 and $s_{y|x}$, tend to favor models that overfit sample datasets, whereas AIC_c prevents such overfitting from occurring. For sound predictive models, the objective is to develop good models from sample datasets that provide reliable predictions for populations. The objective is not merely to describe sample datasets. As the number of covariates increases, models tend to more closely fit sample datasets and reduce the bias. However, as the number of covariates increases and the number of samples in the dataset remains fixed, the precision of the covariate coefficient parameter estimates tends to decrease. The AIC_c criterion moderates this process, producing an optimal compromise between reduced bias and maximal precision. It determines the most parsimonious model, the one with the combined least amount of bias, closest to the sample dataset, and most amount of precision, with the lowest amount of sampling error, for the coefficient estimates relative to the reduced bias.

The AIC_c criterion measures the relative amount of noise, or **entropy**, in the sample data, separating it from the **signal** or **information**. It is a relative measure, since the reality is unknown; the absolute measure of entropy is the calculated AIC_c plus a constant. The constant remains unknown, but since each model has the same constant, AIC_cs may be compared to determine the best fitting models. As AIC_c provides comparative measures of fit between models only, goodness-of-fit tests must also be used in analysis to assess how well-fitting are the best-fitting models.

In general, for any probabilistic statistical model for a sample dataset with a likelihood function $\mathcal{L}$ (see Chapters 2–4, and 6, and 7), AIC and AIC_c are defined using the deviance=$D = -2 \cdot \log(\mathcal{L})$:

$$\begin{aligned} \text{AIC} &= D + 2 \cdot k \\ &= -2 \cdot \log(\mathcal{L}) + 2 \cdot k, \\ \text{AIC}_c &= D + 2 \cdot k + 2 \cdot \frac{k \cdot (k+1)}{n-k-1} \\ &= -2 \cdot \log(\mathcal{L}) + 2 \cdot k + 2 \cdot \frac{k \cdot (k+1)}{n-k-1}. \end{aligned}$$

A.3.8 Bayesian Information Criterion (BIC)

The AIC_c criterion should be applied to models of realities that are complex and infinite- or high-dimensional as are most biological populations (Burnham and Anderson 1998). For such complex realities, finite-dimensional models will necessarily be inaccurate and, at best, approximations. For realities that are finite-dimensional, of fairly low dimension such as $k = 1–5$, with k fixed as the sample size n increases, so-called dimension-consistent criteria such as the Bayesian information criterion (BIC) should be applied.

The **Bayesian information criterion**, developed by Schwarz (1978), also uses a formula based on the deviance or log likelihood and "penalizes" models for the overuse of covariates

$$\begin{aligned} \text{BIC} &= D + k \cdot \log(n) \\ &= -2 \cdot \log(\mathcal{L}) + k \cdot \log(n). \end{aligned}$$

For multiple linear regression models, BIC is given by

$$\begin{aligned}\text{BIC} &= n \cdot \log\left(s_{y|x}^2 \cdot \frac{n-p-1}{n}\right) + k \cdot \log(n) \\ &= n \cdot \log\left(s_{y|x}^2 \cdot \frac{n-k+1}{n}\right) + k \cdot \log(n)\end{aligned}$$

and is derived using Bayesian assumptions of equal priors on each model and vague priors on the parameters (Burnham and Anderson 1998), with the objective of predicting rather than understanding the process of a system. It penalizes more heavily for increases in the number of parameters and hence sometimes tends to select models that are underfit with excessive bias. For natural resource modeling, most realities are likely complex and infinite-dimensional; hence, AIC_c is a more appropriate criteria for comparing statistical models.

A.4 STEPWISE MULTIPLE LINEAR REGRESSION METHODS

In this section we will briefly describe methods for model selection that are applicable to the multiple linear regression case where there are p covariates $\{x_1, x_2, \ldots, x_p\}$ along with the response y with n samples. We shall discuss two exploratory, descriptive, "data dredging" methods for model selection, ways of selecting from among collections of models consisting of linear combinations of covariates. For example, suppose that we are interested in developing a habitat selection model (Manly et al. 1995, 2004) using a collection of habitat covariates as predictor variables and an animal abundance or presence–absence response variable. We could use any of the following criteria as a basis for model selection:

1. Most significant coefficient estimate (i.e., lowest p value)
2. Lowest residual standard error $s_{y|x}$,
3. Highest coefficient of determination R^2
4. Highest adjusted R^2
5. Lowest Mallows C_p, or one closest to p
6. Lowest AIC
7. Lowest BIC

With **stepwise selection multiple linear regression**, we can choose a "best" model according to the covariate coefficient estimate p values, using the following iterative procedure:

1. Choose type I error bounds α_e for the entering covariate coefficients and α_s for the staying covariate coefficients with $\alpha_e < \alpha_s$.
2. Start with the null model with no covariates $y_i = \beta_0 + e_i$.

3. Consider all single-covariate models fitted to the sample dataset and select the covariate, say, x_{c_1}, with coefficient estimate $\hat{\beta}_{c_1}$ having the lowest p value below α_e, obtaining the single covariate model $y_i = \beta_0 + \beta_{c_1} \cdot x_{c_1,i} + e_i$.
4. Consider the addition of another covariate, $x_{c_{j+1}}$, to the currently selected covariate model with $x_{c_1}, x_{c_2}, \ldots, x_{c_j}$ fitted to the sample dataset and add the covariate, say, $x_{c_{j+1}}$, with coefficient estimate $\hat{\beta}_{c_{j+1}}$ having the lowest p value below α_e, if there is one, obtaining the $(k+1)$ covariate model $y_i = \beta_0 + \beta_{c_1} \cdot x_{c_1,i} + + \beta_{c_2} \cdot x_{c_2,i} + \cdots + \beta_{c_{j+1}} \cdot x_{c_{j+1},i} + e_i$; or otherwise stop.
5. Consider the p values of the current model covariate coefficient estimates and drop the covariates with p values above α_s, if there are any.
6. Continue iterating steps 4 and 5, adding and dropping covariates as appropriate until the process stops.

Note that it is important that $\alpha_e < \alpha_s$, or otherwise the process could continue indefinitely, with covariates added and dropped repeatedly.

Forward selection multiple linear regression adds covariates, without dropping any, until the process stops, whereas **backward elimination multiple linear regression** begins with the full model consisting of all covariates and drops covariates with the highest p values, without adding any, until the process stops. Stepwise selection multiple linear regression combines both forward selection and backward elimination multiple linear regression. As mentioned earlier, other optimization criteria besides p values for the coefficient estimates could be used for the covariate selection, such as lowest residual standard error $s_{y|x}$, highest R^2, highest adjusted R^2, lowest Mallows C_p, lowest AIC, or lowest BIC.

Stepwise selection, forward selection, and backward elimination multiple linear regression methods provide convenient methods for choosing models with multivariate sample datasets. However, these methods tend to overfit sample data because of the compounding of type I error caused by the multiple testing of hypotheses in the selection criteria. Therefore, these "data dredging" methods should be viewed as exploratory and descriptive. Results should be interpreted tentatively. These methods are most suitable for formulating model hypotheses that can be tested with additional sample datasets. Only with sufficient goodness-of-fit testing of the inferences, preferably with additional sample datasets, should the results of these methods be used for prediction.

A.5 BEST-SUBSETS SELECTION MULTIPLE LINEAR REGRESSION

Best-subsets selection multiple linear regression selects best-fitting models from the collection of all models that are linear combinations of covariates using some of the following criteria:

1. Highest R^2
2. Highest adjusted R^2

3. Lowest Mallows C_p, or C_p closest to p
4. Lowest AIC
5. Lowest BIC

As with stepwise selection multiple linear regression, best-subsets selection multiple regression is prone to type I error and tends to overfit sample data. As such, results should be viewed as tentative, subject to further study with additional sample datasets, if inferences are to be used for prediction.

A.6 GOODNESS OF FIT

The discussion so far has focused on the interpretation of statistics evaluating the significance and competitiveness of models at fitting sample data. The significance of the model coefficient estimates tests the null hypothesis that the coefficients are equal to 0. The significance of the F test addresses the hypothesis of whether the model differs from the null model $y_i = \beta_0 + e_i$. Other statistics, such as R^2, $s_{y|x}$, the adjusted R^2, Mallows C_p, and, most importantly, AIC_c or BIC, evaluate the fit of the model and serve as comparative statistics useful in evaluating the relative competitiveness of models at differing from the null model and fitting the sample dataset. None of these tests or statistics, however, serve adequately in evaluating model goodness of fit to the reality represented by the sample dataset.

So it is very important to examine goodness of fit of the best-fitting models as part of the overall analysis process. In the following sections, we will briefly address this issue.

A.6.1 Residual Analysis

An important part of goodness-of-fit analysis is always to examine the residuals of a linear regression or multiple linear regression model

$$\hat{e}_i = y_i - \hat{y}_i,$$

where $\hat{y}_i = \hat{\beta}_0 + \sum_{j=1}^{p} \hat{\beta}_j \cdot x_j$, to determine whether the assumptions of the model might be violated. Recall that the linear regression and multiple linear regression models are based on the following assumptions for the population residuals e_i:

1. Independence
2. Normality
3. Homoscedasticity (σ^2 constant)

The residuals $\hat{e}_i$ are estimates of the population residuals e_i and should be examined for apparent violations of these three assumptions. Residual plots that graph the residuals as a function of model fit are readily available in most statistical software. They should be examined for any apparent dependences among the residuals, lack of normality, or heteroscedasticity.

The residuals should occur "randomly," both positively and negatively above and below the model fit axis, without any apparent dependence relationships among them. They should also be normally distributed, with a standard deviation estimated by $s_{y|x}$. To check for normalcy, a normal plot of the residuals may be examined for linearity and a test for normality such as the Anderson–Darling test may be applied. The homoscedasticity assumption of constant variance of the residuals should be evaluated visually in the graph of the residuals, and tested, with Bartlett's test or other tests for homoscedasticity such as Levene's test or the F test. Outliers should be scrutinized for their size and quantity. If the confidence level is 95%, the proportion of outliers beyond the 95% confidence band should not greatly exceed the expected 5%. With small sample sizes, residual analysis can be more of an art than a science in practical application. So a biological explanation may provide the best rationale for the validity of the assumptions. Interested readers may consult Cook (1998) and Cook and Weisberg (1999) to examine this topic of residual analysis in more detail.

A.6.2 Confidence Intervals

Confidence intervals for the predicted mean $\hat{y}|x = \hat{\beta}_1 + \hat{\beta}_1 \cdot x$ may be calculated using the standard error for the mean given by

$$\mathrm{se}_{\hat{y}|x} = s_{y|x} \cdot \sqrt{\frac{1}{n} + \frac{(x - \bar{x})^2}{(n-1) \cdot s_x^2}},$$

where $s_{y|x}$ is the standard error, $\bar{x}$ is the estimated mean of the $\{x_i\}$, s_x^2 is the estimated variance of the $\{x_i\}$, and n is the sample size. Confidence intervals $\mathrm{CI}_{\hat{y}|x}$ may then be calculated and graphed using the formula

$$\mathrm{CI}_{\hat{y}|x} = \hat{y}|x \pm t_{\mathrm{d}f,1-\alpha/2} \cdot \mathrm{se}_{\hat{y}|x},$$

where the degrees of freedom $\mathrm{df} = n - k$ with k parameters in the model and $P = 1 - \alpha$ is the confidence level. The frequentist interpretation of this confidence interval with confidence level P is that, with repeated sampling, P of the sample confidence intervals will on average contain the population mean $\mu|x$ at each x.

A.6.3 Prediction Intervals

Prediction intervals for the predicted $y|x$ may also be calculated and graphed. The expected proportions of the (x_i, y_i) points in the developmental and test datasets should fall within these prediction intervals, for example, 95% of the points within the 95% prediction intervals. Prediction intervals can be calculated using the formula for the standard error for the predicted $y|x$ given by

$$\mathrm{se}_{y|x} = s_{y|x} \cdot \sqrt{1 + \frac{1}{n} + \frac{(x - \bar{x})^2}{(n-1) \cdot s_x^2}},$$

where $s_{y|x}$ is the standard error, $\bar{x}$ is the sample mean of the $\{x_i\}$, s_x^2 is the variance of the $\{x_i\}$, and n is the sample size. Note that there are three components of error in this standard error formula, indicated by the three terms in the square root, the first (i.e., 1) due to variation of the y for a fixed x, and the latter two $[1/n$ and $(x - \bar{x})^2/(n-1) \cdot s_x^2]$ due to variation of the estimates for the linear regression. Using the degrees of freedom, $\mathrm{df} = n - k$ with k parameters in the model, prediction intervals $\mathrm{PI}_{y|x}$ with $P = 1 - \alpha$ level of confidence can be calculated:

$$\mathrm{PI}_{y|x} = \hat{y}|x \pm t_{\mathrm{df},1-\alpha/2} \cdot \mathrm{se}_{y|x}.$$

A.6.4 Cross-Validation and Testing Techniques

For a linear or multiple linear regression model to provide a predictive tool, additional confirmation of the reliability of the model may be examined using either cross-validation with the developmental dataset or testing techniques with additional test datasets. If additional test datasets are available, randomly sampled from the population, goodness of fit can be examined by evaluating the predictive performance of the prediction intervals on the test datasets. For example, 95% prediction intervals can be examined on the test datasets to determine whether they perform as expected, with approximately 95% of the samples in the test dataset within the prediction interval. Alternatively, if additional test datasets are unavailable, the next best alternative is to perform cross-validation analysis on the developmental dataset. The most common method of cross-validation consists of successively omitting individual points, fitting the model to the remaining sample dataset, and examining the deleted point with respect to the prediction interval. The overall predictive accuracy of the deleted points falling within the prediction intervals is an indication of the predictive capabilities of the fitted model.

APPENDIX B
Answers to Problems

Problem 1.1

```
> d
 [1] 29.6 22.7 28.6 23.1 27.3 28.3 23.7 28.4 25.0 24.5 27.7 22.7
[13] 27.6 22.7 27.0 21.2 26.8 28.4 21.2 23.6 24.6 24.0 22.0 28.2
[25] 27.0 22.8 28.4 27.7 31.9 18.9
> dbar <- mean(d)
> s <- stdev(d)
> n <- length(d)
> se <- s/sqrt(n)
```

```
> #   95% confidence
> t <- qt(.975,n-1)
> E <- t*se
> left.lim <- dbar-E
> right.lim <- dbar+E
> c(dbar,s,n,se,t,E,left.lim,right.lim)
[1] 25.520000  3.061260 30.000000 0.558907 2.045230 1.143093
[7] 24.376907 26.663093
```

```
> #   80% confidence
> t <- qt(.90,n-1)
> E <- t*se
> left.lim <- dbar-E
> right.lim <- dbar+E
> c(dbar,s,n,se,t,E,left.lim,right.lim)
[1] 25.5200000 3.0612596 30.0000000 0.5589070 1.3114336
[6] 0.7329694 24.7870306 26.2529694
```

```
> #   67% confidence
> t <- qt(.835,n-1)
```

```
> E <- t*se
> left.lim <- dbar-E
> right.lim <- dbar+E
> c(dbar,s,n,se,t,E,left.lim,right.lim)
[1] 25.5200000 3.0612596 30.0000000 0.5589070 0.9907557
[6] 0.5537403 24.9662597 26.0737403
> #   50% confidence
> t <- qt(.75,n-1)
> E <- t*se
> left.lim <- dbar-E
> right.lim <- dbar+E
> c(dbar,s,n,se,t,E,left.lim,right.lim)
[1] 25.5200000 3.0612596 30.0000000 0.5589070 0.6830439
[6] 0.3817580 25.1382420 25.9017580

> t.test(d,mu=26,conf.level=.95)

        One-sample t-Test

data: d
t = -0.8588, df = 29, p-value = 0.3975
alternative hypothesis: true mean is not equal to 26
95 percent confidence interval:
 24.37691 26.66309
sample estimates:
 mean of x
     25.52

> t.test(d,mu=26,conf.level=.80)

     One-sample t-Test

data: d
t = -0.8588, df = 29, p-value = 0.3975
alternative hypothesis: true mean is not equal to 26
80 percent confidence interval:
 24.78703 26.25297
sample estimates:
 mean of x
     25.52

> t.test(d,mu=26,conf.level=.67)

        One-sample t-Test
```

```
data: d
t = -0.8588, df = 29, p-value = 0.3975
alternative hypothesis: true mean is not equal to 26
67 percent confidence interval:
 24.96626 26.07374
sample estimates:
 mean of x
     25.52

> t.test(d,mu=26,conf.level=.50)

          One-sample t-Test

data: d
t = -0.8588, df = 29, p-value = 0.3975
alternative hypothesis: true mean is not equal to 26
50 percent confidence interval:
 25.13824 25.90176
sample estimates:
 mean of x
     25.52
```

> # With a confidence of 95%, 80%, and 67%, the estimate is below the threshold but the CI overlaps the threshold, creating a conflict between the decisionmaking options. However, if the confidence is lowered to 50%, the estimate and the CI are below the threshold, creating no conflict between the decisionmaking options. This is caused by the lowering of the level of confidence in the decisionmaking. The confidence levels of 95% and 80% are commonly used in natural resource science. A confidence level of 67% provides a t value approximately equal to 1. A confidence level of 50% provides a decision that is correct half of the time.
> # The histogram of $\{d_i\}$ does not suggest that there are any apparent problems with normality of the population based the sample data.

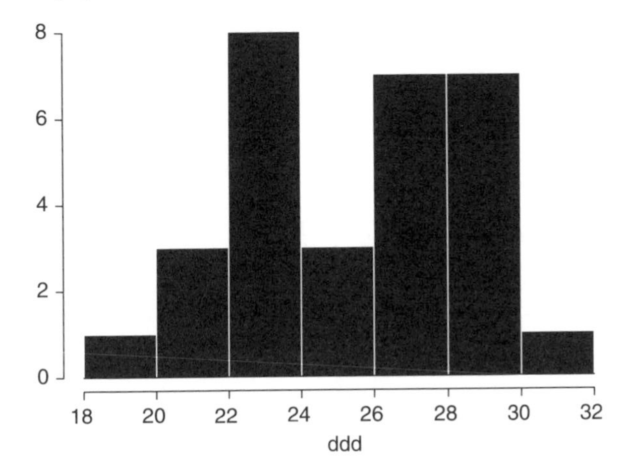

Problem 1.2

```
> phat <- 34/60
> phat
[1] 0.5666667
> se <- sqrt(phat*(1-phat)/59)
> se
[1] 0.06451324
> t <- qt(.975,59)
> t
[1] 2.000995
> E <- t*se
> E
[1] 0.1290907
> phat+E
[1] 0.6957574
> t <- qt(.90,59)
> t
[1] 1.296066
> E <- t*se
> phat+E
[1] 0.6502801
> t <- qt(.75,59)
> E <- t*se
> phat+E
[1] 0.6104499
```

> # With 95% and 80% confidence, the estimate is below the threshold but the CI overlaps the threshold, creating a conflict between the decisionmaking options. However, with the lower 50% confidence, both the estimate and the CI are below the threshold with no conflict created between the decisionmaking options. This disparity is caused by the reduced level of confidence in the decisionmaking.

Problem 1.3

```
> t.test(d,mu=26,alternative=''less'')

      One-sample t-Test

data: d
t = -0.8588, df = 29, p-value = 0.1987
alternative hypothesis: true mean is less than 26
95 percent confidence interval:
      NA 26.46965
```

```
sample estimates:
 mean of x
     25.52
```

> # For confidence levels of 95% and 80% we fail to reject the null hypothesis that the mean is at or above the threshold. However, for confidence levels of 67% and 50%, we reject the null hypothesis that the mean is at or above the threshold. These decision results differ from those based on the CIs in Problem 1.1 because this test is one-tailed whereas the CIs are two-tailed.

Problem 1.4

```
> 1-pnorm(26.0,25.5,3.1)
[1] 0.4359324
```

> There is a risk of 0.4359324 that a decision that the threshold has been exceeded will be incorrect.

Problem 1.5

```
> counts
 [1] 2 1 0 0 2 0 2 1 2 1 1 0 5 1 0 1 4 1 0 4 0 1 2 2 0 4
[27] 2 2 1 0 2 0 3 0 1 0 1 2 1 1 2 3 0 0 2 2 1 1 1 2
> chat <- mean(counts)
> chat
[1] 1.34
> s <- stdev(counts)
> s
[1] 1.22241
> n <- length(counts)
> n
[1] 50
> se <- s/sqrt(n)
> se
[1] 0.1728749
> t <- qt(.975,n-1)
> t
[1] 2.009575
> E <- t*se
> chat-E
[1] 0.992595
> chat+E
[1] 1.687405
```

> # The 95% confidence interval is [0.992595, 1.687405].

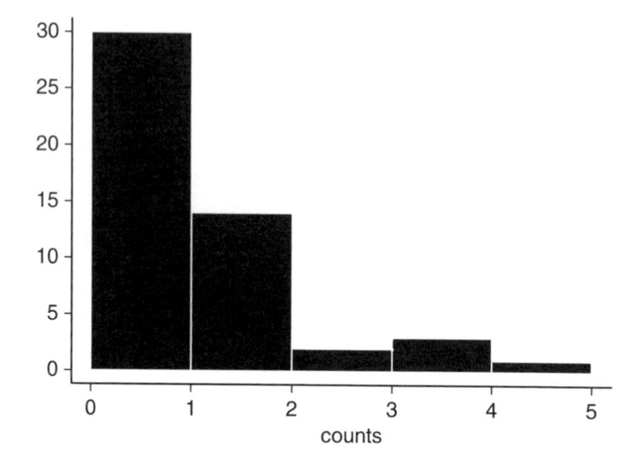

> # The data are quite skewed and nonnormal, so the analysis results may be inaccurate.

Problem 1.6

```
> data
      x      y
 1  14.9   51.9
 2  35.4   72.3
 3  17.4   68.4
 4  30.0   62.3
 5  83.6  141.8
 6  38.2   82.7
 7  65.2   89.4
 8  25.4   69.6
 9   5.9   28.3
10  51.5   94.2
11  31.6   54.9
12  13.1   45.9
13  31.4   72.7
14  98.9  156.0
15  30.4   52.5
16  38.2   77.4
17  94.6  151.6
18  21.8   67.8
19  17.9   48.8
20  45.4   58.3
21  76.5  121.0
```

```
22  80.2   129.2
23  10.6    37.9
24  98.5   147.3
25  66.1   131.1
26   1.8    40.6
27  29.0    59.1
28  98.4   131.6
29  99.4   168.1
30  59.9   108.6
31  46.1    64.9
32  48.3    99.2
33  58.8    93.6
34  10.9    48.5
35   1.1    34.7
36  90.7   146.4
37  28.0    56.0
38  89.1   121.7
39  32.1    92.9
40  77.1   111.8
41  99.2   149.1
42  14.4    60.8
43  67.1    90.2
44  19.2    52.7
45  99.1   140.7
46  38.1    78.9
47   4.3    25.8
48  31.1    75.5
49  94.3   146.1
50   8.7    47.9
> model <- lm(y~x,data)
> summary(model)

Call: lm(formula = y ~ x, data = data)
Residuals:
    Min      1Q Median     3Q    Max
 -26.55 -7.145  1.471 7.312 23.66

Coefficients:
              Value Std. Error t value Pr(>|t|)
(Intercept) 31.5491  2.8722    10.9845  0.0000
          x  1.1741  0.0502    23.3686  0.0000

Residual standard error: 11.37 on 48 degrees of freedom
```

```
Multiple R-Squared: 0.9192
F-statistic: 546.1 on 1 and 48 degrees of freedom, the p
-value is 0

Correlation of Coefficients:
  (Intercept)
x -0.8288
> AIC(model)
[1] 388.9069
```

> # The estimates for β_0 and β_1 are both statistically significant at the 95% level of confidence, since their CIs do not include 0 and their p values are both less than 5%.

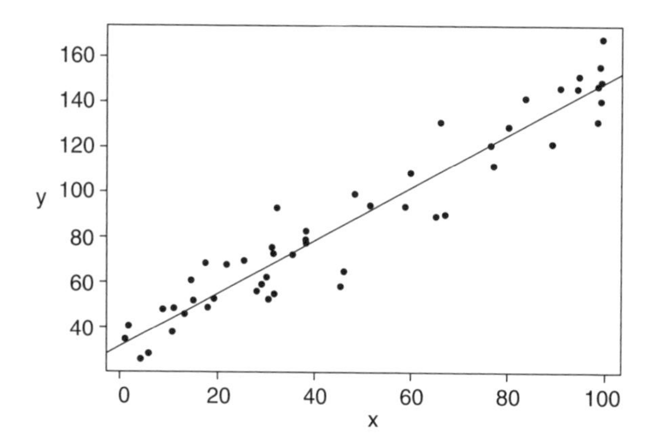

> # The scatterplot suggests that the linear regression model provides a good fit.

Problem 2.1

```
> p <- .50
> 1-(1-p)^2
[1] 0.75
> 1-(1-p)^3
[1] 0.875
> 1-(1-p)^4
[1] 0.9375
> 1-(1-p)^5
[1] 0.96875
> 1-(1-p)^6
[1] 0.984375
> (1-p)^5*p/(1-p)^5
[1] 0.5
```

1. $m = 5$ with P = 0.96875.
2. $m + 1 = 6$ with P = 0.984375, so probability is increased by 0.015625.
3. Bayes probability = P($a \cap b$)/P(b) = 0.03125*0.5/0.03125 = 0.5, where $b = m + 1$ visits with no detections in first m visits and $a = m + 1$ visits with no detections in first m visits and a detection in the $(m + 1)$st visit. This conditional probability equals the unconditional probability because the visit events are independent.

Problem 2.2

The probability of choosing the correct occupied site for each of the three strategies all depend on your initial choice, with three possible ordered combinations of nests (O = occupied, N = nonoccupied), ONN, NON, and NNO, each with probability $\frac{1}{3}$ of occurring (first position – your initial choice, second or third position can be identified by the wildlife officer):

1. ONN—win, NON—lose, NNO—lose: probability $= \frac{1}{3}$
2. ONN—lose, NON—win, NNO—win: probability $= \frac{2}{3}$
3. ONN—win, NON—lose, NNO—lose: probability $= \frac{1}{3}$.

Conclusion: the second option has the unique maximum probability of correctly choosing the occupied site.

Problem 2.3

> # (i) Posteriors are BE(32, 70), BE(31.5, 69.5), BE(33, 71)
> # means:

```
> 32/102
[1] 0.3137255
> 31.5/101
[1] 0.3118812
> 33/104
[1] 0.3173077
> (.3173077-.3118812)/.3118812
[1] 0.01739925
```

> # medians:

```
> qbeta(.5,32,70)
[1] 0.3125037
> qbeta(.5,31.5,69.5)
[1] 0.3106351
> qbeta(.5,33,71)
[1] 0.3161325
```

```
> (.3161325-.3106351)/.3106351
[1] 0.01769729
```

> # CI limits:

```
> qbeta(.025,32,70)
[1] 0.2278132
> qbeta(.025,31.5,69.5)
[1] 0.2257483
> qbeta(.025,33,71)
[1] 0.231841
> (.231841-.2257483)/.2257483
[1] 0.02698891
> qbeta(.975,32,70)
[1] 0.4065568
> qbeta(.975,31.5,69.5)
[1] 0.4050706
> qbeta(.975,33,71)
[1] 0.40943
> (.40943-.4050706)/.4050706
[1] 0.01076207
```

> # All relative differences for means, medians, and CIs are less than 3%.
> # risks:

```
> pbeta(.25,32,70)
[1] 0.07795848
> pbeta(.25,31.5,69.5)
[1] 0.08524389
> pbeta(.25,33,71)
[1] 0.0650023
```

> # The risks are all well below 50%, regardless of prior, based on analysis of the first year's dataset.
> # (ii) Posteriors are BE (31.5, 69.5), BE (44.5, 106.5), BE (66.5, 184.5) and BE (87.5, 263.5) for the 4 years of sequential analysis

```
> pbeta(0.25,31.5,69.5)
[1] 0.0852439
> pbeta(0.25,44.5,106.5)
[1] 0.1114486
> pbeta(0.25,66.5,184.5)
[1] 0.3015056
> pbeta(0.25,87.5,263.5)
[1] 0.5205103
```

> # All risks are below 50% except for the final (fourth), year's analysis. There is an upward trend in risk that can be analyzed by looking at the change in odds ratios.

```
> 0.0852439/(1-0.0852439)
[1] 0.09318757
> 0.1114486/(1-0.1114486)
[1] 0.1254273
> 0.3015056/(1-0.3015056)
[1] 0.4316507
> 0.5205103/(1-0.5205103)
[1] 1.085551
```

> # The odds ratios are increasing, quadrupling and doubling for the last 2 years, respectively.
> # (iii) For the beta prior of the first year's analysis with mean $\mu = 0.333$ and $\sigma = 0.0875$ and the parameters α and β are given by

```
> sqrt((1/3)*(2/3)/29)
[1] 0.08753762
> (0.333^2-0.333^3-(0.333)*(0.0875)^2)/(0.0875)^2
[1] 9.327469 # alpha
> (1/0.0875^2)*0.333*(1.0-0.333)^2+(0.333-1.0)
[1] 18.68295 # beta
> pbeta(0.25,40.327469,87.68295)
[1] 0.0516386
> pbeta(0.25,53.327469,124.68295)
[1] 0.07038743
> pbeta(0.25,75.327469,202.68295)
[1] 0.2182356
> pbeta(0.25,97.327469,280.68295)
[1] 0.3763185
```

> # All risks are now below 50% if the logging company's earlier data are included in the analysis for the first year's prior. The odds of risk for each year is increasing, quadrupling and doubling, for the last 2 years, respectively.

```
> 0.0516386/(1-0.0516386)
[1] 0.05445034
> 0.07038743/(1-0.07038743)
[1] 0.07571695
> 0.2182356/(1-0.2182356)
[1] 0.2791578
> 0.3763185/(1-0.3763185)
[1] 0.6033825
```

Problem 2.4

```
> pgamma(3,0.001+400,0.001+100)
[1] 2.208268e-08
> 1-pgamma(3,0.001+400,0.001+100)
[1] 1
```

```
> pgamma(3,0.001+400,0.001+100)/(1-pgamma(3,0.001+400,0.001+100))
[1] 2.208268e-08
> pgamma(3,0.001+400+260,0.001+100+75)
[1] 8.031383e-09
> 1-pgamma(3,0.001+400+260,0.001+100+75)
[1] 1
> pgamma(3,0.001+400+260,0.001+100+75)/(1-
pgamma(3,0.001+400+260,0.001+100+75))
[1] 8.031383e-09
> pgamma(3,0.001+400+260+140,0.001+100+75+50)
[1] 1.572358e-06
> 1-pgamma(3,0.001+400+260+140,0.001+100+75+50)
[1] 0.9999984
> pgamma(3,0.001+400+260+140,0.001+100+75+50)/(1-
pgamma(3,0.001+400+260+140,0.001+100+75+50))
[1] 1.57236e-06
```

> # The probabilities of exceeding the threshold for the 3 years are 1.0, 1.0, and 0.9999984, approximately.
> # The risks for the 3 years are 2.2×10^{-8}, 8.0×10^{-9}, and 1.6×10^{-6}.
> # The odds for the risks for the 3 years are 2.2×10^{-8}, 8.0×10^{-9}, and 1.6×10^{-6}.
> # The odds for the risks are increasing by a factor of $1600/8 = 200$ between years 2 and 3, so there is some indication of possible trouble ahead.

Problem 2.5

Use the formulas for the mean and precision of the posterior, expressed as a function of the mean and precision of the prior and the dataset found in Section 2.4.1. The precision of the posterior is the sum of the precisions of the prior and the dataset; the mean of the posterior is the weighted average of the means of the prior and the dataset, weighted by the respective precisions. If the prior is noninformative, $\tau_{\text{prior}} \cong 0$ and the posterior mean $\cong$ the sample mean. If the sample size is doubled, the precision will be doubled, the variance will be reduced by $\frac{1}{2}$, and the standard deviation will be reduced by $1/\sqrt{2} = 0.707$. If the sample size is quadrupled, the precision will be quadrupled, the variance will be reduced by $\frac{1}{4}$, and the standard deviation will be reduced by $\frac{1}{2}$. Therefore the sample size needs to be quadrupled to half the standard deviation.

Problem 3.1

```
> 1-pbeta(.75,61,41)
[1] 0.0004146221
> 1-pbeta(.75,135,67)
[1] 0.004968324
```

```
> (1-pbeta(.75,135,67))/pbeta(.75,135,67)
[1] 0.004993132
> (1-pbeta(.75,61,41))/pbeta(.75,61,41)
[1] 0.0004147941
> 4993132/414794
[1] 12.03762
> (1-pbeta(.75,135,67))/pbeta(.75,135,67)/((1-pbeta(.75,61,41))/
pbeta(.75,61,41))
[1] 12.03761
```

> # The Bayes factor for the year 2 analysis is 12.03762, providing strong evidence in favor of H_1: p exceeds 75%, even though the posterior probability for H_1 is only 0.005.

Problem 3.2

> # $\Sigma y_i = 30$, 48, and 55 with $n = 50$.

```
> (1-pgamma(1,0.001,0.001)) # prior probability above 1
[1] 0.006312353
> (1-pgamma(1,0.001,0.001))/pgamma(1,0.001,0.001) # prior odds
[1] 0.006352452
> (1-pgamma(1,30.001,50.001)) # 1st posterior probability above 1
[1] 0.0009169443
> (1-pgamma(1,30.001,50.001))/pgamma(1,30.001,50.001) # 1st posterior
odds
[1] 0.0009177859
> 0.0009177859/0.006352452 # 1st Bayes factor
[1] 0.1444774
> (1-pgamma(1,78.001,100.001)) # 2nd posterior probability above 1
[1] 0.01000833
> (1-pgamma(1,78.001,100.001))/pgamma(1,78.001,100.001) # 2nd
posterior odds
[1] 0.01010951
> 0.01010951/0.009177859 # 2nd Bayes factor
[1] 1.101511
> (1-pgamma(1,133.001,150.001)) # 3rd posterior probability above 1
[1] 0.07434242
> (1-pgamma(1,133.001,150.001))/pgamma(1,133.001,150.001) # 3rd
posterior odds
[1] 0.0803131
> 0.0803131/0.01010951 # 3rd Bayes factor
[1] 7.944312
```

> # None of the posterior probabilities above 1 are substantial.
> # Bayes factors indicate substantial evidence for below 1, minimal evidence for above 1, and substantial evidence for above 1.

Problem 3.3

```
> # sen = .80, spec = .95
> # + test with cancer
> # + test, with cancer: sen/(1 − spec)
> .80/.05
[1] 16
> # + test, without cancer: (1 − spec)/sen
> .05/.80
[1] 0.0625
> # − test, with cancer: (1 − sen)/spec
> .20/.95
[1] 0.2105263
> # − test, without cancer: spec/(1 − sen)
> .95/.20
[1] 4.75
```

> # T^+ increases the odds that she has the disease by 16 if she has the disease and decreases the odds that she has the disease by 0.0625 if she doesn't have the disease. > # T^- increases the odds that she doesn't have the disease by 4.75 if she doesn't have the disease and decreases the odds that she doesn't have the disease by 0.2105263 if she doesn't have the disease.

Problem 3.4

The Bayes risks are:

d1: $L_0^+ \cdot (1\text{-specificity}) \cdot \pi_{prior}(D_0) + L_1^- \cdot (1\text{-sensitivity}) \cdot \pi_{prior}(D_1)$

d2: $L_1^- \cdot \pi_{prior}(D_1)$

d3: $L_0^+ \cdot \pi_{prior}(D_0)$

The Bayes rule is to choose the decision d that minimizes risk:

Let $I = [L_0^+/L_1^-] \cdot [\pi(D_0)/\pi(D_1)]$

If (sensitivity+specificity) ≥ 1

  Choose d3 if

    $I \leq (1\text{-sensitivity})/\text{specificity}$

  Choose d1 if

    $(1\text{-sensitivity})/\text{specificity} < I < \text{sensitivity}/(1\text{-specificity})$

  Choose d2 if

    $I \geq \text{sensitivity}/(1\text{-specificity})$

If (sensitivity+specificity) < 1

  Choose d3 if

    $I \leq 1$

  Choose d2 if

    $I > 1$

Problem 3.5

```
> 3000*.05*.90+20000*.20*.10
[1] 535
> 20000*.10
[1] 2000
> >3000*.90
[1] 2700
```

Bayes risks are d1: 535, d2: 2000, and d3: 2700
> # Best decision is d1: take the test and decide on the basis of the test
Note: I = [3000/20000].[.90/.10] = 1.35
    $(1-\text{sensitivity})/\text{specificity} = .20/.95 = 0.2105263$
    $\text{sensitivity}/(1-\text{specificity}) = .80/.05 = 16$
so choose d1

Problem 3.6

Sensitivity is the probability of deciding whether the threshold has not been exceeded, based on the sample monitoring, if the threshold has not been exceeded.

Specificity is the probability of deciding, whether the threshold has been exceeded, based on the sample monitoring, if the threshold has been exceeded.

The decision can be based on the risk of not exceeding the threshold, the posterior cumulative probability of $p \leq p_c$. If this risk is $< 50\%$, the decision is that the threshold has been exceeded. Otherwise, if this risk $\geq 50\%$, the decision is that the threshold has not been exceeded.

Assuming D_0 is the state of exceeding the threshold (i.e., local viability of the owl) and D_1 is the state of not exceeding the threshold (i.e., local extinction of the owl), the loss L_0^+ is the opportunity cost of the timber not harvested caused by reduced harvesting decided from the monitoring results that conclude the threshold has not been exceeded, when the threshold really has been exceeded and the local owl population is viable. The loss L_1^- is the cost of the loss of the local owl population caused by the failure to reduce the harvesting decided from the monitoring results that conclude the threshold has been exceeded, when the threshold really has not been exceeded.

The Bayes rule is to choose the decision d that minimizes risk, as follows: let $I = [L_0^+/L_1^-] \cdot [\pi(D_0)/\pi(D_1)]$.

```
If (sensitivity+specificity) ≥ 1
  Choose d3 if
     I ≤ (1-sensitivity)/specificity
```

```
  Choose d1 if
    (1-sensitivity)/specificity < I < sensitivity/(1-specificity)
  Choose d2 if
     I ≥ sensitivity/(1-specificity)
If (sensitivity+specificity) < 1
  Choose d3 if
     I ≤ 1
  Choose d2 if
     I > 1
```

The Bayes rule suggests that in certain instances the monitoring is not necessary, depending on the relative magnitudes of $I = [L_0^+/L_1^-] \cdot [\pi(D_0)/\pi(D_1)]$ [i.e., ratios (cost of timber)/(loss of owl) and probability (exceeding threshold)/probability (not exceeding threshold)].

Problem 4.1

```
Program code:


          Bayesian Poisson Model
           Non-informative prior
            y ~ Pois(lambda=3.0)
1. Program
model
{
for(i in 1:n){y[i] ~ dpois(lambda)}
lambda ~ dgamma(0.001,0.001)
}

2. Data
list(y = c(4, 2, 2, 4, 5, 1, 2,
2, 2, 4, 5, 6, 0, 0, 1, 3, 3, 4,
3, 2, 3, 3, 4, 1, 4, 6, 4, 4, 1, 3),
n = 30)

3. Initial values
list(lambda = 1)
```

Outputs:

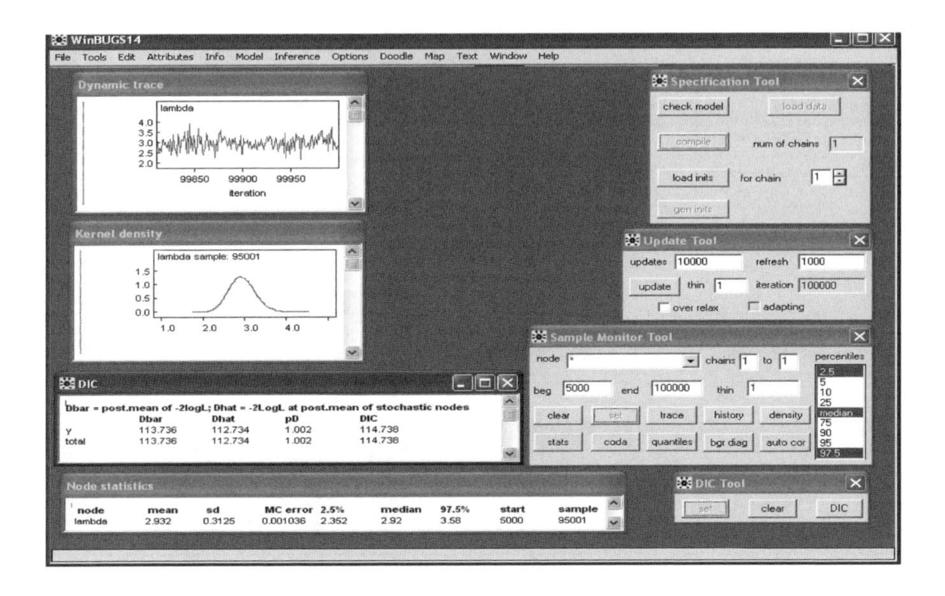
WinBUGS14
File Tools Edit Attributes Info Model Inference Options Doodle Map Text Window Help
Dynamic trace
lambda
99850 99900 99950
iteration
Kernel density
lambda sample: 95001
Specification Tool
check model
load data
compile
num of chains
load inits
for chain
gen inits
Update Tool
updates 10000
refresh 1000
update
thin
iteration 100000
over relax
adapting
Sample Monitor Tool
node
chains 1 to 1
percentiles
beg 5000
end 100000
thin 1
clear set trace history density
stats coda quantiles bgr diag auto cor
DIC
Dbar = post.mean of -2logL; Dhat = -2LogL at post.mean of stochastic nodes
Dbar Dhat pD DIC
y 113.736 112.734 1.002 114.738
total 113.736 112.734 1.002 114.738
Node statistics
node mean sd MC error 2.5% median 97.5% start sample
lambda 2.932 0.3125 0.001036 2.352 2.92 3.58 5000 95001
DIC Tool
set clear DIC

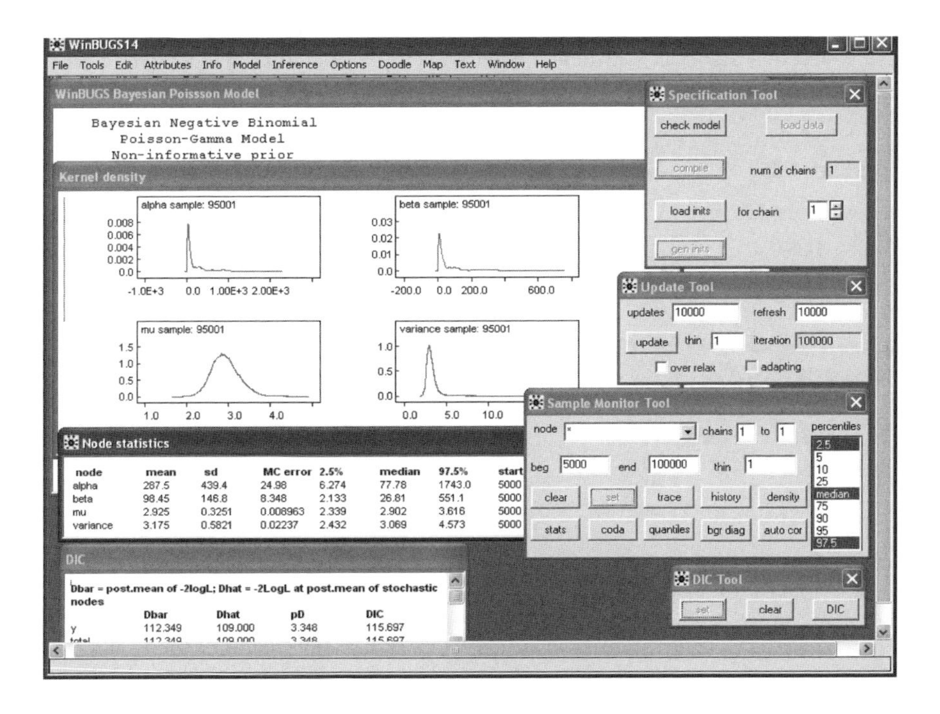
WinBUGS14
File Tools Edit Attributes Info Model Inference Options Doodle Map Text Window Help
WinBUGS Bayesian Poissson Model
Bayesian Negative Binomial
Poisson-Gamma Model
Non-informative prior
Kernel density
alpha sample: 95001
beta sample: 95001
mu sample: 95001
variance sample: 95001
Specification Tool
check model
load data
compile
num of chains
load inits
for chain
gen inits
Update Tool
updates 10000
refresh 10000
update
thin
iteration 100000
over relax
adapting
Sample Monitor Tool
node
chains 1 to 1
percentiles
beg 5000
end 100000
thin 1
clear set trace history density
stats coda quantiles bgr diag auto cor
Node statistics
node mean sd MC error 2.5% median 97.5% start
alpha 287.5 439.4 24.98 6.274 77.78 1743.0 5000
beta 98.45 146.8 8.348 2.133 26.81 551.1 5000
mu 2.925 0.3251 0.008963 2.339 2.902 3.616 5000
variance 3.175 0.5821 0.02237 2.432 3.069 4.573 5000
DIC
Dbar = post.mean of -2logL; Dhat = -2LogL at post.mean of stochastic nodes
Dbar Dhat pD DIC
y 112.349 109.000 3.348 115.697
DIC Tool
set clear DIC

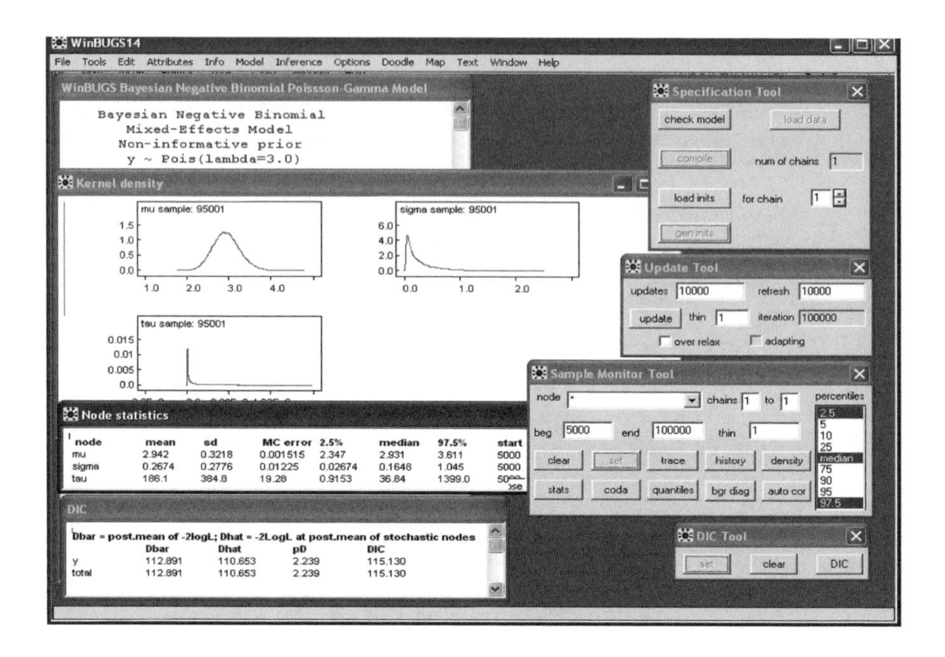

Note that the mean and variance are statistically equal in the Poisson–gamma model results and that sigma is statistically equal to 0 in the mixed-effects model.

Problem 4.2

Program code:

1. Program

```
model
{
for(i in 1:n)
{
y[i] ~ dpois(lambda[i])
log(lambda[i]) <- mu+e[i]
e[i]~ dnorm(0,tau)
}
mu ~ dgamma(0.001,0.001)
tau ~ dgamma(0.001,0.001)
sigma <- 1/sqrt(tau)
}
```

```
Output:
```

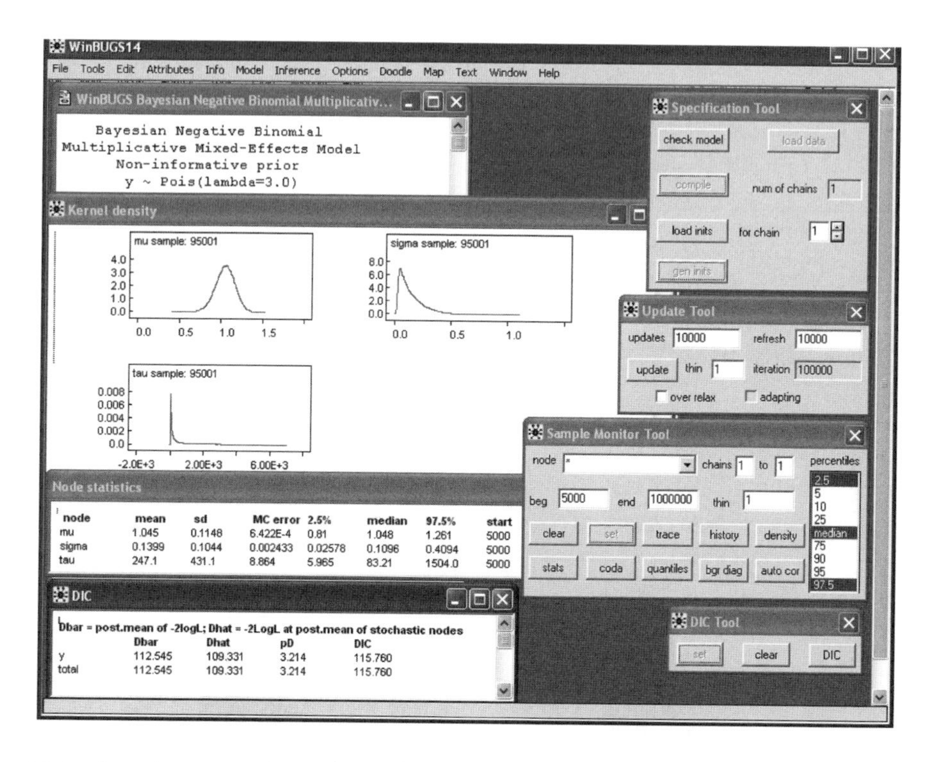

Note that the error sigma is statistically equivalent to 0 in the multiplicative mixed-effects model.

Problem 4.3

```
function(y, k, m, initlambda1, initlambda2)
# Program Gibbs
{
     n <- length(y)
     y1 <- sum(y[1:k])
     y2 <- sum(y[(k+1):n])
     lambda1 <- rep(NA, m)
     lambda2 <- rep(NA, m)
     lambda1[1] <- initlambda1
     lambda2[1] <- initlambda2
     for(i in 2:m) {
          lambda1[i] <- rgamma(1, 0.001 + y1, k)
          lambda2[i] <- rgamma(1, 0.001 + y2, (n - k))
     }
     return(lambda1, lambda2)
```

```
> y
 [1] 4 2 2 4 5 1 2 2 2 4 5 6 0 0 1 3 3 4 3 2 3 3 4 1 4 6 4 4 1 3
> output <- Gibbs(y,20,100000,1,1)
> mean(output[[1]])
[1] 2.749199
> stdev(output[[1]])
[1] 0.3708539
> mean(output[[2]])
[1] 3.298029
> stdev(output[[2]])
[1] 0.5749896
> sum(y[1:20])/20
[1] 2.75
> sum(y[21:30])/10
[1] 3.3
> hist(output[[1]])
> hist(output[[2]])
```

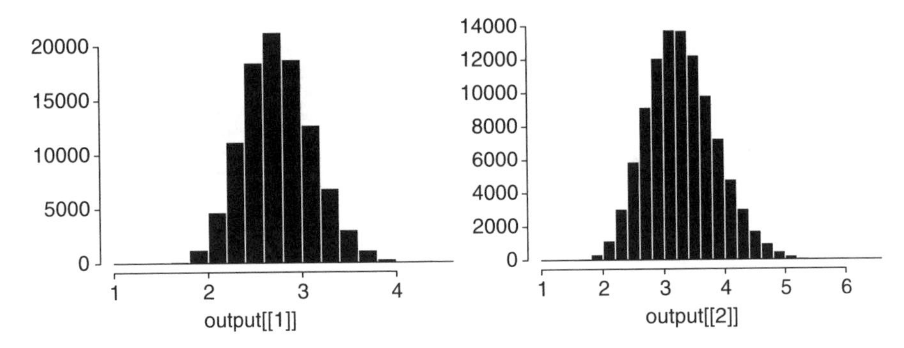

Problem 4.4

```
> dbeta(0,3,2)
[1] 0
> dbeta(.34,3,2)
[1] 0.915552
> # choose p1 = .34
> dbeta(.67,3,2)
[1] 1.777644
> # choose p2 = .67
> dbeta(.52,3,2)
[1] 1.557504
> dbeta(.52,3,2)/dbeta(.67,3,2)
[1] .8761619
> # .91> .87 so choose p3 = p2 = .67
```

Problem 4.5

```
> y <- rnorm(100,15,2)
> y
  [1] 16.85654 19.59268 16.17593 16.44335 15.13185 17.14970
15.69704 17.20139 11.28192
 [10] 16.94269 11.79007 14.35370 17.00237 14.01508 14.81110
15.38694 18.01536 15.59951
 [19] 18.15602 11.71837 18.65446 12.77382 12.47099 12.92969
16.92157 16.96831 14.88682
 [28] 15.96130 15.82643 16.54236 12.81923 14.12794 16.09028
16.81438 14.44691 15.05185
 [37] 19.17883 14.67723 14.62800 14.89441 13.80578 14.68562
15.46704 12.75604 16.55266
 [46] 15.44572 16.95620 14.67055 14.27893 16.83547 19.83804
14.97178 14.31355 17.30323
 [55] 17.13452 14.30720 10.30112 14.03593 15.19265 11.11533
13.54244 17.65803 17.88774
 [64] 16.37850 12.75012 16.72239 14.50407 16.16680 14.70039
12.92720 19.98371 13.56213
 [73] 19.32531 13.71453 13.58275 13.30632 13.07786 14.90847
15.15850 13.49658 12.90805
 [82] 12.28615 14.45550 12.52626 14.45088 12.50655 11.95776
12.53209 19.57162 14.65868
 [91] 13.36388 15.84884 15.96699 14.93516 12.78427 12.97294
14.26660 14.45177 12.60170
[100] 14.89966
> mean(y)
[1] 15.03253
> var(y)
[1] 4.366266
> 100/4 # tau data
[1] 25
> 1/4 # tau prior
[1] 0.25
> 25+.25 # tau posterior - theoretical
[1] 25.25
> (.25/25.25)*10+(25/25.25)*15.03253 # mean posterior -
theoretical
[1] 14.9827
> output1_MCMC(10,2,2,y,.24,15,5000,500000)
> output1[[2]] # rejection rate
[1] 0.3459367
> mean(output1[[1]]) # mean posterior - empirical mean
[1] 14.98291
```

```
> 1/var(output1[[1]]) # tau posterior - empirical precision
[1] 25.56797
```

Problem 4.6

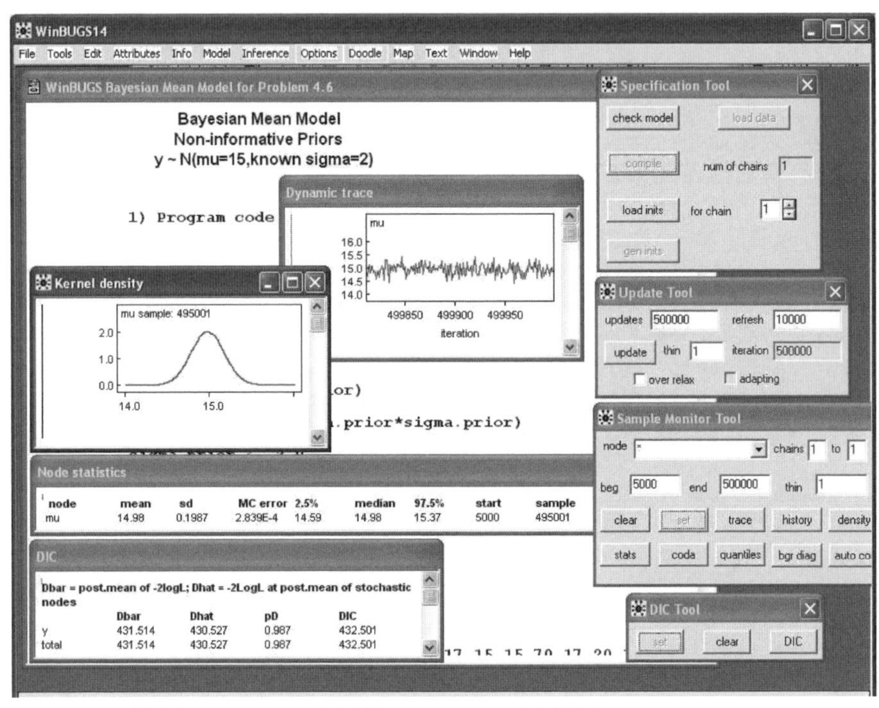

$\mu_{\text{posterior}} = 14.98$, $\sigma_{\text{posterior}} = 0.1987$, $\tau_{\text{posterior}} = 25.3282$.

Problem 5.1

```
> data2
   sample aspect species old.growth rock moss temp moist response
1       1      0       0      0.812    1    1 22.5  12.1     33.6
2       2      1       0      0.564    0    0 22.9  35.2     45.5
3       3      0       0      0.455    0    1 29.7  52.3     67.6
4       4      1       1      0.808    0    0 15.2  15.2     26.9
5       5      0       1      0.225    0    0 17.5  38.7     53.1
6       6      0       0      0.765    0    0 29.8  93.0    105.5
7       7      1       0      0.059    0    0 17.4   2.0     15.6
8       8      1       1      0.038    0    0 20.6  80.5     84.3
9       9      1       0      0.277    0    0 27.9  90.9    102.1
10     10      1       0      0.417    0    1 16.1  63.6     65.9
```

```
11      11      0       0       0.757   0       0 27.8  51.4    62.5
12      12      1       0       0.806   0       0 21.9  96.8   108.7
13      13      1       1       0.015   0       1 21.6  77.9    90.7
14      14      0       0       0.050   0       1 25.3   3.5    20.3
15      15      0       0       0.006   0       1 16.8  49.2    50.8
16      16      0       0       0.646   0       1 15.5  59.6    68.0
17      17      0       1       0.800   0       1 19.9  40.5    48.3
18      18      1       1       0.358   1       0 26.7  81.5    90.8
19      19      1       1       0.151   0       1 18.6  20.3    37.9
20      20      0       0       0.745   0       0 29.1  11.3    25.2
21      21      1       0       0.905   0       1 19.7  93.4   100.2
22      22      0       0       0.206   0       1 16.6  25.0    29.1
23      23      1       1       0.576   0       0 18.4  34.1    40.4
24      24      1       0       0.644   0       1 25.4  46.3    57.7
25      25      1       0       0.501   0       1 22.8  43.0    58.5
26      26      1       0       0.204   0       0 22.0  96.0   105.6
27      27      0       0       0.575   0       1 15.0  33.8    37.7
28      28      1       0       0.777   0       0 25.5  62.5    76.6
29      29      0       0       0.484   0       1 25.0  57.0    67.3
30      30      0       0       0.231   0       1 23.9   4.2    12.9
31      31      1       0       0.489   0       1 20.7  94.5   102.5
32      32      1       1       0.012   0       1 25.4  36.5    55.3
33      33      1       1       0.205   0       0 17.6   7.5    25.6
34      34      0       0       0.238   0       1 22.8  84.6    88.0
35      35      0       0       0.144   0       0 19.7  22.9    26.8
36      36      1       0       0.887   0       0 17.6  54.8    62.9
37      37      0       0       0.001   0       0 15.7  66.8    78.4
38      38      1       1       0.388   0       0 29.1  42.5    57.6
39      39      0       0       0.396   0       1 17.7  72.4    77.1
40      40      0       0       0.751   0       0 18.4   3.8    19.0
41      41      0       0       0.687   0       1 20.9  14.8    27.6
42      42      1       1       0.514   0       0 21.9  52.8    60.3
43      43      1       0       0.371   1       1 27.8  64.2    71.7
44      44      0       0       0.494   0       0 21.5  92.4    95.1
45      45      1       0       0.205   1       0 21.5  22.3    36.3
46      46      0       0       0.919   0       1 29.4  57.9    74.7
47      47      0       0       0.746   1       0 27.6  90.0    98.7
48      48      1       0       0.197   0       0 22.1   5.7    21.1
49      49      1       1       0.655   1       1 26.4  68.1    77.2
50      50      0       0       0.666   1       0 24.8  29.5    44.6
```

```
> output1 <- lm(response~aspect+species+old.growth,data=data2)
>  summary(output1)
Call: lm(formula = response ~ aspect + species + old.growth, data
= data2)
```

```
Residuals:
    Min      1Q    Median     3Q    Max
 -46.32 -22.96  -0.2043 24.86 46.09

Coefficients:
               Value Std. Error  t value Pr(>|t|)
(Intercept)  47.7477    9.2131     5.1826   0.0000
     aspect  13.2737    8.5660     1.5496   0.1281
    species  -6.9754    9.8722    -0.7066   0.4834
 old.growth  15.2475   14.3799     1.0603   0.2945

Residual standard error: 27.87 on 46 degrees of freedom
Multiple R-Squared: 0.07033
F-statistic: 1.16 on 3 and 46 degrees of freedom, the p-value is
0.3353

Correlation of Coefficients:
           (Intercept)   aspect species
    aspect -0.4156
   species -0.2125      -0.3733
old.growth -0.7817       0.0507  0.1606

> output2 <- lm(response~aspect+species+rock,data=data2)
>  summary(output2)

Call: lm(formula = response ~ aspect + species + rock, data =
data2)
Residuals:
    Min      1Q Median     3Q    Max
 -51.81 -24.71 -1.017 23.34 50.76

Coefficients:
               Value Std. Error  t value Pr(>|t|)
(Intercept)  54.7364    5.9762     9.1591   0.0000
     aspect  12.6755    8.6450     1.4662   0.1494
    species  -8.6932    9.8408    -0.8834   0.3816
       rock   5.2043   11.4807     0.4533   0.6525

Residual standard error: 28.14 on 46 degrees of freedom
Multiple R-Squared: 0.05184

F-statistic: 0.8383 on 3 and 46 degrees of freedom, the p-value
is 0.4799
```

```
Correlation of Coefficients:
        (Intercept)  aspect species
 aspect -0.5773
species -0.1352     -0.3864
   rock -0.2390     -0.0351 -0.0082

> output3 <- lm(response~aspect+species+moss,data=data2)
>  summary(output3)

Call: lm(formula = response ~ aspect + species + moss, data =
data2)
Residuals:
    Min     1Q  Median    3Q   Max
 -52.38 -22.36 -0.3957 22.88 50.44

Coefficients:
               Value Std. Error  t value Pr(>|t|)
(Intercept)  55.0630   7.5336     7.3090   0.0000
     aspect  12.9127   8.7850     1.4699   0.1484
    species  -8.6300   9.8699    -0.8744   0.3865
       moss   0.5463   8.1530     0.0670   0.9469

Residual standard error: 28.2 on 46 degrees of freedom
Multiple R-Squared: 0.0477
F-statistic: 0.768 on 3 and 46 degrees of freedom, the p-value is
0.5179

Correlation of Coefficients:
        (Intercept)  aspect species
 aspect -0.5667
species -0.1345     -0.3743
   moss -0.6357      0.1691  0.0401

> output4 <- lm(response~aspect+species+temp,data=data2)
>  summary(output4)

Call: lm(formula = response ~ aspect + species + temp, data =
data2)
Residuals:
    Min     1Q  Median     3Q   Max
 -46.49 -20.13 -0.5598  24.63 41.47
```

```
Coefficients:
                 Value Std. Error  t value Pr(>|t|)
(Intercept)  15.3126  20.4509      0.7488  0.4578
     aspect  12.5388   8.2943      1.5117  0.1374
    species  -6.9314   9.4837     -0.7309  0.4686
       temp   1.7982   0.8830      2.0364  0.0475

Residual standard error: 27.01 on 46 degrees of freedom
Multiple R-Squared: 0.1264
F-statistic: 2.218 on 3 and 46 degrees of freedom, the p-value is
0.09875

Correlation of Coefficients:
        (Intercept)  aspect species
 aspect -0.1488
species -0.1243      -0.3868
   temp -0.9622      -0.0162  0.0893

> output5 <- lm(response~aspect+species+moist,data=data2)
>  summary(output5)

Call: lm(formula = response ~ aspect + species + moist, data =
data2)
Residuals:
    Min      1Q  Median     3Q    Max
 -8.233 -3.005  0.5891  3.31 8.938

Coefficients:
                Value Std. Error  t value  Pr(>|t|)
(Intercept) 13.4517  1.3370      10.0608   0.0000
     aspect  1.4473  1.3807       1.0483   0.3000
    species  0.4119  1.5577       0.2644   0.7926
      moist  0.9264  0.0216      42.8091   0.0000

Residual standard error: 4.414 on 46 degrees of freedom
Multiple R-Squared: 0.9767
F-statistic: 642.2 on 3 and 46 degrees of freedom, the p-value is
0

Correlation of Coefficients:
        (Intercept)  aspect species
 aspect -0.2623
species -0.1949      -0.4024
  moist -0.7326      -0.1923  0.1360
```

```
> output6 <- lm(response~aspect+old.growth+temp,data=data2)
>  summary(output6)

Call: lm(formula = response ~ aspect + old.growth + temp, data =
data2)
Residuals:
    Min     1Q Median     3Q    Max
 -44.11 -19.66   -3.4   25.7  43.62

Coefficients:
               Value Std. Error  t value Pr(>|t|)
(Intercept)  11.1951 20.4572      0.5472  0.5869
     aspect  10.9374  7.6936      1.4216  0.1619
 old.growth  11.3723 14.0432      0.8098  0.4222
       temp   1.7058  0.8977      1.9003  0.0637

Residual standard error: 26.98 on 46 degrees of freedom
Multiple R-Squared: 0.1286
F-statistic: 2.264 on 3 and 46 degrees of freedom, the p-value is
0.09363

Correlation of Coefficients:
           (Intercept)  aspect old.growth
    aspect -0.2279
old.growth -0.1364      0.1193
      temp -0.9047     -0.0053 -0.2064

> output7 <- lm(response~species+old.growth+temp,data=data2)
>  summary(output7)

Call: lm(formula = response ~ species + old.growth + temp, data =
data2)
Residuals:
    Min     1Q Median    3Q   Max
 -49.21 -20.02 -2.366 25.05 48.14

Coefficients:
               Value Std. Error  t value Pr(>|t|)
(Intercept)  18.0175  20.8655     0.8635   0.3923
    species  -0.3860   9.0714    -0.0426   0.9662
 old.growth   8.8792  14.4813     0.6131   0.5428
       temp   1.7105   0.9184     1.8625   0.0689

Residual standard error: 27.57 on 46 degrees of freedom
Multiple R-Squared: 0.09039
```

```
F-statistic: 1.524 on 3 and 46 degrees of freedom, the p-value is
0.221

Correlation of Coefficients:
           (Intercept) species old.growth
   species -0.2206
old.growth -0.1480      0.1797
      temp -0.9177      0.0520 -0.1942

> output8 <- lm(response~moss+temp+moist,data=data2)
>  summary(output8)

Call: lm(formula = response ~ moss + temp + moist, data = data2)
Residuals:
    Min      1Q Median    3Q   Max
 -8.174 -2.975 0.3753 2.285 7.776

Coefficients:
               Value Std. Error  t value Pr(>|t|)
(Intercept)   4.1456   2.9379     1.4111   0.1649
       moss  -0.6874   1.0988    -0.6256   0.5347
       temp   0.5026   0.1294     3.8856   0.0003
      moist   0.9138   0.0191    47.9490   0.0000

Residual standard error: 3.877 on 46 degrees of freedom
Multiple R-Squared: 0.982
F-statistic: 836.7 on 3 and 46 degrees of freedom, the p-value is
0

Correlation of Coefficients:
      (Intercept)    moss    temp
 moss -0.2248
 temp -0.9142      0.0483
moist -0.1091     -0.0056 -0.2141

> output9 <- lm(response~rock+temp,data=data2)
>  summary(output9)

Call: lm(formula = response ~ rock + temp, data = data2)
Residuals:
   Min     1Q Median   3Q   Max
 -50.4 -19.93 -1.012 24.9 49.13
```

```
Coefficients:
               Value Std. Error  t value Pr(>|t|)
(Intercept)  18.7850  20.5933     0.9122   0.3663
       rock  -1.2595  11.6948    -0.1077   0.9147
       temp   1.8625   0.9340     1.9941   0.0520

Residual standard error: 27.39 on 47 degrees of freedom
Multiple R-Squared: 0.08271
F-statistic: 2.119 on 2 and 47 degrees of freedom, the p-value is
0.1315

Correlation of Coefficients:
     (Intercept)    rock
rock  0.2199
temp -0.9792     -0.2985

> output10 <- lm(response~temp+moist,data=data2)
>  summary(output10)

Call: lm(formula = response ~ temp + moist, data = data2)
Residuals:
    Min     1Q Median    3Q   Max
 -7.836 -2.709 0.4237 2.346 7.414

Coefficients:
              Value Std. Error t value Pr(>|t|)
(Intercept)  3.7324  2.8441      1.3123  0.1958
       temp  0.5065  0.1284      3.9460  0.0003
      moist  0.9138  0.0189     48.2598  0.0000

Residual standard error: 3.852 on 47 degrees of freedom
Multiple R-Squared: 0.9819
F-statistic: 1271 on 2 and 47 degrees of freedom, the p-value is
0

Correlation of Coefficients:
      (Intercept)    temp
 temp -0.9281
moist -0.1132     -0.2140

> output11 <- lm(response~rock+moss+temp+moist,data=data2)
>  summary(output11)

Call: lm(formula = response ~ rock + moss + temp + moist, data =
data2)
```

```
Residuals:
    Min      1Q Median    3Q   Max
 -8.159 -2.948  0.395 2.29 7.649

Coefficients:
               Value Std. Error  t value Pr(>|t|)
(Intercept)   4.2009   3.0349     1.3842   0.1731
       rock   0.1485   1.6749     0.0886   0.9298
       moss  -0.6846   1.1113    -0.6160   0.5410
       temp   0.4990   0.1368     3.6472   0.0007
      moist   0.9139   0.0193    47.4231   0.0000

Residual standard error: 3.92 on 45 degrees of freedom
Multiple R-Squared: 0.982
F-statistic: 614 on 4 and 45 degrees of freedom, the p-value is 0

Correlation of Coefficients:
      (Intercept)    rock    moss    temp
 rock  0.2055
 moss -0.2141      0.0284
 temp -0.9155     -0.2942  0.0378
moist -0.1030      0.0180 -0.0051 -0.2098

> output12 <-
lm(response~aspect+species+old.growth+rock+moss+temp+moist,
data=data2)
>  summary(output12)
Call: lm(formula = response ~ aspect + species + old.growth +
rock + moss + temp + moist, data = data2)
Residuals:
    Min      1Q Median    3Q    Max
 -6.644 -2.945 0.1736 2.689 8.034

Coefficients:
               Value Std. Error  t value Pr(>|t|)
(Intercept)   2.4261   3.2115     0.7554   0.4542
     aspect   1.5790   1.2498     1.2634   0.2134
    species   0.8601   1.4014     0.6138   0.5427
 old.growth   1.3029   2.0826     0.6256   0.5349
       rock  -0.0954   1.6831    -0.0567   0.9551
       moss  -0.2597   1.1378    -0.2282   0.8206
       temp   0.5084   0.1388     3.6616   0.0007
      moist   0.9089   0.0198    46.0020   0.0000
```

```
Residual standard error: 3.918 on 42 degrees of freedom
Multiple R-Squared: 0.9832
F-statistic: 351.6 on 7 and 42 degrees of freedom, the p-value is
0

Correlation of Coefficients:
           (Intercept)  aspect species old.growth    rock    moss
temp
    aspect -0.1812
   species -0.1645     -0.3679
old.growth -0.1734      0.0855  0.1408
      rock  0.2240     -0.0390 -0.0447 -0.0845
      moss -0.2632      0.1758  0.0492  0.0672   0.0108
      temp -0.8561      0.0310  0.0501 -0.1380  -0.2826  0.0452
     moist -0.0555     -0.2032  0.1021 -0.0965   0.0296 -0.0395
-0.1924

>
AIC(output1,output2,output3,output4,output5,output6,output7,
output8,output9,output10,output11,output12)
         df      AIC
 output1  5 480.4718
 output2  5 481.4564
 output3  5 481.6743
 output4  5 477.3635
 output5  5 296.1967
 output6  5 477.2330
 output7  5 479.3809
 output8  5 283.2389
 output9  4 477.8016
output10  4 281.6625
output11  6 285.2302
output12  9 287.7390
```

See Fig. B5.9 for a summary of the AIC and AIC weights of the comparative models.

Conclusion: Model 10, with `temp` and `moist`, is the best-fitting. The correct model is model 10 with {`temp`,`moist`}, best fitting the simulated "reality"

$$\texttt{response} \sim \texttt{0.7*temp + 0.9*moist + N(0,4)}.$$

Model 10 analysis provides statistics that are compatible with the "reality."

Figure B5.9. AIC, AIC weight, DIC, and DIC weights for Problems 5.1, 5.3, and 5.4.

Model	Covariates	k	p_D	AIC		Akaike Weights	DIC		DIC Weights
1	aspect, species, old.growth	4	3.914	82.056		<0.00005%	81.550		<0.00005%
2	aspect, species, rock	4	4.104	144.059		<0.00005%	143.858		<0.00005%
3	aspect, species, moss	4	4.086	143.247		<0.00005%	143.007		<0.00005%
4	aspect, species, moist	4	4.067	133.115		<0.00005%	132.863		<0.00005%
5	aspect, species, temp	4	4.083	144.237		<0.00005%	144.008		<0.00005%
6	aspect, old.growth, temp	4	3.791	44.732		29.1853%	44.420		29.5940%
7	rock, temp, moist	4	4.078	132.743		<0.00005%	132.518		<0.00005%
8	moss, temp, moist	4	4.062	132.798		<0.00005%	132.536		<0.00005%
9	old.growth, temp	3	2.846	42.985		69.9167%	42.781		67.1595%
10	rock, moist	3	3.066	142.755		<0.00005%	142.644		<0.00005%
11	rock, moss, temp, moist	5	5.111	134.408		<0.00005%	134.044		<0.00005%
12	aspect, species, old.growth, rock, moss, temp, moist	8	7.329	51.695		0.8981%	48.840		3.2465%
				Total:		100.0000%	*Total:*		100.0000%

Note: k = number of parameters

p_D = Bayesian number of parameters

Problem 5.2

```
> #  Stepwise multiple linear regression analysis in S-Plus and R
> object0 <- lm(response~1,data2)
> scope <- ".~.+aspect+species+old.growth+rock+moss+temp+moist"
> step(object0,scope,direction="forward")
Start:  AIC= 39994.97
 response ~ 1

Single term additions

Model:
response ~ 1

scale:  784.2151

           Df Sum of Sq      RSS       Cp
   <none>                38426.54 39994.97
    aspect  1   1216.28 37210.26 40347.12
   species  1     87.11 38339.43 41476.29
old.growth  1    822.78 37603.76 40740.62
      rock  1    195.75 38230.79 41367.65
      moss  1     16.86 38409.68 41546.54
      temp  1   3169.38 35257.16 38394.02
     moist  1  37498.06   928.48  4065.34

Step:  AIC= 4065.339
 response ~ moist

Single term additions

Model:
response ~ moist

scale:  784.2151

           Df Sum of Sq      RSS       Cp
   <none>                928.4784 4065.339
    aspect  1   30.9901 897.4883 5602.779
   species  1   10.9459 917.5324 5622.823
old.growth  1   16.9201 911.5583 5616.849
      rock  1   23.9238 904.5545 5609.845
      moss  1    9.9676 918.5108 5623.801
      temp  1  231.0531 697.4253 5402.716
Call:
lm(formula = response ~ moist, data = data2)
```

```
Coefficients:
 (Intercept)     moist
    14.14828 0.9297645
Degrees of freedom: 50 total; 48 residual
Residual standard error (on weighted scale): 4.398102
```

> # The stepwise mlr best fitting model is {moist}.
> # Best subsets multiple linear regression in S-Plus and R

```
> models <- list(rep(NA,length(covariates)))
> for (i in 1:length(covariates)) {models[[i]] <-
lm(paste("response~",covariates[i]),data2)}
> aic <- rep(NA,length(covariates))
> for (i in 1:length(covariates)) {aic[i] <-
AIC(lm(paste("response~",covariates[i]),data2))}
> min(aic)
[1] 280.6349
> for (i in 1:length(covariates)) {if (aic[i]==min(aic)) k <- i}
> k
[1] 43
> covariates[k]
[1] "aspect+temp+moist"
> summary(models[[k]])

Call: lm(formula = paste("response~", covariates[i]), data =
data2)
Residuals:
    Min     1Q  Median    3Q   Max
 -6.987 -2.947 0.01178 2.677 8.176

Coefficients:
              Value Std. Error t value Pr(>|t|)
(Intercept) 2.7712  2.8462      0.9736  0.3353
     aspect 1.8359  1.0835      1.6945  0.0969
       temp 0.5181  0.1261      4.1098  0.0002
      moist 0.9087  0.0188     48.3091  0.0000

Residual standard error: 3.778 on 46 degrees of freedom
Multiple R-Squared: 0.9829
F-statistic: 882.2 on 3 and 46 degrees of freedom, the p-value is
0

Correlation of Coefficients:
       (Intercept)  aspect    temp
aspect -0.1993
  temp -0.9190      0.0543
 moist -0.0777     -0.1597 -0.2197
```

> # The best subsets selection mlr best fitting model is {aspect, temp, moist}.
> # Conclusion:
(i) stepwise mlr - model with {moist} is best fitting
(ii) best subsets mlr - model with {aspect, temp, and moist} is best fitting

```
Note:
> outputa <- lm(response~moist,data2)
> outputb <- lm(response~aspect+temp+moist,data2)
> AIC(outputa,outputb,output10)
         df      AIC
 outputa  3 293.9701
 outputb  5 280.6349
output10  4 281.6625
```

> # The correct model is Model #10 with {`temp`,`moist`}, fitting the "reality:"

```
response ~ 0.7*temp + 0.9*moist + N(0,4)
```

> # Best subsets selection slightly overfit the sample dataset, with an additional specious covariate aspect.
> # AIC_c might provide the proper ranking of the models.
> # Note that {`aspect`, `temp`, `moist`} was not among the parsimonious collection of models considered with the a priori strategy of model selection and inference used in Problem 5.1.

Problem 5.3

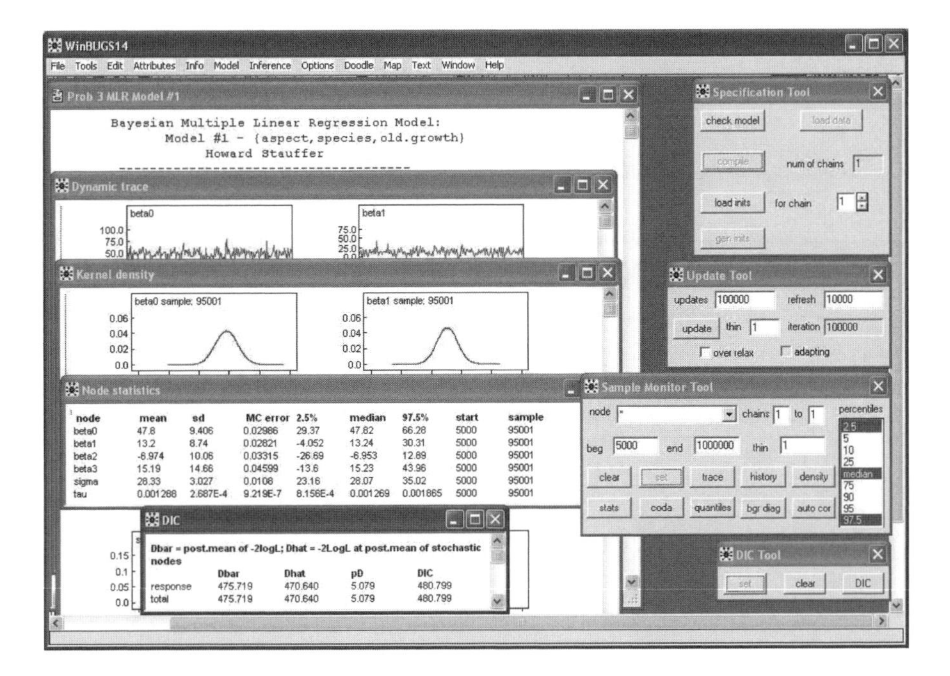

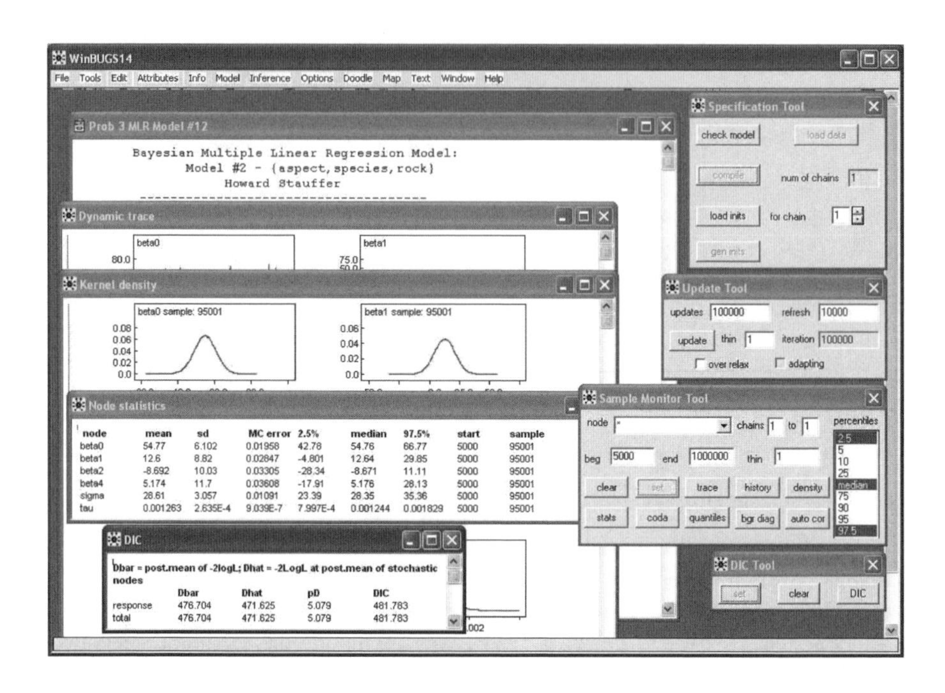
WinBUGS14
File Tools Edit Attributes Info Model Inference Options Doodle Map Text Window Help
Prob 3 MLR Model #12
Bayesian Multiple Linear Regression Model:
Model #2 - (aspect, species, rock)
Howard Stauffer
Dynamic trace
Kernel density
beta0 sample: 95001
beta1 sample: 95001
Node statistics
node mean sd MC error 2.5% median 97.5% start sample
beta0 54.77 6.102 0.01958 42.78 54.76 66.77 5000 95001
beta1 12.6 8.82 0.02847 -4.801 12.64 29.85 5000 95001
beta2 -8.692 10.03 0.03305 -28.34 -8.671 11.11 5000 95001
beta4 5.174 11.7 0.03608 -17.91 5.176 28.13 5000 95001
sigma 28.61 3.057 0.01091 23.39 28.35 35.36 5000 95001
tau 0.001263 2.635E-4 9.039E-7 7.997E-4 0.001244 0.001829 5000 95001
DIC
Dbar = post.mean of -2logL; Dhat = -2LogL at post.mean of stochastic nodes
Dbar Dhat pD DIC
response 476.704 471.625 5.079 481.783
total 476.704 471.625 5.079 481.783
Specification Tool
check model
load data
compile
num of chains 1
load inits
for chain 1
gen inits
Update Tool
updates 100000
refresh 10000
update
thin 1
iteration 100000
over relax
adapting
Sample Monitor Tool
node *
chains 1 to 1
percentiles
beg 5000
end 1000000
thin 1
clear
set
trace
history
density
stats
coda
quantiles
bgr diag
auto cor
DIC Tool
set
clear
DIC

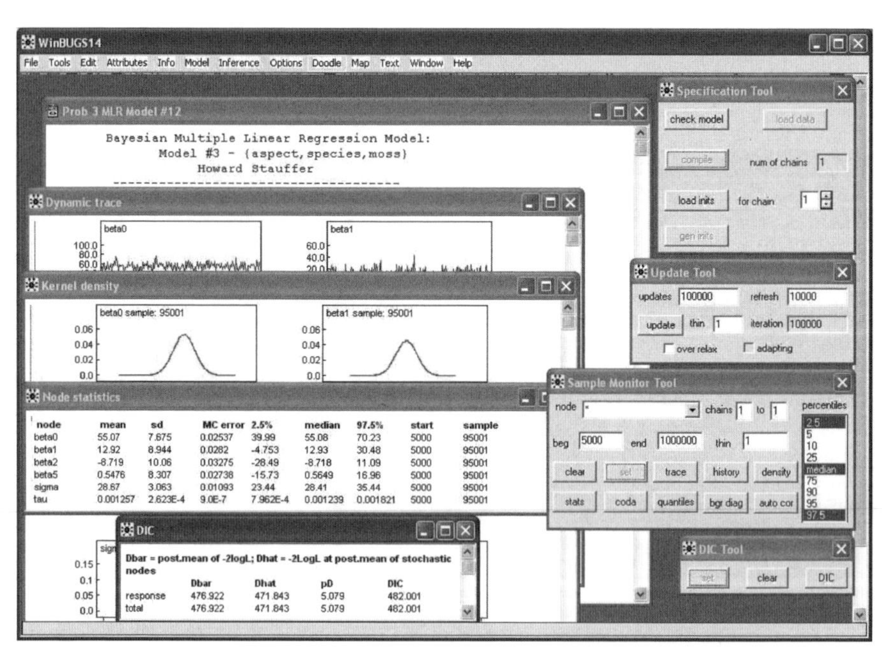
WinBUGS14
File Tools Edit Attributes Info Model Inference Options Doodle Map Text Window Help
Prob 3 MLR Model #12
Bayesian Multiple Linear Regression Model:
Model #3 - (aspect, species, moss)
Howard Stauffer
Dynamic trace
Kernel density
beta0 sample: 95001
beta1 sample: 95001
Node statistics
node mean sd MC error 2.5% median 97.5% start sample
beta0 55.07 7.675 0.02537 39.99 55.08 70.23 5000 95001
beta1 12.92 8.944 0.0282 -4.753 12.93 30.48 5000 95001
beta2 -8.719 10.06 0.03275 -28.49 -8.718 11.09 5000 95001
beta5 0.5476 8.307 0.02738 -15.73 0.5649 16.96 5000 95001
sigma 28.67 3.063 0.01093 23.44 28.41 35.44 5000 95001
tau 0.001257 2.623E-4 9.0E-7 7.962E-4 0.001239 0.001821 5000 95001
DIC
Dbar = post.mean of -2logL; Dhat = -2LogL at post.mean of stochastic nodes
Dbar Dhat pD DIC
response 476.922 471.843 5.079 482.001
total 476.922 471.843 5.079 482.001
Specification Tool
check model
load data
compile
num of chains 1
load inits
for chain 1
gen inits
Update Tool
updates 100000
refresh 10000
update
thin 1
iteration 100000
over relax
adapting
Sample Monitor Tool
node *
chains 1 to 1
percentiles
beg 5000
end 1000000
thin 1
clear
set
trace
history
density
stats
coda
quantiles
bgr diag
auto cor
DIC Tool
set
clear
DIC

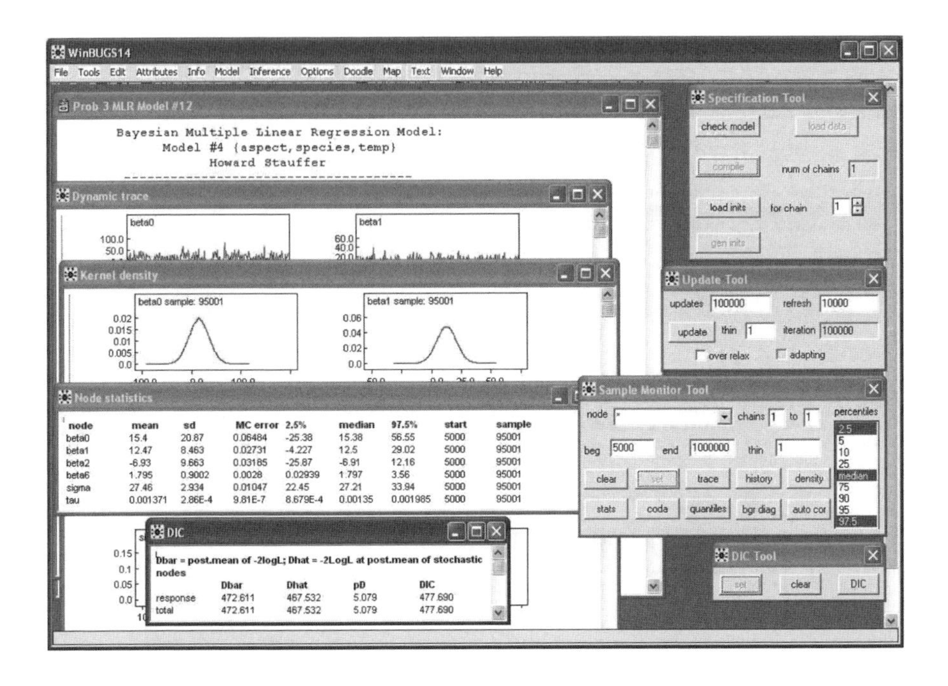
WinBUGS14
File Tools Edit Attributes Info Model Inference Options Doodle Map Text Window Help
Prob 3 MLR Model #12
Bayesian Multiple Linear Regression Model:
Model #4 {aspect,species,temp}
Howard Stauffer
Dynamic trace
Kernel density
beta0 sample: 95001
beta1 sample: 95001
Node statistics
node mean sd MC error 2.5% median 97.5% start sample
beta0 15.4 20.87 0.06484 -25.38 15.38 56.55 5000 95001
beta1 12.47 8.463 0.02731 -4.227 12.5 29.02 5000 95001
beta2 -6.93 9.663 0.03185 -25.87 -6.91 12.16 5000 95001
beta6 1.795 0.9002 0.0028 0.02939 1.797 3.56 5000 95001
sigma 27.46 2.934 0.01047 22.45 27.21 33.94 5000 95001
tau 0.001371 2.86E-4 9.81E-7 8.679E-4 0.00135 0.001985 5000 95001
DIC
Dbar = post.mean of -2logL; Dhat = -2LogL at post.mean of stochastic nodes
Dbar Dhat pD DIC
response 472.611 467.532 5.079 477.690
total 472.611 467.532 5.079 477.690
Specification Tool
check model
load data
compile
num of chains 1
load inits
for chain 1
gen inits
Update Tool
updates 100000
refresh 10000
update
thin 1
iteration 100000
over relax
adapting
Sample Monitor Tool
node *
chains 1 to 1
percentiles
beg 5000
end 1000000
thin 1
clear set trace history density
stats coda quantiles bgr diag auto cor
DIC Tool
set clear DIC

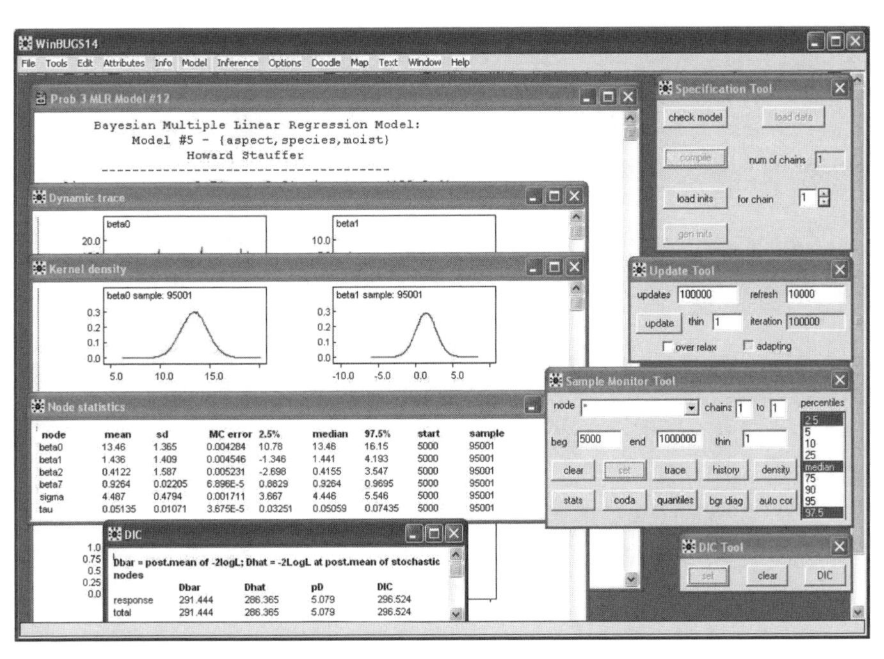
WinBUGS14
File Tools Edit Attributes Info Model Inference Options Doodle Map Text Window Help
Prob 3 MLR Model #12
Bayesian Multiple Linear Regression Model:
Model #5 - {aspect,species,moist}
Howard Stauffer
Dynamic trace
Kernel density
beta0 sample: 95001
beta1 sample: 95001
Node statistics
node mean sd MC error 2.5% median 97.5% start sample
beta0 13.46 1.365 0.004284 10.78 13.46 16.15 5000 95001
beta1 1.436 1.409 0.004546 -1.346 1.441 4.193 5000 95001
beta2 0.4122 1.587 0.005231 -2.698 0.4155 3.547 5000 95001
beta7 0.9264 0.02205 6.896E-5 0.8829 0.9264 0.9695 5000 95001
sigma 4.487 0.4794 0.001711 3.667 4.446 5.546 5000 95001
tau 0.05135 0.01071 3.675E-5 0.03251 0.05059 0.07435 5000 95001
DIC
Dbar = post.mean of -2logL; Dhat = -2LogL at post.mean of stochastic nodes
Dbar Dhat pD DIC
response 291.444 286.365 5.079 296.524
total 291.444 286.365 5.079 296.524
Specification Tool
check model
load data
compile
num of chains 1
load inits
for chain 1
gen inits
Update Tool
updates 100000
refresh 10000
update
thin 1
iteration 100000
over relax
adapting
Sample Monitor Tool
node *
chains 1 to 1
percentiles
beg 5000
end 1000000
thin 1
clear set trace history density
stats coda quantiles bgr diag auto cor
DIC Tool
set clear DIC

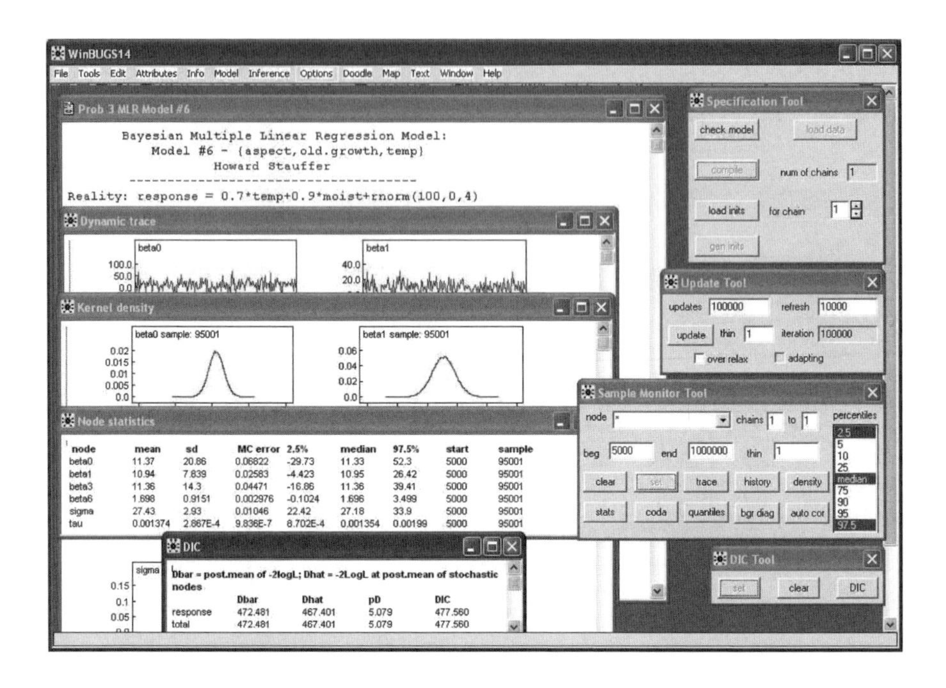
WinBUGS14
File Tools Edit Attributes Info Model Inference Options Doodle Map Text Window Help
Prob 3 MLR Model #6
Bayesian Multiple Linear Regression Model:
Model #6 - {aspect,old.growth,temp}
Howard Stauffer
Reality: response = 0.7*temp+0.9*moist+rnorm(100,0,4)
Dynamic trace
Kernel density
beta0 sample: 95001
beta1 sample: 95001
Node statistics
node mean sd MC error 2.5% median 97.5% start sample
beta0 11.37 20.86 0.06822 -29.73 11.33 52.3 5000 95001
beta1 10.94 7.839 0.02583 -4.423 10.95 26.42 5000 95001
beta3 11.36 14.3 0.04471 -16.86 11.36 39.41 5000 95001
beta6 1.698 0.9151 0.002976 -0.1024 1.696 3.499 5000 95001
sigma 27.43 2.93 0.01046 22.42 27.18 33.9 5000 95001
tau 0.001374 2.867E-4 9.836E-7 8.702E-4 0.001354 0.00199 5000 95001
DIC
Dbar = post.mean of -2logL; Dhat = -2LogL at post.mean of stochastic nodes
Dbar Dhat pD DIC
response 472.481 467.401 5.079 477.560
total 472.481 467.401 5.079 477.560
Specification Tool
check model
load data
compile
num of chains
load inits
for chain
gen inits
Update Tool
updates 100000 refresh 10000
update thin 1 iteration 100000
over relax adapting
Sample Monitor Tool
node * chains 1 to 1 percentiles
beg 5000 end 1000000 thin 1
clear set trace history density
stats coda quantiles bgr diag auto cor
DIC Tool
set clear DIC

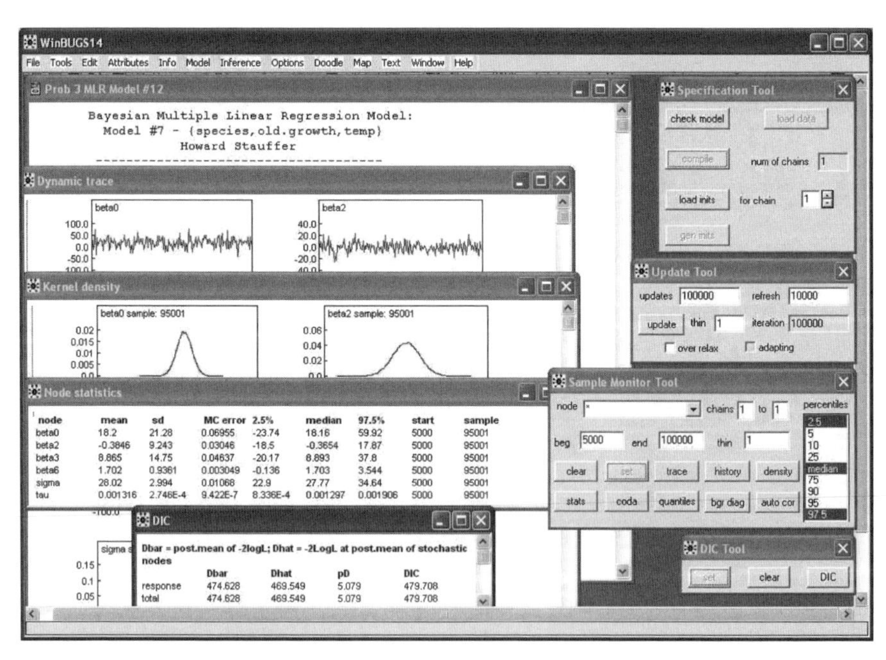
WinBUGS14
File Tools Edit Attributes Info Model Inference Options Doodle Map Text Window Help
Prob 3 MLR Model #12
Bayesian Multiple Linear Regression Model:
Model #7 - {species,old.growth,temp}
Howard Stauffer
Dynamic trace
Kernel density
beta0 sample: 95001
beta2 sample: 95001
Node statistics
node mean sd MC error 2.5% median 97.5% start sample
beta0 18.2 21.28 0.06955 -23.74 18.16 59.92 5000 95001
beta2 -0.3846 9.243 0.03046 -18.5 -0.3654 17.87 5000 95001
beta3 8.865 14.75 0.04637 -20.17 8.893 37.8 5000 95001
beta6 1.702 0.9361 0.003049 -0.136 1.703 3.544 5000 95001
sigma 28.02 2.994 0.01068 22.9 27.77 34.64 5000 95001
tau 0.001316 2.746E-4 9.422E-7 8.336E-4 0.001297 0.001906 5000 95001
DIC
Dbar = post.mean of -2logL; Dhat = -2LogL at post.mean of stochastic nodes
Dbar Dhat pD DIC
response 474.628 469.549 5.079 479.708
total 474.628 469.549 5.079 479.708
Specification Tool
check model
load data
compile
num of chains
load inits
for chain
gen inits
Update Tool
updates 100000 refresh 10000
update thin 1 iteration 100000
over relax adapting
Sample Monitor Tool
node * chains 1 to 1 percentiles
beg 5000 end 100000 thin 1
clear set trace history density
stats coda quantiles bgr diag auto cor
DIC Tool
set clear DIC

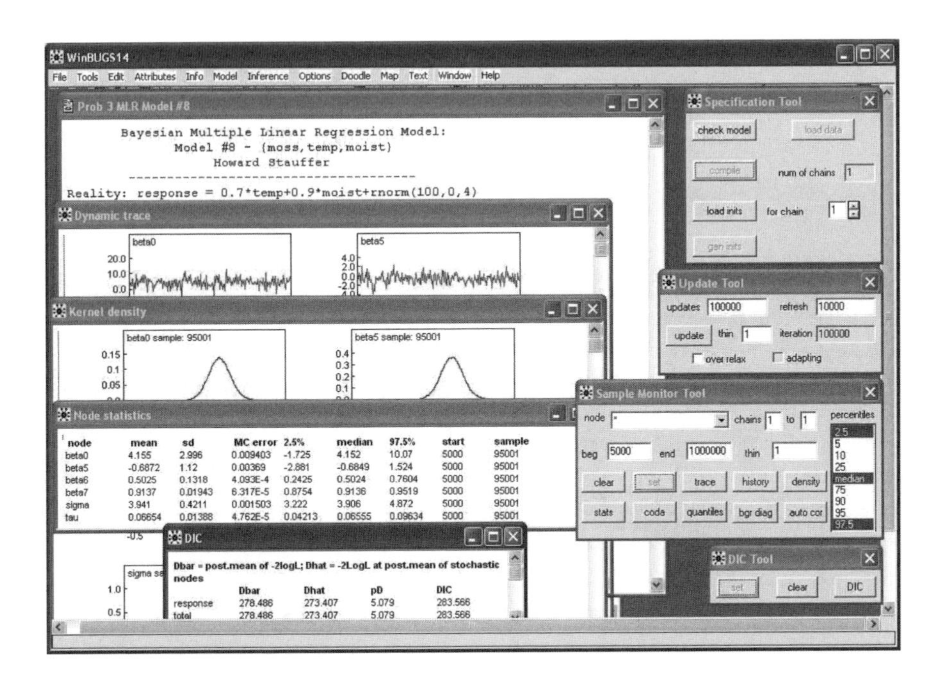
WinBUGS14
File Tools Edit Attributes Info Model Inference Options Doodle Map Text Window Help
Prob 3 MLR Model #8
Bayesian Multiple Linear Regression Model:
Model #8 - {moss,temp,moist}
Howard Stauffer
Reality: response = 0.7*temp+0.9*moist+rnorm(100,0,4)
Dynamic trace
beta0
beta5
Kernel density
beta0 sample: 95001
beta5 sample: 95001
Node statistics
node mean sd MC error 2.5% median 97.5% start sample
beta0 4.155 2.996 0.009403 -1.725 4.152 10.07 5000 95001
beta5 -0.6872 1.12 0.00369 -2.881 -0.6849 1.524 5000 95001
beta6 0.5025 0.1318 4.093E-4 0.2425 0.5024 0.7604 5000 95001
beta7 0.9137 0.01943 6.317E-5 0.8754 0.9136 0.9519 5000 95001
sigma 3.941 0.4211 0.001503 3.222 3.906 4.872 5000 95001
tau 0.06654 0.01388 4.762E-5 0.04213 0.06555 0.09634 5000 95001
DIC
Dbar = post.mean of -2logL; Dhat = -2LogL at post.mean of stochastic nodes
Dbar Dhat pD DIC
response 278.486 273.407 5.079 283.566
total 278.486 273.407 5.079 283.566
Specification Tool
check model
load data
compile
num of chains 1
load inits
for chain 1
gen inits
Update Tool
updates 100000
refresh 10000
update
thin 1
iteration 100000
over relax
adapting
Sample Monitor Tool
node *
chains 1 to 1
percentiles
beg 5000
end 1000000
thin 1
clear set trace history density
stats coda quantiles bgr diag auto cor
2.5 5 10 25 median 75 90 95 97.5
DIC Tool
set clear DIC

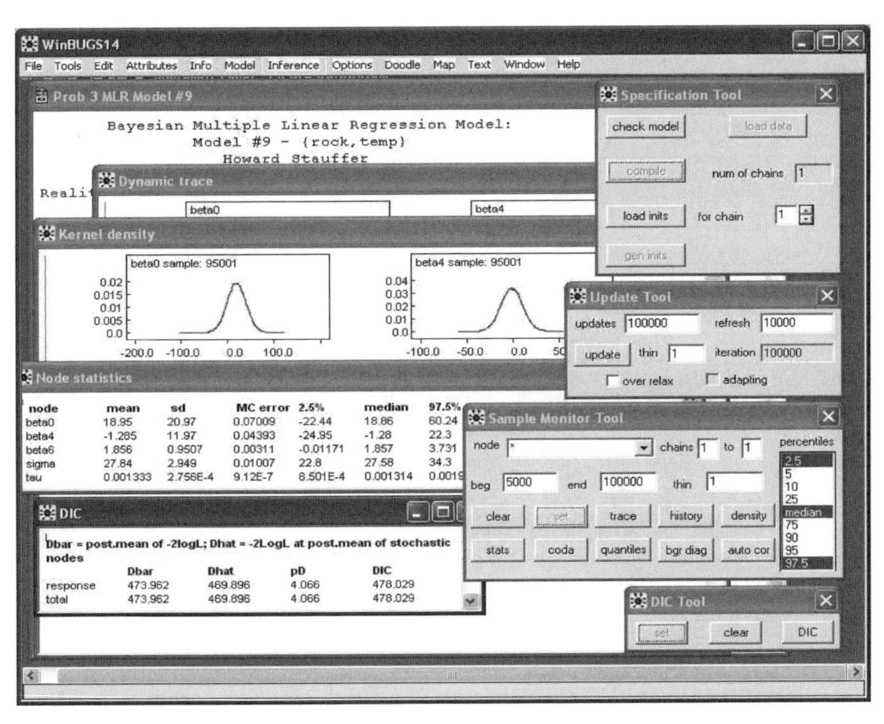
WinBUGS14
File Tools Edit Attributes Info Model Inference Options Doodle Map Text Window Help
Prob 3 MLR Model #9
Bayesian Multiple Linear Regression Model:
Model #9 - {rock,temp}
Howard Stauffer
Dynamic trace
beta0
beta4
Kernel density
beta0 sample: 95001
beta4 sample: 95001
Node statistics
node mean sd MC error 2.5% median 97.5%
beta0 18.95 20.97 0.07009 -22.44 18.86 60.24
beta4 -1.285 11.97 0.04393 -24.95 -1.28 22.3
beta6 1.856 0.9507 0.00311 -0.01171 1.857 3.731
sigma 27.84 2.949 0.01007 22.8 27.58 34.3
tau 0.001333 2.756E-4 9.12E-7 8.501E-4 0.001314
DIC
Dbar = post.mean of -2logL; Dhat = -2LogL at post.mean of stochastic nodes
Dbar Dhat pD DIC
response 473.962 469.896 4.066 478.029
total 473.962 469.896 4.066 478.029
Specification Tool
check model
load data
compile
num of chains 1
load inits
for chain 1
gen inits
Update Tool
updates 100000
refresh 10000
update
thin 1
iteration 100000
over relax
adapting
Sample Monitor Tool
node *
chains 1 to 1
percentiles
beg 5000
end 100000
thin 1
clear set trace history density
stats coda quantiles bgr diag auto cor
2.5 5 10 25 median 75 90 95 97.5
DIC Tool
set clear DIC

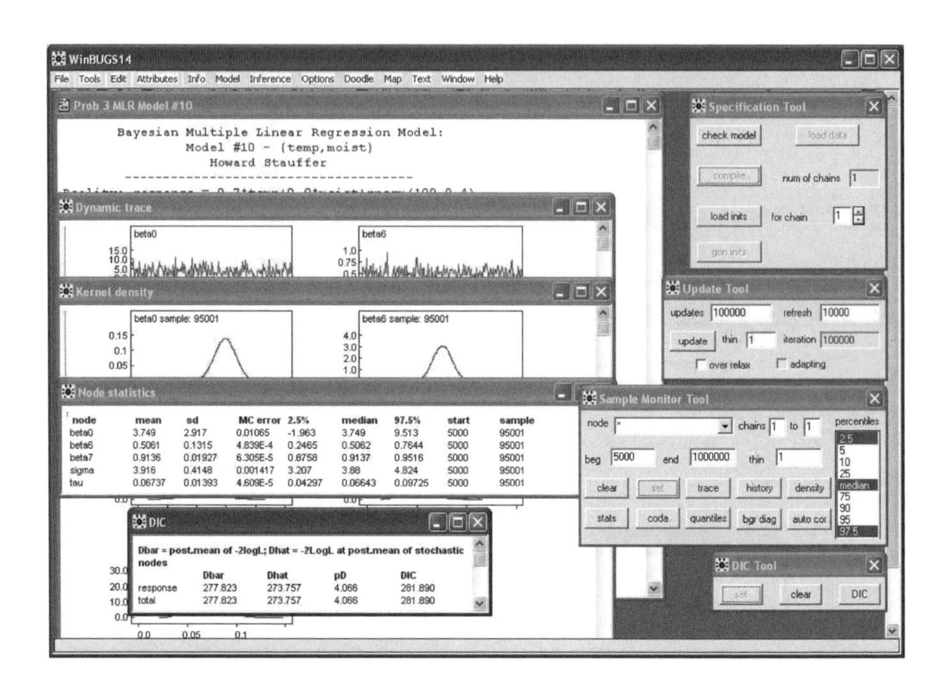

WinBUGS14
File Tools Edit Attributes Info Model Inference Options Doodle Map Text Window Help
Prob 3 MLR Model #10
Bayesian Multiple Linear Regression Model:
Model #10 - (temp,moist)
Howard Stauffer
Dynamic trace
beta0
beta6
Kernel density
beta0 sample: 95001
beta6 sample: 95001
Node statistics
node mean sd MC error 2.5% median 97.5% start sample
beta0 3.749 2.917 0.01065 -1.963 3.749 9.513 5000 95001
beta6 0.5061 0.1315 4.839E-4 0.2465 0.5062 0.7644 5000 95001
beta7 0.9136 0.01927 6.305E-5 0.8758 0.9137 0.9516 5000 95001
sigma 3.916 0.4148 0.001417 3.207 3.88 4.824 5000 95001
tau 0.06737 0.01393 4.609E-5 0.04297 0.06643 0.09725 5000 95001
DIC
Dbar = post.mean of -2logL; Dhat = -2LogL at post.mean of stochastic nodes
Dbar Dhat pD DIC
response 277.823 273.757 4.066 281.890
total 277.823 273.757 4.066 281.890
Specification Tool
check model
load data
compile
num of chains 1
load inits
for chain 1
gen inits
Update Tool
updates 100000
refresh 10000
update
thin 1
iteration 100000
over relax
adapting
Sample Monitor Tool
node *
chains 1 to 1
percentiles
beg 5000
end 1000000
thin 1
clear set trace history density
stats coda quantiles bgr diag auto cor
2.5 5 10 25 median 75 90 95 97.5
DIC Tool
set clear DIC

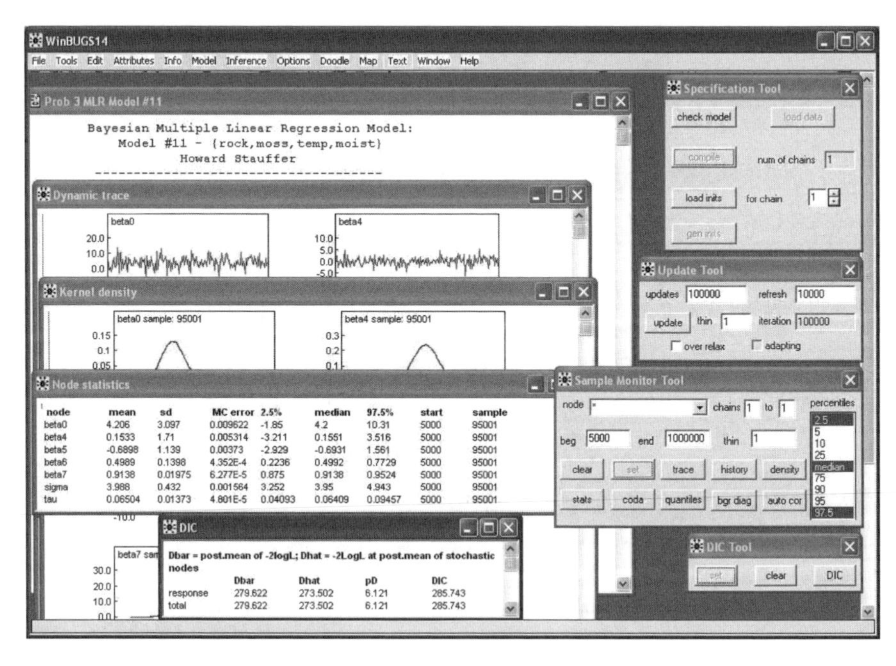

WinBUGS14
File Tools Edit Attributes Info Model Inference Options Doodle Map Text Window Help
Prob 3 MLR Model #11
Bayesian Multiple Linear Regression Model:
Model #11 - (rock,moss,temp,moist)
Howard Stauffer
Dynamic trace
beta0
beta4
Kernel density
beta0 sample: 95001
beta4 sample: 95001
Node statistics
node mean sd MC error 2.5% median 97.5% start sample
beta0 4.206 3.097 0.009622 -1.85 4.2 10.31 5000 95001
beta4 0.1533 1.71 0.005314 -3.211 0.1551 3.516 5000 95001
beta5 -0.6898 1.139 0.00373 -2.929 -0.6931 1.561 5000 95001
beta6 0.4989 0.1398 4.352E-4 0.2236 0.4992 0.7729 5000 95001
beta7 0.9138 0.01975 6.277E-5 0.875 0.9138 0.9524 5000 95001
sigma 3.988 0.432 0.001564 3.252 3.95 4.943 5000 95001
tau 0.06504 0.01373 4.801E-5 0.04093 0.06409 0.09457 5000 95001
DIC
Dbar = post.mean of -2logL; Dhat = -2LogL at post.mean of stochastic nodes
Dbar Dhat pD DIC
response 279.622 273.502 6.121 285.743
total 279.622 273.502 6.121 285.743
Specification Tool
check model
load data
compile
num of chains 1
load inits
for chain 1
gen inits
Update Tool
updates 100000
refresh 10000
update
thin 1
iteration 100000
over relax
adapting
Sample Monitor Tool
node *
chains 1 to 1
percentiles
beg 5000
end 1000000
thin 1
clear set trace history density
stats coda quantiles bgr diag auto cor
2.5 5 10 25 median 75 90 95 97.5
DIC Tool
set clear DIC

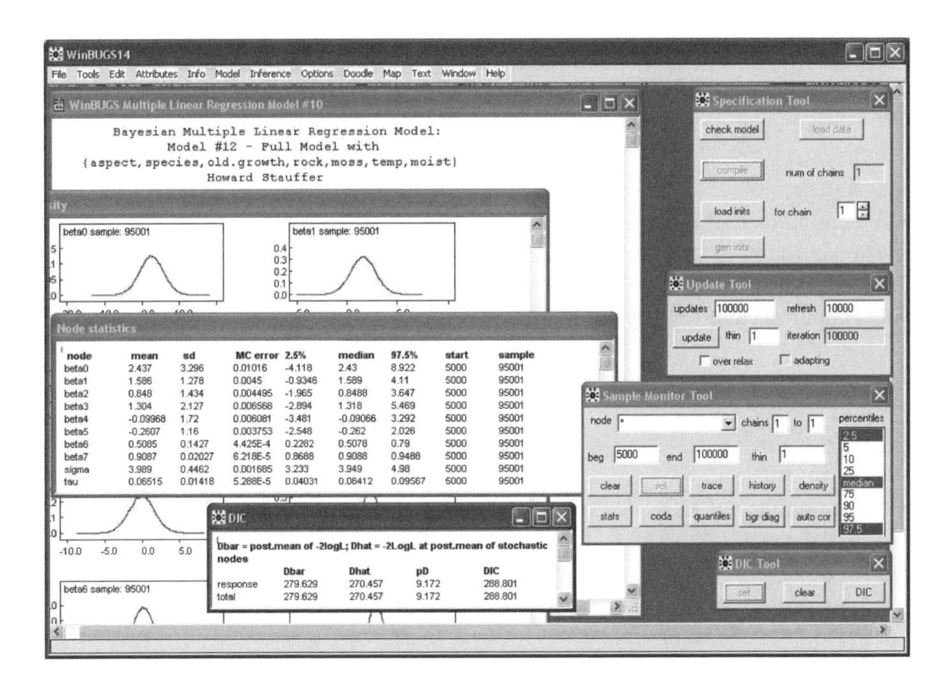

See Fig. B5.9 for a summary of the DIC and DIC weights of the comparative models.

Conclusion: The Bayesian DIC values are similar to the frequentist AIC values in Problem 5.1 with the same model ranking. Model 10 is the best-fitting model with the Bayesian statistical analysis with DIC = 281.890.

Problem 5.4

> # The unconditional shrinkage estimates of the coefficients of temp (beta6) and moist (beta7), using AIC weights, are:

> # unconditional shrinkage estimates of beta6 using AIC weights

```
> .2720*.5026+.5983*.5065+.1005*.4990+.0287*.5084
[1] 0.5044867
```

> # unconditional shrinkage estimates of beta6 se using AIC weights

```
> .2720*.1294+.5983*.1284+.1005*.1368+.0287*.1388
[1] 0.1297505
```

> # unconditional shrinkage estimates of beta7 using AIC weights

```
> .0004*.9264+.2720*.9138+.5983*.9138+.1005*.9139+.0287*.9089
[1] 0.913583
```

> # unconditional shrinkage estimates of beta7 se using AIC weights

```
> .0004*.0216+.2720*.0191+.5983*.0189+.1005*.0193+.0287*.0198
[1] 0.01901962
```

> # The results without shrinkage are very similar to those with shrinkage since the AIC weights for the models without temp and moist are negligible.

> # The results, using DIC weights, are very similar to those using AIC weights since the weights are similar and the model coefficient estimates are similar.
> # The importance of the covariates, using AIC weights are:
> # importance of aspect

```
> .0004+.0287
[1] 0.0291
```

> # importance of species

```
> .0004+.0287
[1] 0.0291
```

> # importance of old.growth

```
> .0287
[1] 0.0287
```

> # importance of rock

```
> .1005+.0287
[1] 0.1292
```

> # importance of moss

```
> .2720+.1005+.0287
[1] 0.4012
```

> # importance of temp

```
> .2720+.5983+.1005+.0287
[1] 0.9995
```

> # importance of moist

```
> .0004+.2720+.5983+.1005+.0287
[1] 0.9999
```

> # The importance of the covariates, using DIC weights are very similar, since the AIC and DIC weights are similar.

Problem 5.5

> # Recall, the simulated "reality" is
response ~ 0.7*temp + 0.9*moist + N(0, 4).

```
> We examine goodness-of-fit for the Correct Model #10
> summary(output10)

Call: lm(formula = response ~ temp + moist, data = data2)
Residuals:
    Min      1Q Median    3Q   Max
 -7.836 -2.709 0.4237 2.346 7.414

Coefficients:
               Value Std. Error t value Pr(>|t|)
(Intercept)   3.7324   2.8441    1.3123  0.1958
       temp   0.5065   0.1284    3.9460  0.0003
      moist   0.9138   0.0189   48.2598  0.0000
```

```
Residual standard  error: 3.852 on 47 degrees of freedom
Multiple R-Squared:  0.9819
F-statistic: 1271 on 2 and 47 degrees of freedom, the p-value is
0

Correlation of Coefficients:
      (Intercept)    temp
 temp -0.9281
moist -0.1132     -0.2140
```

> # Histogram of the Model #10 residuals

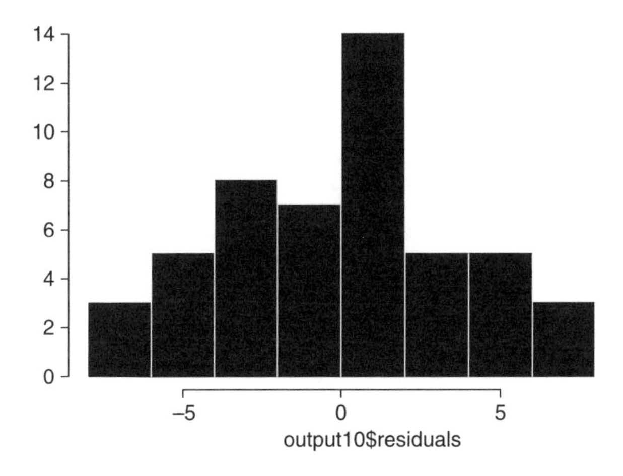

> # Graph of model 10 residuals/fitted values shows no indications of dependence or outliers.

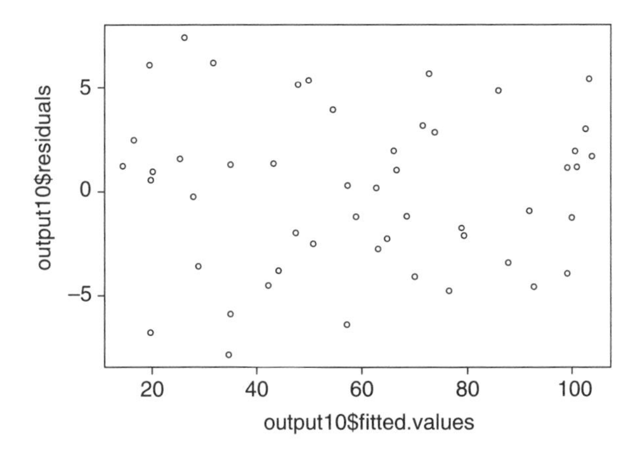

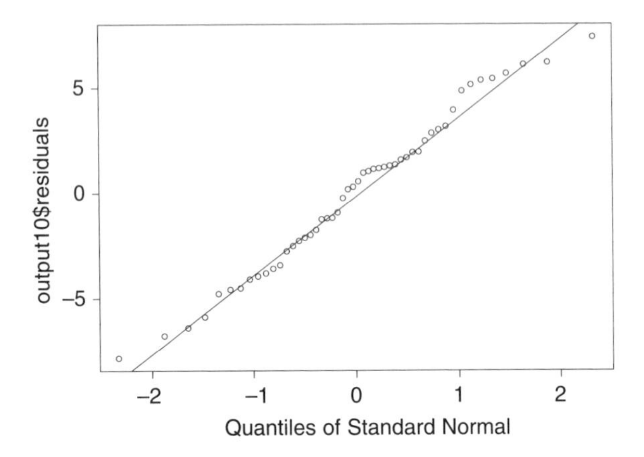

> # Normal plot of residuals shows no abnormalities of note

> # Cross validation program:

```
> cross.validation
function(model, j, k, data1)
{
# model is of the form response~sum of 2 covariates
# j,k are the data1 covariate data1 columns in the model
        n <- dim(data1)[1]
        residuals <- rep(NA, n)
        for(i in 1:n) {
                output <- lm(model, data = data1[ - i, ])
                residuals[i] <- data1$response[i] - sum(output$
                    coefficients * c(1, data1[i, j], data1[i, k]))
        }
        return(residuals)
}
> output <- cross.validation("response~temp+moist",7,8,data2)
> output
 [1]  7.8395715 -2.0498825  1.1267600  1.7248288  5.3709663
1.8987201
 [7]  1.3372920 -3.6043921  1.2794691 -4.4184753 -2.4144195
5.8661706
[13]  5.0531451  0.6105728 -6.7461355  2.1169389 -2.5867780
-0.9862320
```

```
[19]  6.5000915 -4.1219727  1.2442463 -6.2503065 -3.9577061
-1.2464063
[25]  4.0140447  3.2404994 -4.8776173  2.9403953 -1.2156973
-7.3480755
[31]  2.0913613  5.5649725  6.5676859 -4.8236869 -8.1508647
0.1866867
[37]  6.1770243  0.3176671 -1.8707143  2.6705973 -0.2543623
-2.8300577
[43] -5.0577898 -4.2416706  1.3498520  3.4305151 -1.3577022
1.0327350
[49] -2.2290007  1.4074559
> hist(output)
]]>
```

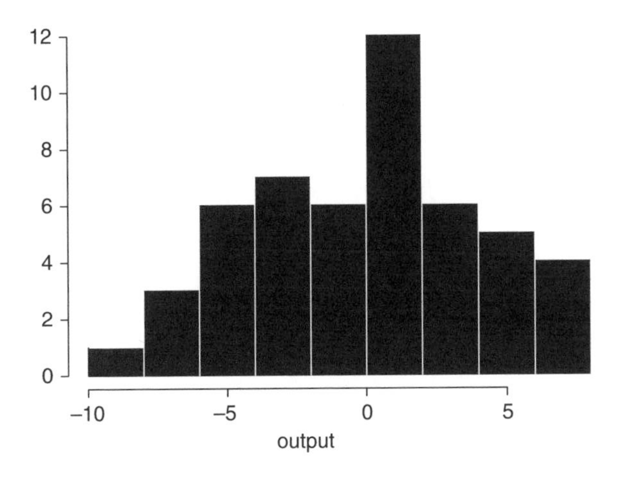

Problem 5.6 (Report)

Problem 6.1

```
> # The "reality" is given by the logit = 3.0*old.growth + 2.0*temp for the standardized
covariates
> data2$logit _ 3*data2$old.growth+2*data2$temp
> data2$prob _ exp(logit)/(1+exp(logit))
> data2$response _ rbinom(100,1,prob)
> data2
   sample aspect species old.growth  rock  moss  temp moist logit         prob   response
 1      1   0.92   -0.48       0.29 -0.38  1.19  0.24  0.27  1.35 0.7941296282        1
 2      2   0.92   -0.48       0.04 -0.38 -0.83  0.47  0.28  1.06 0.7426905453        0
 3      3  -1.08   -0.48      -0.81 -0.38  1.19  0.86 -0.87 -0.71 0.3295988402        0
 4      4   0.92   -0.48      -0.35 -0.38 -0.83 -0.48  0.63 -2.01 0.1181569778        0
 5      5  -1.08    2.05       1.02 -0.38 -0.83  0.86 -0.40  4.78 0.9916739069        1
 6      6  -1.08   -0.48       0.69 -0.38  1.19  1.66 -0.37  5.39 0.9954587438        1
 7      7   0.92   -0.48      -1.34 -0.38  1.19  0.05 -1.55 -3.92 0.0194550846        0
 8      8   0.92   -0.48      -0.25 -0.38 -0.83  0.36  0.05 -0.03 0.4925005624        1
 9      9   0.92    2.05      -1.28  2.57 -0.83  1.46  0.95 -0.92 0.2849578943        0
10     10  -1.08    2.05      -0.51 -0.38  1.19  1.50 -1.30  1.47 0.8130573860        1
11     11  -1.08    2.05      -0.37 -0.38 -0.83 -1.60 -0.95 -4.31 0.0132554814        0
12     12   0.92   -0.48       1.35 -0.38  1.19 -0.88 -1.40  2.29 0.9080454501        1
13     13  -1.08    2.05       0.53 -0.38 -0.83 -1.65 -1.74 -1.71 0.1531637158        0
14     14  -1.08   -0.48       1.28 -0.38 -0.83 -1.30  0.82  1.24 0.7755640143        1
15     15  -1.08   -0.48      -0.72 -0.38 -0.83 -0.62 -0.84 -3.40 0.0322954647        0
16     16  -1.08   -0.48      -0.32 -0.38 -0.83 -0.50 -1.73 -1.96 0.1234670476        0
17     17  -1.08   -0.48       1.03  2.57 -0.83  1.64 -1.80  6.37 0.9982907673        1
18     18   0.92   -0.48      -0.37 -0.38  1.19 -0.30 -0.80 -1.71 0.1531637158        0
19     19   0.92   -0.48       0.27 -0.38  1.19 -0.37  1.77  0.07 0.5174928577        1
20     20   0.92   -0.48       0.36 -0.38  1.19 -0.70  1.03 -0.32 0.4206757479        0
21     21   0.92   -0.48       0.15  2.57 -0.83 -0.70  1.34 -0.95 0.2788848220        0
22     22  -1.08   -0.48       0.14 -0.38 -0.83 -1.05  0.41 -1.68 0.1570954689        0
23     23   0.92   -0.48       1.79 -0.38  1.19 -0.78  0.79  3.81 0.9783317337        1
24     24   0.92   -0.48      -0.28 -0.38 -0.83 -1.57 -1.84 -3.98 0.0183428907        0
25     25  -1.08   -0.48       1.67 -0.38 -0.83  0.41 -1.60  5.83 0.9970705300        1
26     26   0.92   -0.48       1.97 -0.38 -0.83 -1.14  1.71  3.63 0.9741687615        1
27     27   0.92   -0.48      -0.65 -0.38 -0.83  0.94  1.26 -0.07 0.4825071423        0
28     28   0.92   -0.48       1.91 -0.38 -0.83 -1.02 -0.93  3.69 0.9756364057        1
29     29  -1.08   -0.48       1.73 -0.38 -0.83  0.91  1.55  7.01 0.9990980057        1
30     30   0.92   -0.48       1.16 -0.38  1.19 -0.20 -0.36  3.08 0.9560601846        1
31     31  -1.08    2.05       0.29  2.57 -0.83  0.02  0.65  0.91 0.7130001628        0
32     32   0.92   -0.48      -0.26 -0.38  1.19 -1.66  0.71 -4.10 0.0163024994        0
33     33   0.92   -0.48      -0.42 -0.38  1.19 -1.63  0.39 -4.52 0.0107717300        0
34     34  -1.08   -0.48       0.00 -0.38  1.19 -0.67  0.52 -1.34 0.2075100586        0
35     35   0.92   -0.48      -0.07 -0.38 -0.83 -0.34  1.52 -0.89 0.2911098274        0
36     36   0.92    2.05      -1.05 -0.38  1.19  0.99  0.00 -1.17 0.2368549843        0
37     37   0.92   -0.48      -0.13 -0.38 -0.83  1.23  1.26  2.07 0.8879529614        1
38     38   0.92   -0.48       0.26 -0.38  1.19  0.35  0.39  1.48 0.8145725807        1
39     39   0.92   -0.48      -0.02 -0.38 -0.83  1.76  0.65  3.46 0.9695279667        1
40     40  -1.08   -0.48      -0.56  2.57  1.19 -0.33  0.36 -2.34 0.0878639148        0
41     41   0.92   -0.48       0.89  2.57 -0.83  1.46 -1.16  5.59 0.9962788707        1
42     42  -1.08   -0.48      -0.06 -0.38  1.19 -1.61 -0.02 -3.40 0.0322954647        0
43     43   0.92   -0.48      -1.23 -0.38 -0.83  1.46  0.12 -0.77 0.3164791063        0
44     44  -1.08   -0.48       0.43 -0.38  1.19  0.73  0.76  2.75 0.9399133498        1
45     45  -1.08   -0.48      -0.24 -0.38 -0.83  1.12 -1.58  1.52 0.8205384806        1
46     46   0.92   -0.48      -1.40 -0.38 -0.83 -0.66  0.48 -5.52 0.0039898651        0
47     47  -1.08   -0.48      -0.57  2.57  1.19  1.61  0.56  1.51 0.8190612068        1
48     48   0.92   -0.48       0.64 -0.38  1.19 -1.70  1.09 -1.48 0.1854274193        0
49     49   0.92   -0.48      -1.15 -0.38  1.19 -0.92 -0.45 -5.29 0.0050164684        0
50     50   0.92    2.05      -0.53 -0.38 -0.83 -0.18  1.53 -1.95 0.1245533582        0
51     51   0.92   -0.48       1.38 -0.38  1.19  0.89  0.63  5.92 0.9973219908        1
52     52   0.92   -0.48       1.98  2.57 -0.83  0.15 -1.57  6.24 0.9980539390        1
53     53  -1.08   -0.48      -1.35 -0.38 -0.83  0.32  0.90 -3.41 0.0319843975        0
54     54   0.92    2.05       0.85 -0.38  1.19  0.76  0.10  4.07 0.9832093519        1
55     55  -1.08   -0.48       0.73 -0.38 -0.83  1.45 -0.42  5.09 0.9938796692        1
```

```
56     56  -1.08    2.05       0.01 -0.38  1.19 -0.69  0.80 -1.35 0.2058703718      0
57     57  -1.08   -0.48       0.56 -0.38 -0.83  1.20 -0.83  4.08 0.9833736439      1
58     58  -1.08   -0.48      -1.34 -0.38 -0.83 -1.59 -0.27 -7.20 0.0007460288      0
59     59  -1.08   -0.48       1.98 -0.38  1.19  0.34 -1.45  6.62 0.9986683447      1
60     60   0.92   -0.48      -1.36  2.57  1.19 -1.48  1.42 -7.04 0.0008753596      0
61     61  -1.08   -0.48       0.73 -0.38 -0.83  0.05  0.25  2.29 0.9080454501      1
62     62   0.92   -0.48      -0.80 -0.38 -0.83  0.40 -0.03 -1.60 0.1679816149      0
63     63   0.92   -0.48      -0.92 -0.38  1.19  0.66  1.38 -1.44 0.1915453486      0
64     64   0.92   -0.48      -1.02 -0.38  1.19 -1.40 -0.78 -5.86 0.0028431372      0
65     65  -1.08   -0.48      -1.35 -0.38  1.19 -0.11  0.12 -4.27 0.0137889885      0
66     66  -1.08    2.05      -0.03 -0.38 -0.83  0.16 -1.79  0.23 0.5572478546      0
67     67   0.92   -0.48       0.20 -0.38 -0.83 -1.75 -0.79 -2.90 0.0521535631      0
68     68   0.92   -0.48      -0.16 -0.38 -0.83  1.04  0.94  1.60 0.8320183851      0
69     69  -1.08    2.05       0.11 -0.38 -0.83  1.69  0.80  3.71 0.9761073106      1
70     70  -1.08   -0.48      -1.10  2.57 -0.83  0.11  0.98 -3.08 0.0439398154      0
71     71   0.92   -0.48       0.00 -0.38 -0.83 -0.22 -1.59 -0.44 0.3917409693      1
72     72  -1.08   -0.48      -0.69  2.57  1.19 -0.46 -0.50 -2.99 0.0478796898      0
73     73   0.92   -0.48       1.53 -0.38 -0.83  0.71  1.22  6.01 0.9975519196      1
74     74   0.92   -0.48      -1.27 -0.38  1.19 -1.17  0.57 -6.15 0.0021289397      0
75     75   0.92   -0.48      -0.17 -0.38  1.19  0.58  0.25  0.65 0.6570104627      0
76     76  -1.08   -0.48      -0.65  2.57 -0.83 -0.94  0.43 -3.83 0.0212483227      0
77     77  -1.08   -0.48      -0.60 -0.38  1.19 -0.24  0.13 -2.28 0.0927929531      0
78     78  -1.08   -0.48      -0.67 -0.38 -0.83  1.10  0.18  0.19 0.5473576181      0
79     79  -1.08   -0.48      -0.76 -0.38 -0.83  1.52 -0.98  0.76 0.6813537338      1
80     80  -1.08    2.05       0.23 -0.38 -0.83 -0.43  0.96 -0.17 0.4576020592      1
81     81   0.92   -0.48      -1.05 -0.38 -0.83 -0.38  0.68 -3.91 0.0196467699      0
82     82   0.92    2.05      -0.95 -0.38 -0.83  0.17  0.28 -2.51 0.0751601095      0
83     83   0.92   -0.48      -1.12 -0.38 -0.83  0.25 -0.84 -2.86 0.0541667005      0
84     84  -1.08   -0.48      -1.16 -0.38 -0.83  1.24  1.29 -1.00 0.2689414214      1
85     85   0.92    2.05      -1.23 -0.38  1.19  0.35  1.53 -2.99 0.0478796898      0
86     86  -1.08   -0.48      -0.98 -0.38 -0.83  1.12 -0.83 -0.70 0.3318122278      0
87     87  -1.08    2.05       1.80 -0.38 -0.83 -1.59 -1.25  2.22 0.9020311957      1
88     88  -1.08   -0.48      -1.32 -0.38  1.19  0.37 -1.43 -3.22 0.0384199855      0
89     89  -1.08   -0.48       1.33 -0.38 -0.83 -0.12  0.19  3.75 0.9770226301      1
90     90  -1.08    2.05       1.13 -0.38  1.19 -0.27  0.77  2.85 0.9453186828      1
91     91   0.92   -0.48      -1.18 -0.38 -0.83  1.03 -1.35 -1.48 0.1854274193      0
92     92   0.92    2.05       1.97 -0.38 -0.83 -1.40  0.24  3.11 0.9573033558      1
93     93  -1.08   -0.48      -1.42 -0.38  1.19  0.91 -0.05 -2.44 0.0801729122      0
94     94   0.92   -0.48      -1.46 -0.38  1.19 -0.96 -1.62 -6.30 0.0018329389      0
95     95   0.92    2.05      -0.12 -0.38  1.19 -0.25 -0.31 -0.86 0.2973393457      0
96     96  -1.08   -0.48       0.08 -0.38 -0.83  0.88  0.32  2.00 0.8807970780      1
97     97  -1.08   -0.48       0.69 -0.38 -0.83 -1.07 -1.49 -0.07 0.4825071423      0
98     98   0.92   -0.48       1.57 -0.38 -0.83 -1.17 -0.43  2.37 0.9145108606      1
99     99   0.92   -0.48       1.43 -0.38  1.19 -0.11  0.65  4.07 0.9832093519      1
100   100   0.92   -0.48       1.04  2.57 -0.83  0.81  0.45  4.74 0.9913370562      1
```

```
> model21 _ glm(response~aspect+species+old.growth,data=data2,
    family=binomial(link=logit))
> summary(model21)

Call: glm(formula = response ~ aspect + species + old.growth,
 family =binomial(link = logit), data = data2)
Deviance Residuals:
       Min          1Q      Median          3Q       Max
 -2.018374 -0.6262515 -0.2038182 0.4335786 2.374611

Coefficients:
                 Value Std. Error   t value
(Intercept) -0.3812985  0.2945342 -1.294581
     aspect -0.3804848  0.3108366 -1.224067
    species -0.2735954  0.3053769 -0.895927
 old.growth  2.5162455  0.4964602  5.068373
```

```
(Dispersion Parameter for Binomial family taken to be 1 )
    Null Deviance: 136.663 on 99 degrees of freedom
Residual Deviance: 73.6352 on 96 degrees of freedom

Number of Fisher Scoring Iterations: 5

 Correlation of Coefficients:
           (Intercept)     aspect    species
    aspect  0.0947743
   species -0.0070858   0.3118578
old.growth -0.0227379  -0.1806913 -0.2075402

> model22 _
glm(response~aspect+species+rock,data=data2,family=binomial
   (link=logit))
> summary(model22)

Call: glm(formula = response ~ aspect + species + rock, family =
binomial(link = logit), data = data2)
Deviance Residuals:
       Min         1Q     Median         3Q        Max
 -1.168238 -1.009788 -0.9664637  1.234161  1.469901

Coefficients:
                  Value Std. Error    t value
(Intercept) -0.28445294  0.2029802 -1.4013828
     aspect -0.19315599  0.2052469 -0.9410909
    species -0.04381164  0.2065612 -0.2121001
       rock -0.08732775  0.2080100 -0.4198247

(Dispersion Parameter for Binomial family taken to be 1 )

    Null Deviance: 136.663 on 99 degrees of freedom

Residual Deviance: 135.6378 on 96 degrees of freedom

Number of Fisher Scoring Iterations: 2

Correlation of Coefficients:
        (Intercept)    aspect    species
 aspect  0.0241944
species  0.0051767  0.1222465
   rock  0.0104506  0.0726275  0.0463164
```

```
> model23 _ glm(response~aspect+species+moss,data=data2,
      family=binomial(link=logit))
> summary(model23)

Call: glm(formula = response ~ aspect + species + moss,
      family = binomial(link = logit), data = data2)

Deviance Residuals:
       Min        1Q     Median          3Q        Max
 -1.213441 -1.07437 -0.9117188  1.284061  1.521494

Coefficients:
                  Value Std. Error    t value
(Intercept) -0.28777002  0.2038046 -1.4119897
     aspect -0.16581842  0.2064232 -0.8032935
    species -0.04631863  0.2072606 -0.2234801
       moss -0.20580646  0.2075097 -0.9917918

(Dispersion Parameter for Binomial family taken to be 1 )

    Null Deviance: 136.663 on 99 degrees of freedom

Residual Deviance: 134.8263 on 96 degrees of freedom

Number of Fisher Scoring Iterations: 2

Correlation of Coefficients:
        (Intercept)    aspect    species
 aspect  0.0204747
species  0.0061224  0.1165137
   moss  0.0329830 -0.0966413  0.0332432

> model24 _ glm(response~aspect+species+temp,data=data2,family=
   binomial(link=logit))
> summary(model24)

Call: glm(formula = response ~ aspect + species + temp, family =
binomial(link = logit), data = data2)
Deviance Residuals:
       Min        1Q     Median          3Q        Max
 -1.470158 -1.00431 -0.6476229  1.078913  1.840854
```

```
Coefficients:
                  Value Std. Error    t value
(Intercept) -0.31972304  0.2158474 -1.4812456
     aspect -0.09995994  0.2172582 -0.4600974
    species -0.02904393  0.2172947 -0.1336615
       temp  0.72868208  0.2315928  3.1463938

(Dispersion Parameter for Binomial family taken to be 1 )

    Null Deviance: 136.663 on 99 degrees of freedom

Residual Deviance: 124.6936 on 96 degrees of freedom

Number of Fisher Scoring Iterations: 3

Correlation of Coefficients:
         (Intercept)    aspect    species
  aspect  0.0013862
 species  0.0007743  0.0889408
    temp -0.1041291  0.1074510  0.0054037

> model25 _ glm(response~aspect+species+moist,data=data2,family=
   binomial(link=logit))
> summary(model25)

Call: glm(formula = response ~ aspect + species + moist,family =
   binomial(link = logit), data = data2)
Deviance Residuals:
       Min         1Q     Median         3Q        Max
 -1.151355 -0.9987872 -0.9974408  1.247247  1.412807

Coefficients:
                   Value Std. Error      t value
(Intercept) -0.284168027  0.2028145 -1.401122896
     aspect -0.187356391  0.2096769 -0.893548046
    species -0.039878178  0.2063326 -0.193271357
      moist  0.001078895  0.2089821  0.005162618

(Dispersion Parameter for Binomial family taken to be 1 )

    Null Deviance: 136.663 on 99 degrees of freedom

Residual Deviance: 135.8159 on 96 degrees of freedom

Number of Fisher Scoring Iterations: 2
```

```
Correlation of Coefficients:
        (Intercept)    aspect    species
 aspect  0.0231497
species  0.0044574  0.1253947
  moist  0.0000237 -0.2203928 -0.0456131

> model26 _ glm(response~aspect+old.growth+temp,data=data2,family
   =binomial(link=logit))
> summary(model26)

Call: glm(formula = response ~ aspect + old.growth + temp, family
= binomial(link = logit), data = data2)

Deviance Residuals:
       Min         1Q       Median        3Q       Max
 -1.818713 -0.2298326 -0.01874344  0.10295  2.162767

Coefficients:
                Value Std. Error    t value
(Intercept) -0.7432215  0.4558625 -1.6303633
     aspect -0.2792430  0.4280498 -0.6523611
 old.growth  4.8563381  1.1088344  4.3796785
       temp  3.0948623  0.7969715  3.8832782

(Dispersion Parameter for Binomial family taken to be 1 )

    Null Deviance: 136.663 on 99 degrees of freedom

Residual Deviance: 36.31113 on 96 degrees of freedom

Number of Fisher Scoring Iterations: 6

Correlation of Coefficients:
          (Intercept)     aspect old.growth
    aspect  0.0775643
old.growth -0.1883359 -0.0578475
      temp -0.3239957 -0.0217232  0.7861620

> model27 _ glm(response~rock+temp+moist,data=data2,
   family=binomial(link=logit))
> summary(model27)

Call: glm(formula = response ~ rock + temp + moist, family =
binomial(link = logit), data = data2)
```

```
Deviance Residuals:
       Min          1Q     Median         3Q       Max
 -1.549069 -0.9826195 -0.6460244  1.054946  1.800411

Coefficients:
                  Value  Std.Error    t value
(Intercept) -0.32192768  0.2164081 -1.4875951
       rock -0.15748363  0.2229356 -0.7064087
       temp  0.76042412  0.2317981  3.2805447
      moist -0.05660421  0.2164813 -0.2614739

(Dispersion Parameter for Binomial family taken to be 1 )

    Null Deviance: 136.663 on 99 degrees of freedom

Residual Deviance: 124.3217 on 96 degrees of freedom

Number of Fisher Scoring Iterations: 3

Correlation of Coefficients:
      (Intercept)       rock        temp
 rock  0.0277688
 temp -0.1102726 -0.1297368
moist  0.0023346 -0.0240196 -0.0314682

> model28 _ glm(response~moss+temp+moist,data=data2,family=
   binomial(link=logit))
> summary(model28)

Call: glm(formula = response ~ moss + temp + moist, family =
   binomial(link = logit), data = data2)
Deviance Residuals:
      Min          1Q     Median         3Q       Max
-1.555659 -0.9738612 -0.6638551  1.079298  1.748465

Coefficients:
                  Value Std. Error    t value
(Intercept) -0.31974716  0.2161358 -1.4793808
       moss -0.14699914  0.2183618 -0.6731908
       temp  0.72619244  0.2309462  3.1444225
      moist -0.05459733  0.2167654 -0.2518729

(Dispersion Parameter for Binomial family taken to be 1 )

    Null Deviance: 136.663 on 99 degrees of freedom
```

```
Residual Deviance: 124.3772 on 96 degrees of freedom

Number of Fisher Scoring Iterations: 3

Correlation of Coefficients:
      (Intercept)       moss       temp
 moss  0.0139125
 temp -0.0983389  0.0868243
moist -0.0003751 -0.0499147 -0.0430762

> model29 _ glm(response~old.growth+temp,data=data2,family=
    binomial(link=logit))
> summary(model29)

Call: glm(formula = response ~ old.growth + temp, family =
    binomial(link = logit), data = data2)
Deviance Residuals:
       Min         1Q      Median          3Q       Max
 -1.949256 -0.250522 -0.02094977  0.09681103  2.275508

Coefficients:
                 Value Std. Error   t value
(Intercept) -0.7291304  0.4517408 -1.614046
 old.growth  4.8729916  1.1152687  4.369343
       temp  3.1216931  0.8002679  3.900810

(Dispersion Parameter for Binomial family taken to be 1 )

    Null Deviance: 136.663 on 99 degrees of freedom

Residual Deviance: 36.7349 on 97 degrees of freedom

Number of Fisher Scoring Iterations: 6

Correlation of Coefficients:
           (Intercept) old.growth
old.growth -0.1809192
      temp -0.3234724   0.7954179

> model30 _ glm(response~rock+moist,data=data2,family=binomial
   (link=logit))
> summary(model30)

Call: glm(formula = response ~ rock + moist, family = binomial
   (link = logit), data = data2)
```

```
Deviance Residuals:
       Min         1Q     Median         3Q       Max
 -1.098287 -1.069386 -0.9886188  1.290709  1.388836

Coefficients:
                  Value Std. Error    t value
(Intercept) -0.28211668  0.2021157 -1.3958176
       rock -0.07089319  0.2069095 -0.3426290
      moist -0.03597933  0.2034385 -0.1768561

(Dispersion Parameter for Binomial family taken to be 1 )

    Null Deviance: 136.663 on 99 degrees of freedom

Residual Deviance: 136.5051 on 97 degrees of freedom

Number of Fisher Scoring Iterations: 2

Correlation of Coefficients:
      (Intercept)      rock
 rock  0.0085030
moist  0.0048325 -0.0607497
> model31 _ glm(response~rock+moss+temp+
moist,data=data2,family=binomial(link=logit))

> summary(model31)

Call: glm(formula = response ~ rock + moss + temp + moist, family
= binomial(link = logit), data = data2)
Deviance Residuals:
       Min        1Q       Median         3Q       Max
 -1.602848 -0.9767413 -0.6729573  1.092539  1.731579

Coefficients:
                 Value Std. Error    t value
(Intercept) -0.3225569  0.2169275 -1.4869338
       rock -0.1734555  0.2249926 -0.7709385
       moss -0.1633909  0.2201368 -0.7422243
       temp  0.7446446  0.2326483  3.2007304
      moist -0.0468392  0.2170510 -0.2157982

(Dispersion Parameter for Binomial family taken to be 1 )

    Null Deviance: 136.663 on 99 degrees of freedom

Residual Deviance: 123.7693 on 95 degrees of freedom
```

```
Number of Fisher Scoring Iterations: 3

Correlation of Coefficients:
      (Intercept)      rock        moss       temp
 rock  0.0293166
 moss  0.0174321  0.1009758
 temp -0.1032482 -0.1282630  0.0670949
moist  0.0073141 -0.0337258 -0.0591562 -0.0411908

> model32 _ glm(response~aspect+species+old.growth+rock+moss+temp
   +moist,data=data2,family=binomial(link=logit))
> summary(model32)

Call: glm(formula = response ~ aspect + species + old.growth +
rock + moss + temp + moist, family = binomial(link = logit), data
= data2)
Deviance Residuals:
       Min          1Q      Median          3Q       Max
 -1.920072 -0.1728077 -0.01371914 0.09407092 2.285338

Coefficients:
                 Value Std. Error    t value
(Intercept) -0.9210613  0.5222776 -1.7635475
     aspect -0.6858480  0.5868230 -1.1687476
    species -0.2351692  0.4874347 -0.4824630
 old.growth  5.2572678  1.3176968  3.9897400
       rock -0.7270474  0.6207932 -1.1711587
       moss  0.1372742  0.4664094  0.2943213
       temp  3.3465016  0.9158027  3.6541732
      moist  0.3283434  0.5127001  0.6404199

(Dispersion Parameter for Binomial family taken to be 1 )

    Null Deviance: 136.663 on 99 degrees of freedom

Residual Deviance: 34.11214 on 92 degrees of freedom

Number of Fisher Scoring Iterations: 7
```

```
Correlation of Coefficients:
           (Intercept)     aspect    species old.growth       rock       moss       temp
    aspect   0.3120590
   species  -0.0601317  0.3005190
old.growth  -0.3098927 -0.3738107 -0.0983157
      rock   0.3635868  0.3579570 -0.1410673 -0.3311216
      moss  -0.0417944 -0.1844661  0.0559806  0.0685948 -0.1298313
      temp  -0.4361095 -0.2503376  0.0753317  0.8131769 -0.3834336  0.0947506
     moist  -0.3174575 -0.4535135  0.0229843  0.2760200 -0.3411515 -0.0497759  0.2569112

> attributes(model21)
$names:
 [1] "coefficients"      "residuals"     "fitted.values"  "effects"    "R"
 [6] "rank"              "assign"        "df.residual"    "weights"    "family"
[11] "linear.predictors" "deviance"      "null.deviance"  "call"       "iter"
[16] "y"                 "contrasts"     "terms"          "formula"    "control"

$class:
[1] "glm" "lm"

> aic21 _ model21$deviance+2*model21$rank
> aic22 _ model22$deviance+2*model22$rank
> aic23 _ model23$deviance+2*model23$rank
> aic24 _ model24$deviance+2*model24$rank
> aic25 _ model25$deviance+2*model25$rank
> aic26 _ model26$deviance+2*model26$rank
> aic27 _ model27$deviance+2*model27$rank
> aic28 _ model28$deviance+2*model28$rank
> aic29 _ model29$deviance+2*model29$rank
> aic30 _ model30$deviance+2*model30$rank
> aic31 _ model31$deviance+2*model31$rank
> aic32 _ model32$deviance+2*model32$rank
> c(aic21,aic22,aic23,aic24,aic25,aic26,aic27,aic28,aic29,aic30,
  aic31,aic32)
 [1]  81.63520
 [2] 143.63784
 [3] 142.82631
 [4] 132.69362
 [5] 143.81594
 [6]  44.31113
 [7] 132.32165
 [8] 132.37720
 [9]  42.73490
[10] 142.50506
```

```
[11] 133.76932
[12]  50.11214
> aicc21 _ aic21+2*model21$rank*(model21$rank+1)/
  (length(model21$residuals)-model21$rank-1)
> aicc22 _ aic22+2*model22$rank*(model22$rank+1)/
  (length(model22$residuals)-model22$rank-1)
> aicc23 _ aic23+2*model23$rank*(model23$rank+1)/
  (length(model23$residuals)-model23$rank-1)
> aicc24 _ aic24+2*model24$rank*(model24$rank+1)/
  (length(model24$residuals)-model24$rank-1)
> aicc25 _ aic25+2*model25$rank*(model25$rank+1)/
  (length(model25$residuals)-model25$rank-1)
> aicc26 _ aic26+2*model26$rank*(model26$rank+1)/
  (length(model26$residuals)-model26$rank-1)
> aicc27 _ aic27+2*model27$rank*(model27$rank+1)/
  (length(model27$residuals)-model27$rank-1)
> aicc28 _ aic28+2*model28$rank*(model28$rank+1)/
  (length(model28$residuals)-model28$rank-1)
> aicc29 _ aic29+2*model29$rank*(model29$rank+1)/
  (length(model29$residuals)-model29$rank-1)
> aicc30 _ aic30+2*model30$rank*(model30$rank+1)/
  (length(model30$residuals)-model30$rank-1)
> aicc31 _ aic31+2*model31$rank*(model31$rank+1)/
  (length(model31$residuals)-model31$rank-1)
> aicc32 _ aic32+2*model32$rank*(model32$rank+1)/
  (length(model32$residuals)-model32$rank-1)
> c(aicc21,aicc22,aicc23,aicc24,aicc25,
  aicc26,aicc27,aicc28,aicc29,aicc30,aicc31,aicc32)
 [1]  82.05625
 [2] 144.05889
 [3] 143.24736
 [4] 133.11467
 [5] 144.23699
 [6]  44.73218
 [7] 132.74270
 [8] 132.79825
 [9]  42.98490
[10] 142.75506
[11] 134.40762
[12]  51.69456
```

> # Model 9, with `old.growth` and `temp`, is the best-fitting model, with statistics that are compatible with the "reality," based on an analysis strategy of a priori parsimonious model selection and inference using AIC.

Problem 6.2

```
> covariates
 [1] "aspect"
 [2] "species"
 [3] "old.growth"
 [4] "rock"
 [5] "moss"
 [6] "temp"
 [7] "moist"
 [8] "aspect+species"
 [9] "aspect+old.growth"
[10] "aspect+rock"
[11] "aspect+moss"
[12] "aspect+temp"
[13] "aspect+moist"
[14] "species+old.growth"
[15] "species+rock"
[16] "species+moss"
[17] "species+temp"
[18] "species+moist"
[19] "old.growth+rock"
[20] "old.growth+moss"
[21] "old.growth+temp"
[22] "old.growth+moist"
[23] "rock+moss"
[24] "rock+temp"
[25] "rock+moist"
[26] "moss+temp"
[27] "moss+moist"
[28] "temp+moist"
[29] "aspect+species+old.growth"
[30] "aspect+species+rock"
[31] "aspect+species+moss"
[32] "aspect+species+temp"
[33] "aspect+species+moist"
[34] "aspect+old.growth+rock"
[35] "aspect+old.growth+moss"
[36] "aspect+old.growth+temp"
[37] "aspect+old.growth+moist"
[38] "aspect+rock+moss"
[39] "aspect+rock+temp"
[40] "aspect+rock+moist"
```

```
[41] "aspect+moss+temp"
[42] "aspect+moss+moist"
[43] "aspect+temp+moist"
[44] "species+old.growth+rock"
[45] "species+old.growth+moss"
[46] "species+old.growth+temp"
[47] "species+old.growth+moist"
[48] "species+rock+moss"
[49] "species+rock+temp"
[50] "species+rock+moist"
[51] "species+moss+temp"
[52] "species+moss+moist"
[53] "species+temp+moist"
[54] "old.growth+rock+moss"
[55] "old.growth+rock+temp"
[56] "old.growth+rock+moist"
[57] "old.growth+moss+temp"
[58] "old.growth+moss+moist"
[59] "old.growth+temp+moist"
[60] "rock+moss+temp"
[61] "rock+moss+moist"
[62] "rock+temp+moist"
[63] "moss+temp+moist"
[64] "aspect+species+old.growth+rock"
[65] "aspect+species+old.growth+moss"
[66] "aspect+species+old.growth+temp"
[67] "aspect+species+old.growth+moist"
[68] "aspect+species+rock+moss"
[69] "aspect+species+rock+temp"
[70] "aspect+species+rock+moist"
[71] "aspect+species+moss+temp"
[72] "aspect+species+moss+moist"
[73] "aspect+species+temp+moist"
[74] "aspect+old.growth+rock+moss"
[75] "aspect+old.growth+rock+temp"
[76] "aspect+old.growth+rock+moist"
[77] "aspect+old.growth+moss+temp"
[78] "aspect+old.growth+moss+moist"
[79] "aspect+old.growth+temp+moist"
[80] "aspect+rock+moss+temp"
[81] "aspect+rock+moss+moist"
[82] "aspect+rock+temp+moist"
[83] "aspect+moss+temp+moist"
[84] "species+old.growth+rock+moss"
```

```
 [85] "species+old.growth+rock+temp"
 [86] "species+old.growth+rock+moist"
 [87] "species+old.growth+moss+temp"
 [88] "species+old.growth+moss+moist"
 [89] "species+old.growth+temp+moist"
 [90] "species+rock+moss+temp"
 [91] "species+rock+moss+moist"
 [92] "species+rock+temp+moist"
 [93] "species+moss+temp+moist"
 [94] "old.growth+rock+moss+temp"
 [95] "old.growth+rock+moss+moist"
 [96] "old.growth+rock+temp+moist"
 [97] "old.growth+moss+temp+moist"
 [98] "rock+moss+temp+moist"
 [99] "aspect+species+old.growth+rock+moss"
[100] "aspect+species+old.growth+rock+temp"
[101] "aspect+species+old.growth+rock+moist"
[102] "aspect+species+old.growth+moss+temp"
[103] "aspect+species+old.growth+moss+moist"
[104] "aspect+species+old.growth+temp+moist"
[105] "aspect+species+rock+moss+temp"
[106] "aspect+species+rock+moss+moist"
[107] "aspect+species+rock+temp+moist"
[108] "aspect+species+moss+temp+moist"
[109] "aspect+old.growth+rock+moss+temp"
[110] "aspect+old.growth+rock+moss+moist"
[111] "aspect+old.growth+rock+temp+moist"
[112] "aspect+old.growth+moss+temp+moist"
[113] "aspect+rock+moss+temp+moist"
[114] "species+old.growth+rock+moss+temp"
[115] "species+old.growth+rock+moss+moist"
[116] "species+old.growth+rock+temp+moist"
[117] "species+old.growth+moss+temp+moist"
[118] "species+rock+moss+temp+moist"
[119] "old.growth+rock+moss+temp+moist"
[120] "species+old.growth+rock+moss+temp+moist"
[121] "aspect+old.growth+rock+moss+temp+moist"
[122] "aspect+species+rock+moss+temp+moist"
[123] "aspect+species+old.growth+moss+temp+moist"
[124] "aspect+species+old.growth+rock+temp+moist"
[125] "aspect+species+old.growth+rock+moss+moist"
[126] "aspect+species+old.growth+rock+moss+temp"
[127] "aspect+species+old.growth+rock+moss+temp+moist"
```

```
> models <- list(rep(NA,length(covariates)))
> for (i in 1:length(covariates)) {models[[i]] <-
 glm(paste("response~",covariates[[i]]),data=data2,family=
binomial(link=logit))}

> aic <- rep(NA,length(covariates))
> for (i in 1:length(covariates)) {aic[i] <- models[[i]]
$deviance+2*models[[i]]$rank}
> aic
  [1] 139.85339 140.65531  79.51750 140.53631 139.48914 128.91362
140.62353 141.81597  80.46451 141.68293 140.87633 130.71150 141.85338
 [14]  81.17793 142.52625 141.47155 130.90486 142.61660  81.40933
81.26398  42.73490  81.49832 141.29015 130.38997 142.50506 130.44062
 [27] 141.47541 130.83112  81.63520 143.63784 142.82631 132.69362
143.81594  82.29365  82.31451  44.31113  82.27267 142.64205 132.15225
 [40] 143.68207 132.28469 142.87395 132.67354  83.06473  82.88769
44.50219  83.16640 143.26811 132.37783 144.49579 132.43030 143.45855
 [53] 132.82489  83.11493  43.73011  83.38248  44.72662  83.20725
44.73473 131.81585 143.28249 132.32165 132.37720 83.41078 83.46074
 [66]  45.72469  83.43748 144.58208 134.12878 145.63625 134.26597
144.82273 134.65905  84.11005  44.94795  84.06180  46.23969  84.05198
 [79]  46.22166 133.63155 144.63436 134.12704 134.25567  84.73434
45.67655  85.04619  46.49698  84.84144  46.49710 133.80105 145.26111
 [92] 134.31255 134.36900  45.71764  85.03685  45.67334  46.72661
133.76932  85.20296  46.65120  85.15489  47.64900  85.17926  47.64217
[105] 135.60616 146.57197 136.10700 136.23973  46.80726  85.78260
46.45420  48.16111 135.61598  47.66746 86.66670  47.64091  48.48968
[118] 135.75705  47.66931  49.63741  48.34955 137.59319  49.57600
48.19900  86.83034  48.53391  50.11214
> min(aic)
[1] 42.7349
> for (i in 1:length(covariates)) {if (aic[i]==min(aic)) k <- i}
> k
[1] 21
> covariates[[k]]
[1] "old.growth+temp"
> summary(models[[k]])

Call: glm(formula = response ~ old.growth + temp, family =
binomial(link = logit), data = data2)
Deviance Residuals:
      Min         1Q      Median          3Q       Max
-1.949256 -0.250522 -0.02094977 0.09681103 2.275508
```

```
Coefficients:
                 Value Std. Error   t value
(Intercept) -0.7291304  0.4517408 -1.614046
 old.growth  4.8729916  1.1152687  4.369343
       temp  3.1216931  0.8002679  3.900810

(Dispersion Parameter for Binomial family taken to be 1 )

    Null Deviance: 136.663 on 99 degrees of freedom

Residual Deviance: 36.7349 on 97 degrees of freedom

Number of Fisher Scoring Iterations: 6

Correlation of Coefficients:
           (Intercept) old.growth
old.growth -0.1809192
      temp -0.3234724   0.7954179
```

> # Model 21, consisting of `old.growth` and `temp`, is the best-fitting model, with statistics that are compatible with the "reality," based on an analysis strategy of a posteriori model selection and inference.

Problem 6.3

```
Model #1

model
{
for(i in 1:n)
   {
      response[i] ~ dbin(p[i],1)
      logit(p[i]) <-beta0+beta1*aspect[i]+beta2*species[i]
                             +beta3*old.growth[i]
          }
beta0 ~ dnorm(0,0.1)
beta1 ~ dnorm(0,0.1)
beta2 ~ dnorm(0,0.1)
beta3 ~ dnorm(0,0.1)
}
```

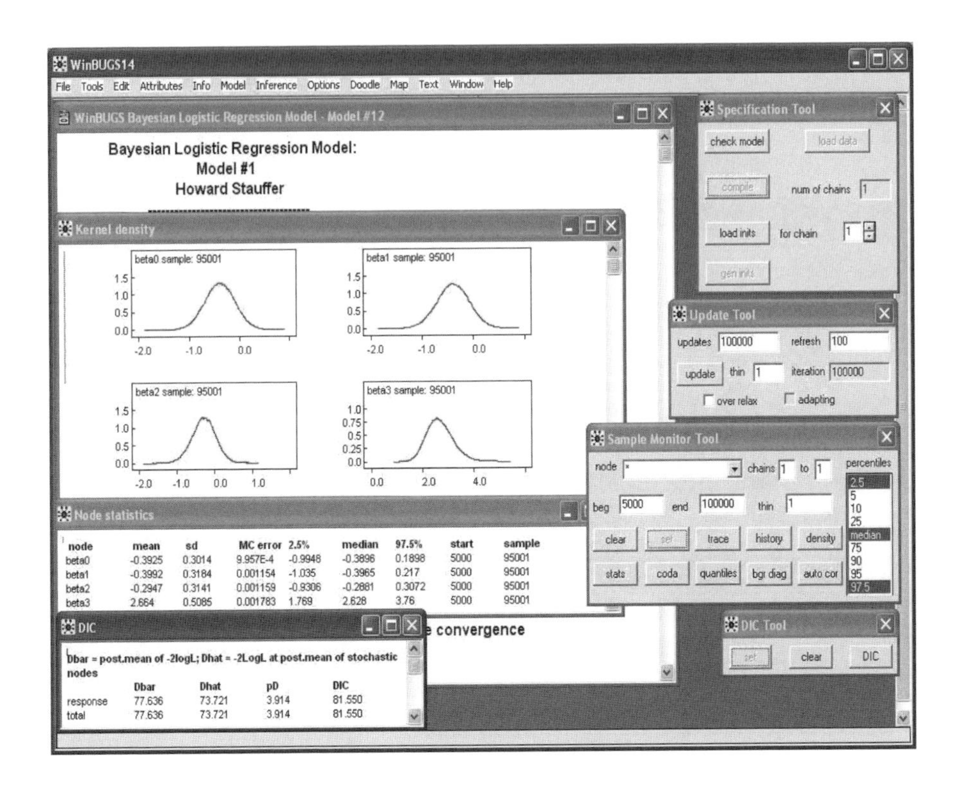

Model #2

```
model
{
for(i in 1:n)
   {
       response[i] ~ dbin(p[i],1)
       logit(p[i]) <- beta0+beta1*aspect[i]+beta2*species[i]
                          +beta4*rock[i]
          }
beta0 ~ dnorm(0,0.000001)
beta1 ~ dnorm(0,0.000001)
beta2 ~ dnorm(0,0.000001)
beta4 ~ dnorm(0,0.000001)
}
```

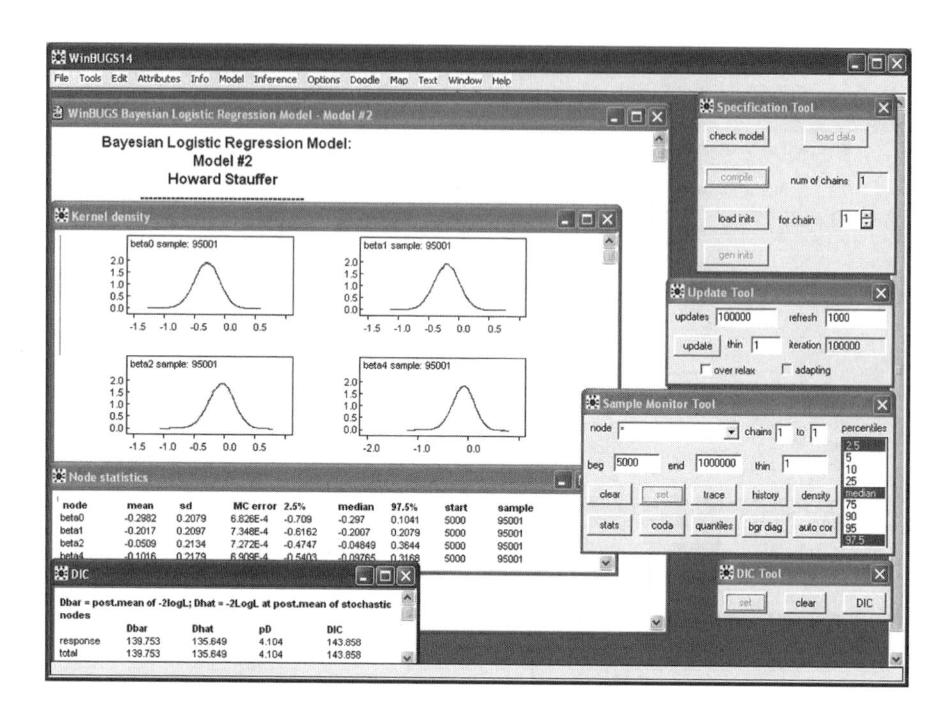

Model #3

```
model
{
for(i in 1:n)
   {
      response[i] ~ dbin(p[i],1)
      logit(p[i]) <- beta0+beta1*aspect[i]+beta2*species[i]
                         +beta5*moss[i]
         }
beta0 ~ dnorm(0,0.000001)
beta1 ~ dnorm(0,0.000001)
beta2 ~ dnorm(0,0.000001)
beta5 ~ dnorm(0,0.000001)
}
```

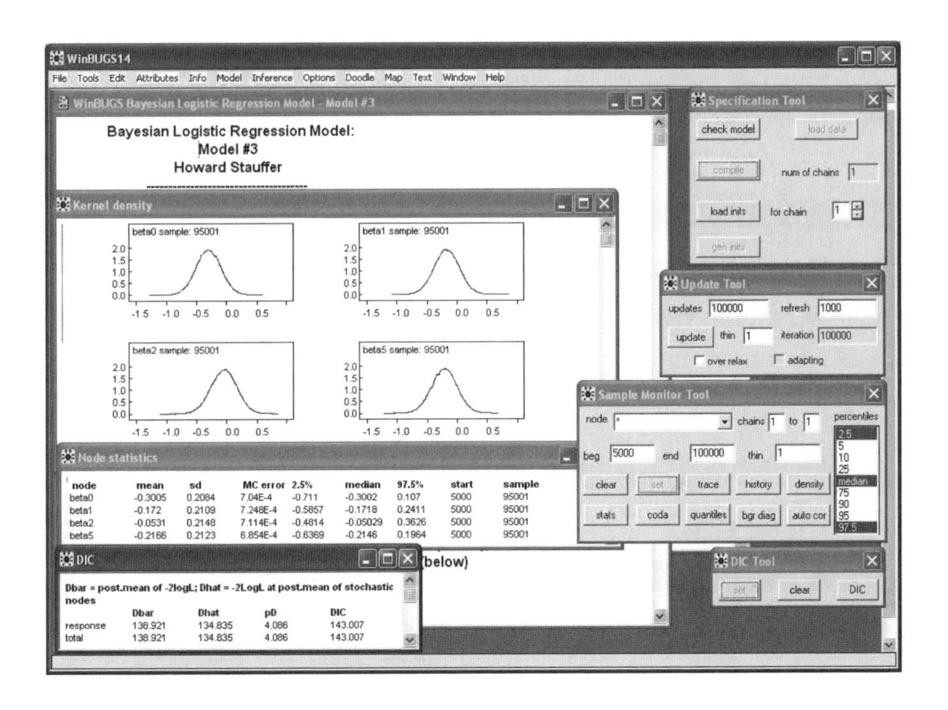

Model #4

```
model
{
for(i in 1:n)
   {
      response[i] ~ dbin(p[i],1)
      logit(p[i]) <- beta0+beta1*aspect[i]+beta2*species[i]
                           +beta6*temp[i]
         }
beta0 ~ dnorm(0,0.000001)
beta1 ~ dnorm(0,0.000001)
beta2 ~ dnorm(0,0.000001)
beta6 ~ dnorm(0,0.000001)
}
```

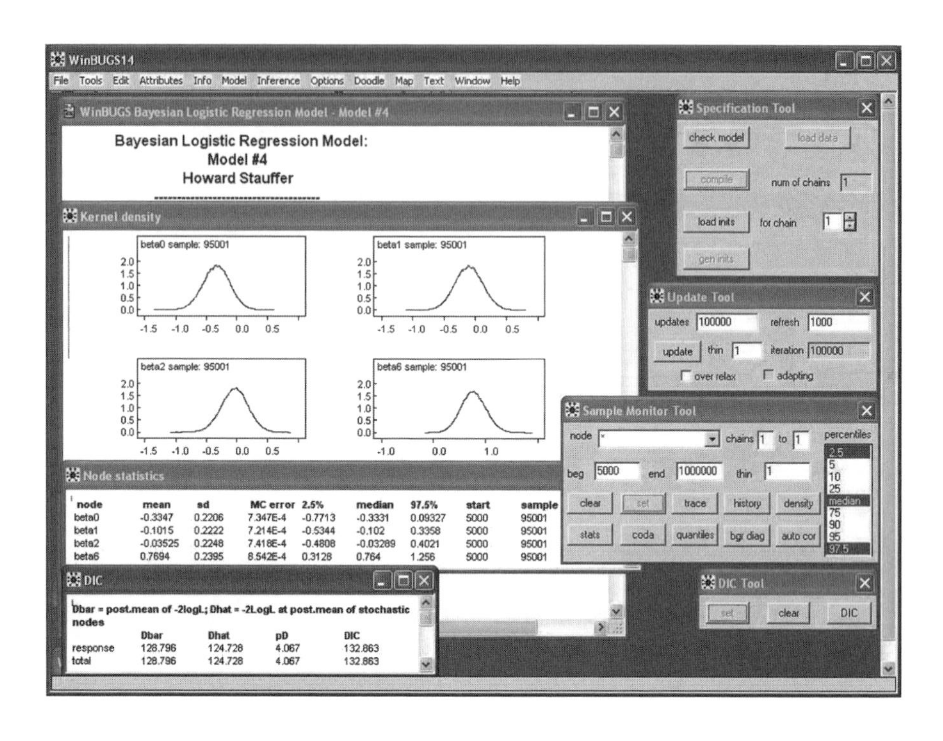

Model #5

```
model
{
for(i in 1:n)
   {
      response[i] ~ dbin(p[i],1)
      logit(p[i]) <- beta0+beta1*aspect[i]+beta2*species[i]
                          +beta7*moist[i]
         }
beta0 ~ dnorm(0,0.000001)
beta1 ~ dnorm(0,0.000001)
beta2 ~ dnorm(0,0.000001)
beta7 ~ dnorm(0,0.000001)
}
```

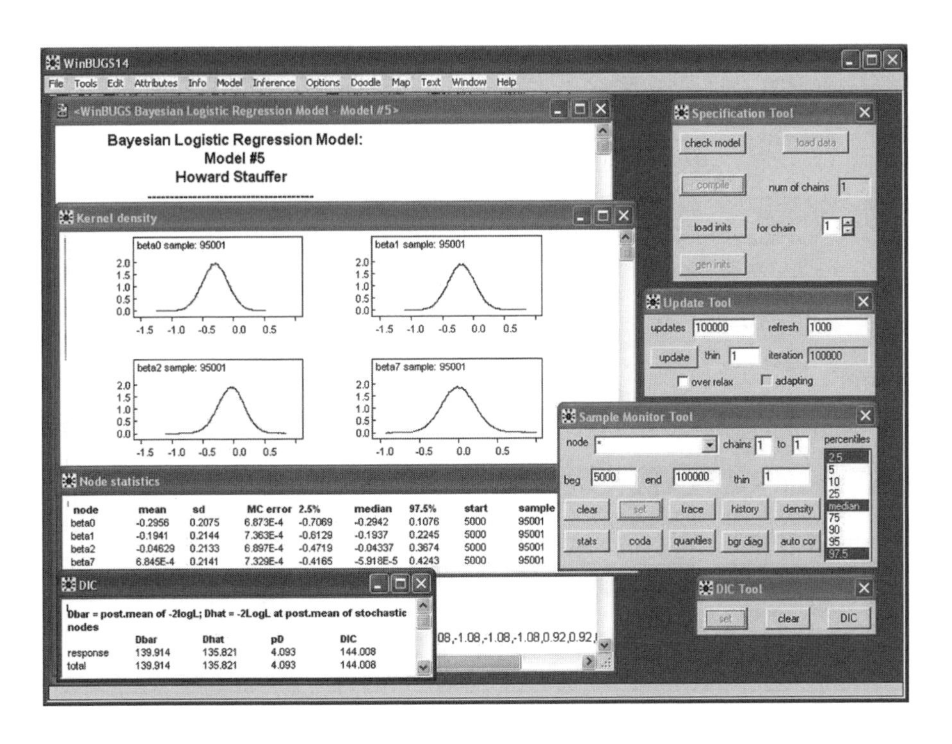

Model #6

```
model
{
for(i in 1:n)
   {
      response[i] ~ dbin(p[i],1)
      logit(p[i]) <- beta0+beta1*aspect[i]
                        +beta3*old.growth[i]
                        +beta6*temp[i]
         }
beta0 ~ dnorm(0,0.000001)
beta1 ~ dnorm(0,0.000001)
beta3 ~ dnorm(0,0.000001)
beta6 ~ dnorm(0,0.000001)
}
```

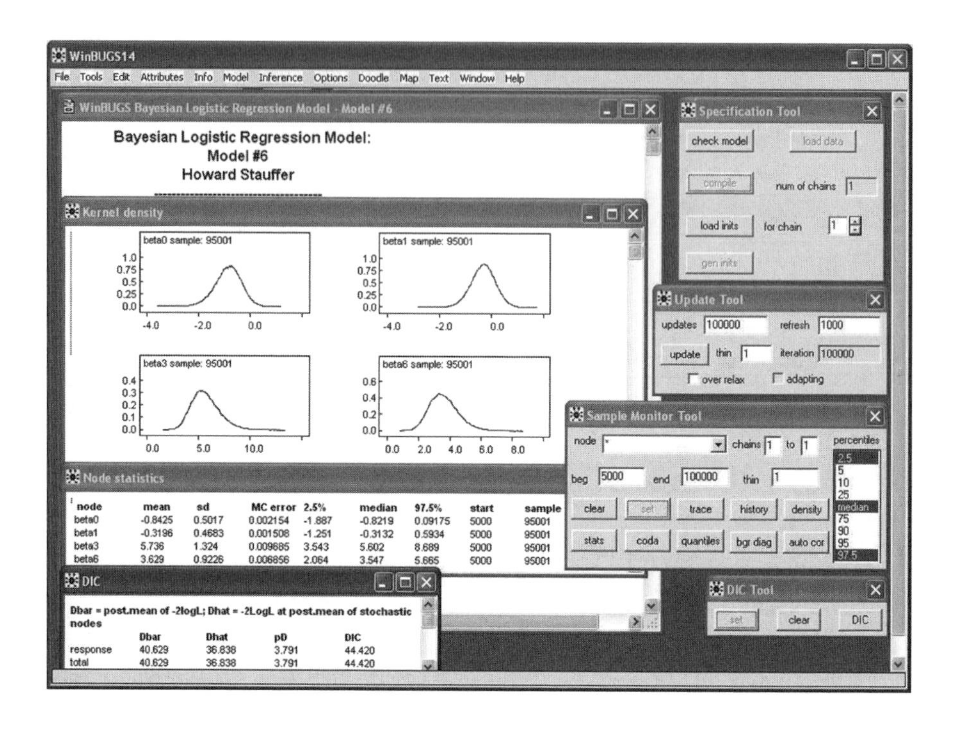

Model #7

```
model
{
for(i in 1:n)
   {
      response[i] ~ dbin(p[i],1)
      logit(p[i]) <- beta0
                        +beta4*rock[i]
                        +beta6*temp[i] + beta7*moist[i]
         }
beta0 ~ dnorm(0,0.000001)
beta4 ~ dnorm(0,0.000001)
beta6 ~ dnorm(0,0.000001)
beta7 ~ dnorm(0,0.000001)
}
```

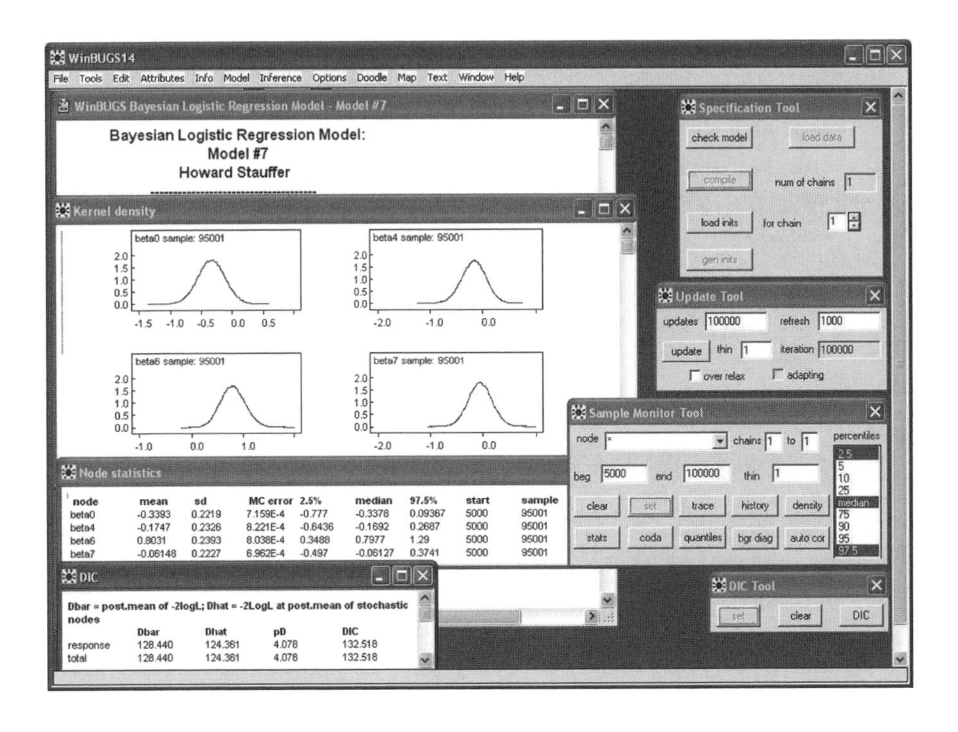

Model #8

```
model
{
for(i in 1:n)
   {
      response[i] ~ dbin(p[i],1)
      logit(p[i]) <- beta0
                        +beta5*moss[i]
                        +beta6*temp[i] + beta7*moist[i]
         }
beta0 ~ dnorm(0,0.000001)
beta5 ~ dnorm(0,0.000001)
beta6 ~ dnorm(0,0.000001)
beta7 ~ dnorm(0,0.000001)
}
```

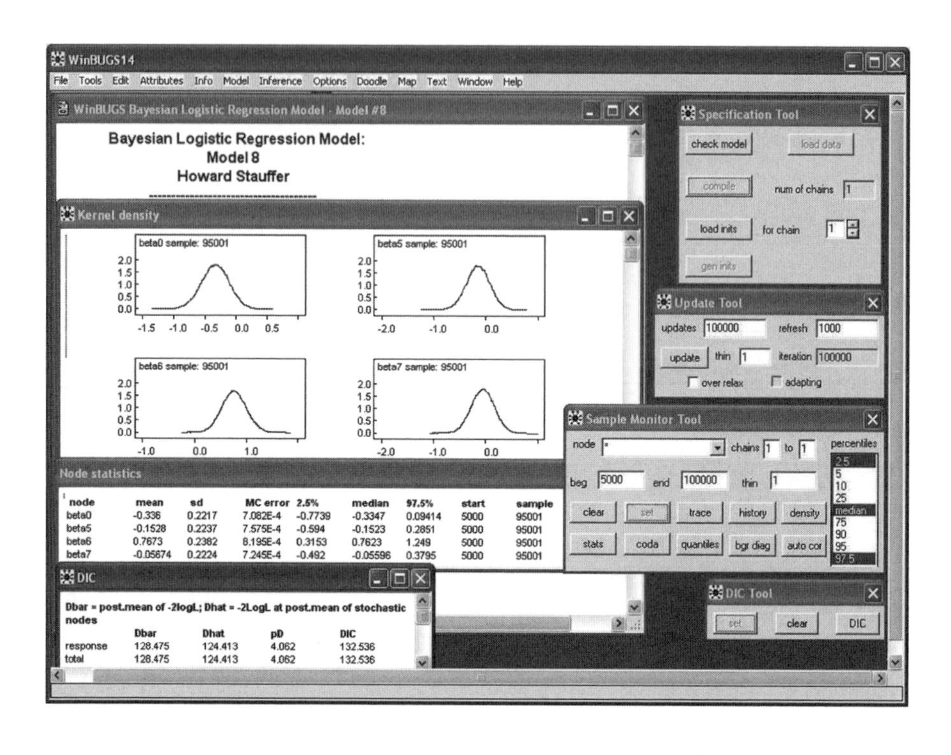

Model #9

```
model
{
for(i in 1:n)
   {
      response[i] ~ dbin(p[i],1)
      logit(p[i]) <- beta0
                          +beta3*old.growth[i]
                          +beta6*temp[i]
         }
beta0 ~ dnorm(0,0.000001)
beta3 ~ dnorm(0,0.000001)
beta6 ~ dnorm(0,0.000001)
}
```

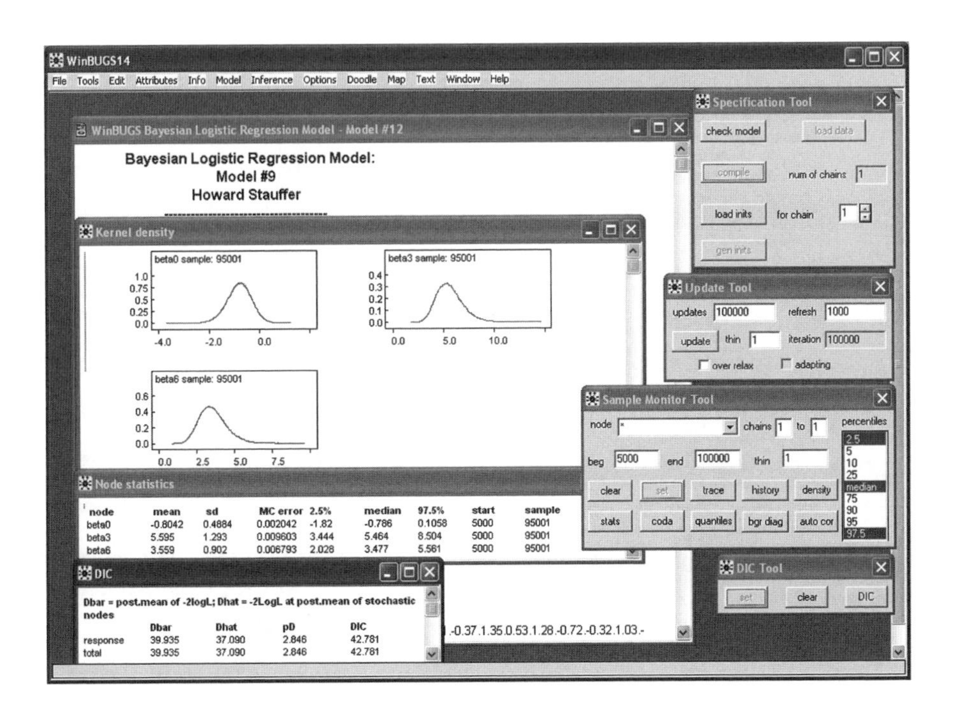


```
Model #10

model
{
for(i in 1:n)
   {
      response[i] ~ dbin(p[i],1)
      logit(p[i]) <- beta0
                        +beta4*rock[i]
                        + beta7*moist[i]
         }
beta0 ~ dnorm(0,0.000001)
beta4 ~ dnorm(0,0.000001)
beta7 ~ dnorm(0,0.000001)
}
```

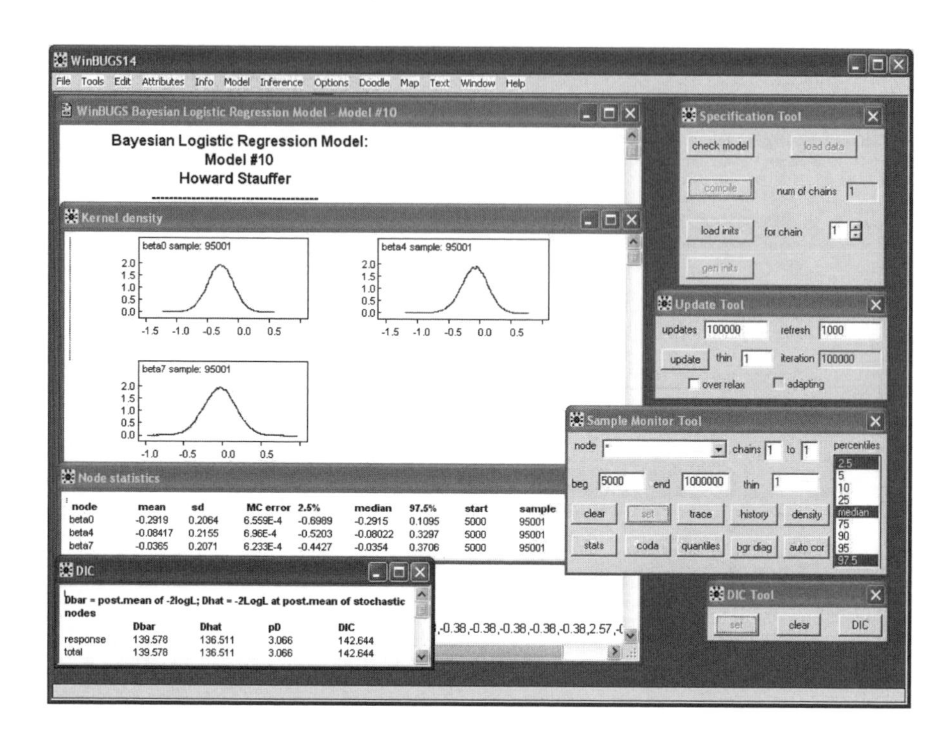

Model #11

```
model
{
for(i in 1:n)
   {
      response[i]  ~ dbin(p[i],1)
      logit(p[i])  <- beta0
                            +beta4*rock[i]+beta5*moss[i]
                            +beta6*temp[i] + beta7*moist[i]
         }
beta0 ~ dnorm(0,0.000001)
beta4 ~ dnorm(0,0.000001)
beta5 ~ dnorm(0,0.000001)
beta6 ~ dnorm(0,0.000001)
beta7 ~ dnorm(0,0.000001)
}
```

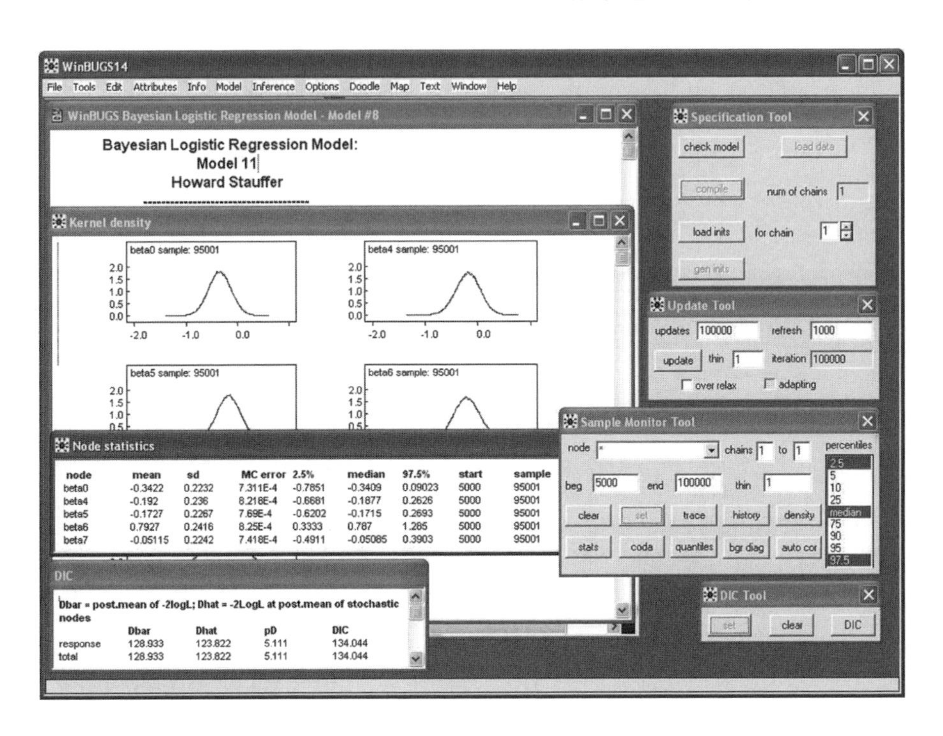


```
Model #12
model
{
for(i in 1:n)
   {
      response[i] ~ dbin(p[i],1)
      logit(p[i]) <- beta0+beta1*aspect[i]+beta2*species[i]
                        +beta3*old.growth[i]+beta4*rock[i]+beta5*moss[i]
                        +beta6*temp[i] +beta7*moist[i]
         }
beta0 ~ dnorm(0,0.1)
beta1 ~ dnorm(0,0.1)
beta2 ~ dnorm(0,0.1)
beta3 ~ dnorm(0,0.1)
beta4 ~ dnorm(0,0.1)
beta5 ~ dnorm(0,0.1)
beta6 ~ dnorm(0,0.1)
beta7 ~ dnorm(0,0.1)
# Note: the priors must be very precise in order to have
#    convergence for this poor fitting model
```

```
# Note: change models by deleting appropriate terms
#    in the description of logit(p[i]) (above), priors for
#    beta (above),
#    data in the list (below), and starting values for beta (below)

}
```

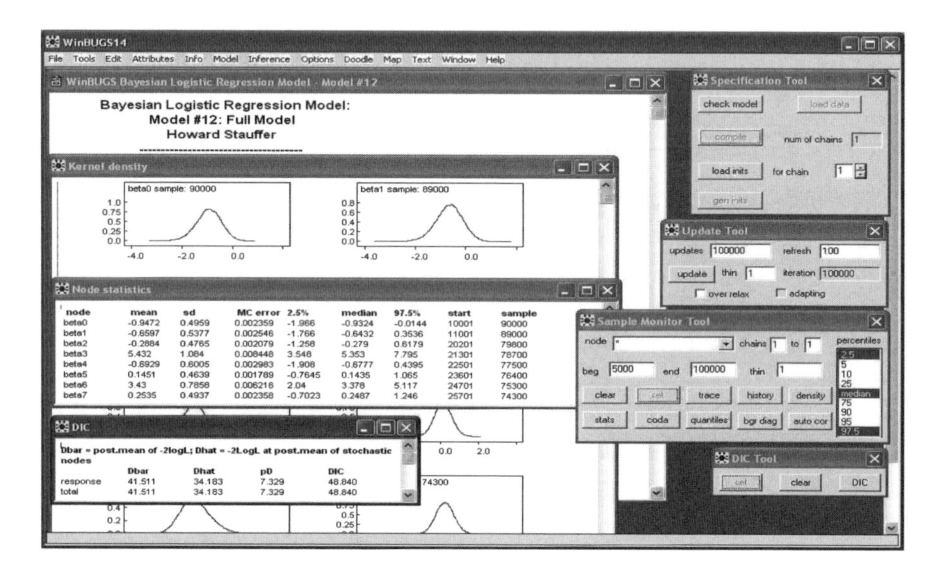

Summary (see Fig. B6.9).

Conclusion: Model 9, with old.growth and temp, is the best-fitting model, based on an analysis strategy of a priori parsimonious model selection and inference using DIC with Bayesian statistical analysis. The Bayesian results with DIC are similar to the frequentist results with AIC in Problem 6.1.

Problem 6.4

Summary (see Fig. B6.10).

Figure B6.9. Answers to Problems 6.1–6.3, statistics for 12 models: AIC and Akaike weights; DIC and DIC weights.

Model	Covariates	k	p_D	AIC		Akaike Weights	DIC		DIC Weights
1	aspect, species, old.growth	4	3.914	82.056		< 0.00005%	81.550		< 0.00005%
2	aspect, species, rock	4	4.104	144.059		< 0.00005%	143.858		< 0.00005%
3	aspect, species, moss	4	4.086	143.247		< 0.00005%	143.007		< 0.00005%
4	aspect, species, moist	4	4.067	133.115		< 0.00005%	132.863		< 0.00005%
5	aspect, species, temp	4	4.083	144.237		< 0.00005%	144.008		< 0.00005%
6	aspect, old.growth, temp	4	3.791	44.732		29.1853%	44.420		29.5940%
7	rock, temp, moist	4	4.078	132.743		< 0.00005%	132.518		< 0.00005%
8	moss, temp, moist	4	4.062	132.798		< 0.00005%	132.536		< 0.00005%
9	old.growth, temp	3	2.846	42.985		69.9167%	42.781		67.1595%
10	rock, moist	3	3.066	142.755		< 0.00005%	142.644		< 0.00005%
11	rock, moss, temp, moist	5	5.111	134.408		< 0.00005%	134.044		< 0.00005%
12	aspect, species, old.growth, rock, moss, temp, moist	8	7.329	51.695		0.8981%	48.840		3.2465%
				Total:		100.0000%	*Total:*		100.0000%

Note: k = number of parameters
p_D = Bayesian number of parameters

Figure B6.10. Answer to Problem 6.4, statistics for 12 models: AIC and Akaike weights; DIC and DIC weights; model averaging results.

Model	Covariates	DIC	DIC Weights	old.growth Estimate	old.growth Standard Error	temp Estimate	temp Standard Error
1	aspect, species,old. growth	81.550	0.0000%	2.6640	0.5085	0.0000	0.0000
2	aspect, species, rock	143.858	0.0000%	0.0000	0.0000	0.0000	0.0000
3	aspect, species, moss	143.007	0.0000%	0.0000	0.0000	0.0000	0.0000
4	aspect, species, temp	132.863	0.0000%	0.0000	0.0000	0.0000	0.0000
5	aspect, species, moist	144.008	0.0000%	0.0000	0.0000	0.0000	0.0000
6	aspect,old.growth, temp	44.420	29.5940%	5.7360	1.3240	3.6290	0.9226
7	rock, temp, moist	132.518	0.0000%	0.0000	0.0000	0.8031	0.2393
8	moss, temp, moist	132.536	0.0000%	0.0000	0.0000	0.7673	0.2382
9	old.growth, temp	42.781	67.1595%	5.5950	1.2930	3.5590	0.9020
10	rock, moist	142.644	0.0000%	0.0000	0.0000	0.0000	0.0000
11	rock, moss, temp, moist	134.044	0.0000%	0.0000	0.0000	0.7927	0.2416
12	aspect, species, old.growth, rock, moss, temp, moist	48.840	3.2465%	5.4320	1.0840	3.4300	0.7858
		Total:	100.0000%	5.6314	1.2954	3.5755	0.9043

Note: k = number of parameters

Conclusion: The unconditional estimates of `old.growth` coefficient (and standard error) are 5.6314 (1.2954) and of `temp` coefficient (and standard error) are 3.5755 (0.9043). The shrinkage estimators are very similar since most of the insignificant weights are very close to 0. These estimates are calculated using DIC weights and means (and standard deviations) of the posteriors for `old.growth` and `temp`. The results for AIC weights will be similar since the coefficient estimates and standard errors and AIC weights for the frequentist statistical analysis are very similar to those for the Bayesian statistical analysis.

```
> # The importance of the covariates, using DIC weights, are:
> # aspect
> .2959+.0325
[1] 0.3284

> # species
> .0325
[1] 0.0325

> # old.growth
> .2959+.6716+.0325
[1] 1

> # rock
> .0325
[1] 0.0325

> # moss
> .0325
[1] 0.0325

> # temp
> .2959+.6716+.0325

> # moist
> .0325
[1] 0.0325

> # The results, using AIC weights, will be similar, since the AIC weights are similar
to the DIC weights.
```

Problem 6.5

The classification results for the frequentist statistical analysis best-fitting model 9 are as follows:

```
function(output, data)
{
      freq00 <- 0
      freq01 <- 0
      freq10 <- 0
      freq11 <- 0
      n <- dim(data)[1]
      sensitivity <- rep(NA, 19)
      specificity <- rep(NA, 19)
      correct <- rep(NA, 19)
      c <- rep(NA,19)
      j <- 1
      for(pcutoff in seq(0.05, 0.95, 0.05)) {
             for(i in 1:n) {
                   if(data$response[i] == 0 & output$fitted.values[i] <= pcutoff)
                         freq00 <- freq00 + 1
                   if(data$response[i] == 0 & output$fitted.values[i] <= pcutoff)
                         freq01 <- freq01 + 1
                   if(data$response[i] == 1 & output$fitted.values[i] <= pcutoff)
                         freq10 <- freq10 + 1
                   if(data$response[i] == 1 & output$fitted.values[i] <= pcutoff)
                         freq11 <- freq11 + 1
            }
            sensitivity[j] <- freq11/(freq10 + freq11)
            specificity[j] <- freq00/(freq00 + freq01)
            correct[j] <- (freq00 + freq11)/(freq00 + freq01 +freq10 + freq11)
            c <- sensitivity/max(1-specificity,0.00001)
            j <- j + 1
 }
 prob.cutoff <- seq(0.05, 0.95, 0.05)
 results <- data.frame(prob.cutoff, sensitivity, specificity,
   correct, c)
 return(results)
}
```

```
> classification(model9,data2)
   prob.cutoff sensitivity specificity   correct        c
 1        0.05   1.0000000   0.6491228 0.8000000 2.850000
 2        0.10   0.9883721   0.6929825 0.8200000 2.816860
 3        0.15   0.9844961   0.7426901 0.8466667 2.805814
 4        0.20   0.9767442   0.7675439 0.8575000 2.783721
 5        0.25   0.9720930   0.7859649 0.8660000 2.770465
 6        0.30   0.9651163   0.8011696 0.8716667 2.750581
 7        0.35   0.9568106   0.8145363 0.8757143 2.726910
 8        0.40   0.9476744   0.8267544 0.8787500 2.700872
 9        0.45   0.9405685   0.8382066 0.8822222 2.680620
10        0.50   0.9348837   0.8473684 0.8850000 2.664419
11        0.55   0.9302326   0.8548644 0.8872727 2.651163
12        0.60   0.9244186   0.8625731 0.8891667 2.634593
13        0.65   0.9194991   0.8690958 0.8907692 2.620572
14        0.70   0.9152824   0.8759398 0.8928571 2.608555
15        0.75   0.9116279   0.8830409 0.8953333 2.598140
16        0.80   0.9084302   0.8892544 0.8975000 2.589026
17        0.85   0.8974008   0.8947368 0.8958824 2.557592
18        0.90   0.8875969   0.9005848 0.8950000 2.529651
19        0.95   0.8763770   0.9058172 0.8931579 2.497674
```

The optimal cutoff point is 0.80, with correct classification 0.8975, sensitivity 0.9084, and specificity 0.8893.

Problem 7.1

```
> Orthodont
Grouped Data: distance ~ age | Subject
     distance age   Subject    Sex
  1      26.0   8       M01   Male
  2      25.0  10       M01   Male
  3      29.0  12       M01   Male
  4      31.0  14       M01   Male
  5      21.5   8       M02   Male
  6      22.5  10       M02   Male
  7      23.0  12       M02   Male
  8      26.5  14       M02   Male
  9      23.0   8       M03   Male
 10      22.5  10       M03   Male
 11      24.0  12       M03   Male
 12      27.5  14       M03   Male
 13      25.5   8       M04   Male
 14      27.5  10       M04   Male
 15      26.5  12       M04   Male
```

```
   distance age Subject  Sex
16     27.0  14     M04 Male
17     20.0   8     M05 Male
18     23.5  10     M05 Male
19     22.5  12     M05 Male
20     26.0  14     M05 Male
21     24.5   8     M06 Male
22     25.5  10     M06 Male
23     27.0  12     M06 Male
24     28.5  14     M06 Male
25     22.0   8     M07 Male
26     22.0  10     M07 Male
27     24.5  12     M07 Male
28     26.5  14     M07 Male
29     24.0   8     M08 Male
30     21.5  10     M08 Male
31     24.5  12     M08 Male
32     25.5  14     M08 Male
33     23.0   8     M09 Male
34     20.5  10     M09 Male
35     31.0  12     M09 Male
36     26.0  14     M09 Male
37     27.5   8     M10 Male
38     28.0  10     M10 Male
39     31.0  12     M10 Male
40     31.5  14     M10 Male
41     23.0   8     M11 Male
42     23.0  10     M11 Male
43     23.5  12     M11 Male
44     25.0  14     M11 Male
45     21.5   8     M12 Male
46     23.5  10     M12 Male
47     24.0  12     M12 Male
48     28.0  14     M12 Male
49     17.0   8     M13 Male
50     24.5  10     M13 Male
51     26.0  12     M13 Male
52     29.5  14     M13 Male
53     22.5   8     M14 Male
54     25.5  10     M14 Male
55     25.5  12     M14 Male
56     26.0  14     M14 Male
57     23.0   8     M15 Male
58     24.5  10     M15 Male
59     26.0  12     M15 Male
```

```
    distance age  Subject    Sex
60      30.0  14      M15   Male
61      22.0   8      M16   Male
62      21.5  10      M16   Male
63      23.5  12      M16   Male
64      25.0  14      M16   Male
65      21.0   8      F01 Female
66      20.0  10      F01 Female
67      21.5  12      F01 Female
68      23.0  14      F01 Female
69      21.0   8      F02 Female
70      21.5  10      F02 Female
71      24.0  12      F02 Female
72      25.5  14      F02 Female
73      20.5   8      F03 Female
74      24.0  10      F03 Female
75      24.5  12      F03 Female
76      26.0  14      F03 Female
77      23.5   8      F04 Female
78      24.5  10      F04 Female
79      25.0  12      F04 Female
80      26.5  14      F04 Female
81      21.5   8      F05 Female
82      23.0  10      F05 Female
83      22.5  12      F05 Female
84      23.5  14      F05 Female
85      20.0   8      F06 Female
86      21.0  10      F06 Female
87      21.0  12      F06 Female
88      22.5  14      F06 Female
89      21.5   8      F07 Female
90      22.5  10      F07 Female
91      23.0  12      F07 Female
92      25.0  14      F07 Female
93      23.0   8      F08 Female
94      23.0  10      F08 Female
95      23.5  12      F08 Female
96      24.0  14      F08 Female
97      20.0   8      F09 Female
98      21.0  10      F09 Female
99      22.0  12      F09 Female
100     21.5  14      F09 Female
101     16.5   8      F10 Female
102     19.0  10      F10 Female
103     19.0  12      F10 Female
```

```
    distance age   Subject    Sex
104      19.5  14       F10 Female
105      24.5   8       F11 Female
106      25.0  10       F11 Female
107      28.0  12       F11 Female
108      28.0  14       F11 Female
names(Orthodont)
[1] "distance" "age" "Subject" "Sex"
> out1 <- gls(distance~1,Orthodont)
> out2 <- gls(distance~age,Orthodont)
> out2a <- gls(distance~age-1,Orthodont)
> out3 <- gls(distance~age+age^2,Orthodont)
> out3a <- gls(distance~age+age^2-1,Orthodont)
> out3b <- gls(distance~age^2-1-age,Orthodont)
> AIC(out1,out2,out2a,out3,out3a,out3b)
      df      AIC
 out1  2 542.2815
 out2  3 515.1695
out2a  2 624.2148
 out3  4 520.6967
out3a  3 532.0595
out3b  2 762.6865
> summary(out1)

Generalized least squares fit by REML
  Model: distance ~ 1
  Data: Orthodont
        AIC       BIC     logLik
  542.2815  547.6272  -269.1408

Coefficients:
                Value Std.Error  t-value p-value
(Intercept)  24.02315 0.2818024 85.24819  <.0001

Standardized residuals:
      Min         Q1         Med        Q3       Max
-2.568875 -0.6908298 -0.09326993 0.6750213 2.553067

Residual standard error: 2.928577
Degrees of freedom: 108 total; 107 residual
> summary(out2)
Generalized least squares fit by REML
  Model: distance ~ age
  Data: Orthodont
        AIC      BIC    logLik
  515.1695 523.1598 -254.5847
```

```
Coefficients:
                Value Std.Error  t-value p-value
(Intercept)  16.76111  1.225560 13.67629  <.0001
        age   0.66019  0.109182  6.04667  <.0001

Correlation:
     (Intr)
age -0.98

Standardized residuals:
       Min         Q1           Med          Q3       Max
 -2.563389  -0.62187  -0.07225954  0.5328229  2.48967

Residual standard error: 2.537151
Degrees of freedom: 108 total; 106 residual
> summary(out2a)
Generalized least squares fit by REML
  Model: distance ~ age - 1
  Data: Orthodont
       AIC       BIC     logLik
  624.2148  629.5605  -310.1074

Coefficients:
       Value  Std.Error  t-value p-value
age 2.123457 0.03599327 58.99595  <.0001

Standardized residuals:
       Min         Q1          Med          Q3       Max
 -2.436066 -0.546902 0.09409067 0.9115034 2.503694

Residual standard error: 4.198734
Degrees of freedom: 108 total; 107 residual
> summary(out3)
Generalized least squares fit by REML
  Model: distance ~ age + age^2
  Data: Orthodont
       AIC       BIC     logLik
  520.6967  531.3125  -256.3483

Coefficients:
               Value Std.Error   t-value p-value
(Intercept) 20.11759  7.211753  2.789557  0.0063
        age  0.02361  1.352152  0.017462  0.9861
   I(age^2)  0.02894  0.061259  0.472340  0.6377
```

```
 Correlation:
         (Intr)    age
     age -0.996
I(age^2)  0.985 -0.997

Standardized residuals:
       Min          Q1         Med         Q3      Max
 -2.599429 -0.6155872 -0.0545411 0.547229 2.52598

Residual standard error: 2.5465
Degrees of freedom: 108 total; 105 residual
> summary(out3a)
Generalized least squares fit by REML
  Model: distance ~ age + age^2 - 1
  Data: Orthodont
       AIC       BIC    logLik
  532.0595 540.0498 -263.0297

Coefficients:
             Value  Std.Error    t-value  p-value
     age   3.779112  0.1299307   29.08559   <.0001
I(age^2)  -0.139447  0.0107778  -12.93837   <.0001

Correlation:
           age
I(age^2) -0.985

Standardized residuals:
       Min          Q1         Med          Q3       Max
 -2.386645 -0.6183548 0.01477009 0.6331542 2.357222

Residual standard error: 2.626696
Degrees of freedom: 108 total; 106 residual
> summary(out3b)
Generalized least squares fit by REML
  Model: distance ~ age^2 - 1 - age
  Data: Orthodont
       AIC       BIC    logLik
  762.6865 768.0321 -379.3432

Coefficients:
             Value    Std.Error   t-value  p-value
I(age^2) 0.1692879 0.005571219  30.38615   <.0001
```

```
Standardized residuals:
       Min         Q1        Med        Q3       Max
 -1.746102 -0.3769319 0.4909963  1.112952  2.127112

Residual standard error: 7.834838
Degrees of freedom: 108 total; 107 residual
```

> # Model 2 is best-fitting, with `distance ~ age.`

Problem 7.2

```
> out4a <- lme(distance~age,Orthodont,random=~1|Sex)
> out4b <- lme(distance~age,Orthodont,random=~age|Sex)
> out4c <- lme(distance~age,Orthodont,random=~age-1|Sex)
> out5a <- lme(distance~age,Orthodont,random=~1|Subject)
> out5b <- lme(distance~age,Orthodont,random=~age|Subject)
> out5c <- lme(distance~age,Orthodont,random=~age-1|Subject)
> out6a <- lme(distance~age,Orthodont,random=~1|Sex/Subject)
> out6b <- lme(distance~age,Orthodont,random=~age|Sex/Subject)
> out6c <- lme(distance~age,Orthodont,random=~age-1|Sex/Subject)
> AIC(out4a,out4b,out4c,out5a,out5b,out5c,out6a,out6b,out6c)
      df      AIC
out4a  4 497.0458
out4b  6 498.7545
out4c  4 494.9654
out5a  4 455.0025
out5b  6 454.6367
out5c  4 453.0857
out6a  5 452.0344
out6b  9 453.2491
out6c  5 449.5305
> summary(out4a)
Linear mixed-effects model fit by REML
 Data: Orthodont
       AIC      BIC    logLik
  497.0458 507.6995 -244.5229

Random effects:
 Formula:  ~ 1 | Sex
        (Intercept) Residual
StdDev:     1.61078 2.271713

Fixed effects: distance ~ age
                Value Std.Error  DF   t-value p-value
(Intercept) 16.55410  1.582118 105  10.46325  <.0001
        age  0.66019  0.097759 105   6.75319  <.0001
```

```
 Correlation:
    (Intr)
age -0.68

Standardized Within-Group Residuals:
       Min          Q1          Med          Q3       Max
 -2.620689 -0.6398049 -0.01048698  0.5397587 2.379617

Number of Observations: 108
Number of Groups: 2
> intervals(out4a)
Approximate 95% confidence intervals

 Fixed effects:
                lower       est.       upper
(Intercept) 13.4170490 16.5540974  19.6911457
        age  0.4663472  0.6601852   0.8540231

 Random Effects:
  Level: Sex
                   lower    est.     upper
sd((Intercept)) 0.3789406 1.61078  6.847011
 Within-group standard error:
    lower     est.    upper
 1.984329 2.271713 2.600718
> summary(out4b)
Linear mixed-effects model fit by REML
 Data: Orthodont
       AIC      BIC    logLik
  498.7545 514.7351 -243.3772

Random effects:
 Formula: ~ age | Sex
 Structure: General positive-definite
              StdDev   Corr
(Intercept) 0.7178830 (Inter
        age 0.2119552 -1
   Residual 2.2461781
Fixed effects: distance ~ age
              Value Std.Error  DF   t-value  p-value
(Intercept) 16.85355  1.198003 105  14.06804   <.0001
        age  0.63289  0.178413 105   3.54734   0.0006
 Correlation:
    (Intr)
age -0.837
```

```
Standardized Within-Group Residuals:
      Min          Q1           Med           Q3         Max
 -2.49162 -0.5691045 -0.08825623    0.5978222   2.351662

Number of Observations: 108
Number of Groups: 2
> intervals(out4b)
Approximate 95% confidence intervals

 Fixed effects:
                lower       est.         upper
(Intercept) 14.4781327 16.8535517    19.2289707
        age  0.2791308  0.6328906     0.9866505

 Random Effects:
  Level: Sex
                                 lower            est.       upper
     sd((Intercept))  0.002676612     0.7178830 192.540401
             sd(age)  0.022181098     0.2119552   2.025374
cor((Intercept),age) -1.000000000    -0.9999710         NA
 Within-group standard error:
    lower     est.    upper
 1.962026 2.246178 2.571483
> summary(out4c)
Linear mixed-effects model fit by REML
> Data: Orthodont
        AIC       BIC    logLik
   494.9654 505.6192 -243.4827

Random effects:
 Formula:  ~ age - 1  | Sex
             age  Residual
StdDev: 0.1492784  2.248511

Fixed effects: distance ~ age
                 Value Std.Error   DF  t-value  p-value
(Intercept) 16.76111  1.086134  105 15.43190   <.0001
        age  0.64097  0.143239  105  4.47481   <.0001
 Correlation:
    (Intr)
age -0.662
```

```
Standardized Within-Group Residuals:
       Min          Q1         Med          Q3       Max
 -2.543496 -0.5491384 -0.0857501   0.5744827 2.357976

Number of Observations: 108
Number of Groups: 2
> intervals(out4c)
Approximate 95% confidence intervals

 Fixed effects:
                lower       est.     upper
(Intercept) 14.6075078 16.7611111 18.914714
        age  0.3569506  0.6409668  0.924983
 Random Effects:
 Level: Sex
           lower      est.     upper
sd(age) 0.0355699 0.1492784 0.6264859

 Within-group standard error:
    lower     est.    upper
 1.964065 2.248511 2.574153
> summary(out5a)
Linear mixed-effects model fit by REML
 Data: Orthodont
       AIC      BIC    logLik
  455.0025 465.6563 -223.5013

Random effects:
 Formula:  ~ 1 | Subject
        (Intercept) Residual
StdDev:    2.114724 1.431592

Fixed effects: distance ~ age
               Value Std.Error DF   t-value   p-value
(Intercept) 16.76111 0.8023952 80  20.88885    <.0001
        age  0.66019 0.0616059 80  10.71626    <.0001
 Correlation:
    (Intr)
age -0.845

Standardized Within-Group Residuals:
       Min          Q1          Med          Q3        Max
 -3.664539 -0.5350798 -0.01289591  0.4874286   3.721785

Number of Observations: 108
```

```
Number of Groups: 27
> intervals(out5a)
Approximate 95% confidence intervals

 Fixed effects:
                lower       est.      upper
(Intercept) 15.1642937 16.7611111 18.3579285
        age  0.5375855  0.6601852  0.7827849

 Random Effects:
 Level: Subject
                  lower     est.   upper
sd((Intercept)) 1.561205 2.114724 2.86449

 Within-group standard error:
    lower     est.    upper
 1.226093 1.431592 1.671535
> summary(out5b)
Linear mixed-effects model fit by REML
 Data: Orthodont
       AIC      BIC    logLik
  454.6367 470.6173 -221.3183

Random effects:
 Formula:  ~ age | Subject
 Structure: General positive-definite
             StdDev   Corr
(Intercept) 2.327037 (Inter
        age 0.226427 -0.609
   Residual 1.310040

Fixed effects: distance ~ age
               Value Std.Error DF  t-value  p-value
(Intercept) 16.76111 0.7752464 80 21.62037   <.0001
        age  0.66019 0.0712532 80  9.26534   <.0001
 Correlation:
    (Intr)
age -0.848

Standardized Within-Group Residuals:
       Min          Q1           Med        Q3       Max
 -3.223106 -0.4937604 0.007316863 0.472151 3.916033

Number of Observations: 108
Number of Groups: 27
```

```
> intervals(out5b)
Approximate 95% confidence intervals

 Fixed effects:
                lower       est.      upper
(Intercept) 15.2183215 16.7611111 18.3039007
        age  0.5183868  0.6601852  0.8019835

 Random Effects:
  Level: Subject
                             lower       est.     upper
     sd((Intercept))  0.9503642  2.3270368 5.6979214
             sd(age)  0.1026589  0.2264270 0.4994129
cor((Intercept),age) -0.9379522 -0.6093317 0.2959052

 Within-group standard error:
    lower    est.    upper
 1.084904 1.31004 1.581896
> summary(out5c)
Linear mixed-effects model fit by REML
 Data: Orthodont
       AIC      BIC    logLik
  453.0857 463.7394 -222.5428

Random effects:
 Formula:  ~ age - 1 | Subject
             age Residual
StdDev: 0.1895482 1.412641

Fixed effects: distance ~ age
               Value Std.Error DF  t-value p-value
(Intercept) 16.76111 0.6823703 80 24.56307  <.0001
        age  0.66019 0.0708954 80  9.31210  <.0001
 Correlation:
    (Intr)
age -0.84

Standardized Within-Group Residuals:
      Min         Q1         Med          Q3       Max
 -3.93173 -0.4520521 0.002636414 0.4834305 3.639918

Number of Observations: 108
Number of Groups: 27
> intervals(out5c)
Approximate 95% confidence intervals
```

```
 Fixed effects:
                  lower       est.      upper
(Intercept) 15.4031509 16.7611111 18.1190713
        age  0.5190989  0.6601852  0.8012715

 Random Effects:
  Level: Subject
             lower      est.     upper
sd(age) 0.1401078 0.1895482 0.2564348

 Within-group standard error:
    lower     est.    upper
 1.209871 1.412641 1.649393
> summary(out6a)
Linear mixed-effects model fit by REML
 Data: Orthodont
       AIC      BIC    logLik
  452.0344 465.3516 -221.0172

Random effects:
 Formula:  ~ 1 | Sex
        (Intercept)
StdDev:     1.55039

 Formula:  ~ 1 | Subject %in% Sex
        (Intercept) Residual
StdDev:    1.807425 1.431592

Fixed effects: distance ~ age
                Value Std.Error DF  t-value p-value
(Intercept) 16.56933  1.343685 80 12.33126  <.0001
        age  0.66019  0.061606 80 10.71626  <.0001
 Correlation:
    (Intr)
age -0.504

Standardized Within-Group Residuals:
       Min         Q1          Med         Q3       Max
 -3.739259 -0.5466211 -0.01599553 0.4519955 3.667103

Number of Observations: 108
Number of Groups:
 Sex Subject %in% Sex
   2               27
```

```
> intervals(out6a)
Approximate 95% confidence intervals

 Fixed effects:
                 lower       est.      upper
(Intercept) 13.8953121 16.5693297 19.2433473
        age  0.5375855  0.6601852  0.7827849

 Random Effects:
  Level: Sex
                    lower    est.    upper
sd((Intercept)) 0.3261313 1.55039 7.370373
  Level: Subject
                   lower     est.   upper
sd((Intercept)) 1.310314 1.807425 2.49313

 Within-group standard error:
    lower     est.    upper
 1.226087 1.431592 1.671542
> summary(out6b)
Linear mixed-effects model fit by REML
 Data: Orthodont
       AIC    BIC    logLik
  453.2491 477.22 -217.6245
Random effects:
 Formula:  ~ age | Sex
 Structure: General positive-definite
              StdDev   Corr
(Intercept) 0.6998300 (Inter
        age 0.2070146 -1

 Formula:  ~ age | Subject %in% Sex
 Structure: General positive-definite
              StdDev   Corr
(Intercept) 2.2754327 (Inter
        age 0.1726564 -0.634
   Residual 1.3099227

Fixed effects: distance ~ age
              Value Std.Error DF  t-value p-value
(Intercept) 16.84921 0.9152379 80 18.40965  <.0001
        age  0.63412 0.1605152 80  3.95054  0.0002
 Correlation:
     (Intr)
age -0.795
```

```
Standardized Within-Group Residuals:
       Min          Q1          Med        Q3       Max
 -3.247803 -0.4087215 0.008375599 0.432053 3.861081

Number of Observations: 108
Number of Groups:
 Sex Subject %in% Sex
   2              27
> intervals(out6b)
Approximate 95% confidence intervals

 Fixed effects:
                 lower       est.      upper
(Intercept) 15.0278323 16.8492139 18.6705955
        age  0.3146863  0.6341218  0.9535573

 Random Effects:
  Level: Sex
                           lower       est.     upper
    sd((Intercept))  0.02486172  0.6998300 19.699442
           sd(age)  0.04045817  0.2070146  1.059243
cor((Intercept),age) -1.00000000 -0.9999519  1.000000
  Level: Subject
                           lower       est.     upper
    sd((Intercept))  0.90309418  2.2754327 5.7331717
           sd(age)  0.05280769  0.1726564 0.5645057
cor((Intercept),age) -0.96232650 -0.6337161 0.4472527

 Within-group standard error:
    lower     est.    upper
 1.084813 1.309923 1.581745
> summary(out6c)
Linear mixed-effects model fit by REML
 Data: Orthodont
      AIC      BIC    logLik
  449.5305 462.8477 -219.7652

Random effects:
 Formula:  ~ age - 1 | Sex
             age
StdDev: 0.1442001

 Formula:  ~ age - 1 | Subject %in% Sex
             age Residual
StdDev: 0.1596937 1.412641
```

```
Fixed effects: distance ~ age
                Value Std.Error DF  t-value p-value
(Intercept) 16.76111 0.6823703 80 24.56307  <.0001
        age  0.64225 0.1227672 80  5.23147  <.0001
 Correlation:
    (Intr)
age -0.485

Standardized Within-Group Residuals:
       Min         Q1          Med        Q3      Max
 -3.977661 -0.4036487 -0.02843643 0.503207 3.582271

Number of Observations: 108
Number of Groups:
 Sex Subject %in% Sex
   2              27
> intervals(out6c)
Approximate 95% confidence intervals

 Fixed effects:
                  lower       est.      upper
(Intercept) 15.4031509 16.7611111 18.1190713
        age  0.3979377  0.6422521  0.8865666

 Random Effects:
  Level: Sex
              lower      est.    upper
sd(age) 0.03101527 0.1442001 0.670433
  Level: Subject
             lower      est.    upper
sd(age) 0.1158322 0.1596937 0.220164

 Within-group standard error:
    lower     est.    upper
 1.209877 1.412641 1.649386
```

Conclusion: Model 6c is best-fitting, with distance $\sim$ age, random = $\sim$age-1|Sex/Subject.

Answers to Problem 7.2 are summarized in Fig. B7.16.

Figure B.7.16. Answer to Problem 7.2: comparative frequentist statistics for linear regression and mixed-effects models.

Model	Covariates	*k*	AIC	AIC Weights
1	Linear regression: 1	2	542.2815	< 0.00005
2	Linear regression: 1, age	3	515.1695	< 0.00005
2a	Linear regression: age	2	624.2148	< 0.00005
3	Linear regression: 1, age, age^2	4	520.6967	< 0.00005
3a	Linear regression: age, age^2	3	532.0595	< 0.00005
3b	Linear regression: age^2	2	762.6865	< 0.00005
4a	Mixed effects on 1: ~1|Sex	4	497.0458	< 0.00005
4b	Mixed effects on 1, age: ~age|Sex	5	498.7545	< 0.00005
4c	Mixed effects on age: ~age-1|Sex	4	494.9654	< 0.00005
5a	Mixed effects on 1: ~1|Subject	4	455.0025	0.0370
5b	Mixed effects on 1, age: ~age|Subject	5	454.6367	0.0444
5c	Mixed effects on age: ~age-1|Subject	4	453.0857	0.0964
6a	Mixed effects on 1: ~1|Sex/Subject	5	452.0344	0.1631
6b	Mixed effects on 1, age: ~age|Sex/Subject	6	453.2491	0.0888
6c	Mixed effects on age: ~age-1|Sex/Subject	5	449.5305	0.5703
				1.0000

Problem 7.3

```
> plot(Orthodont)
```

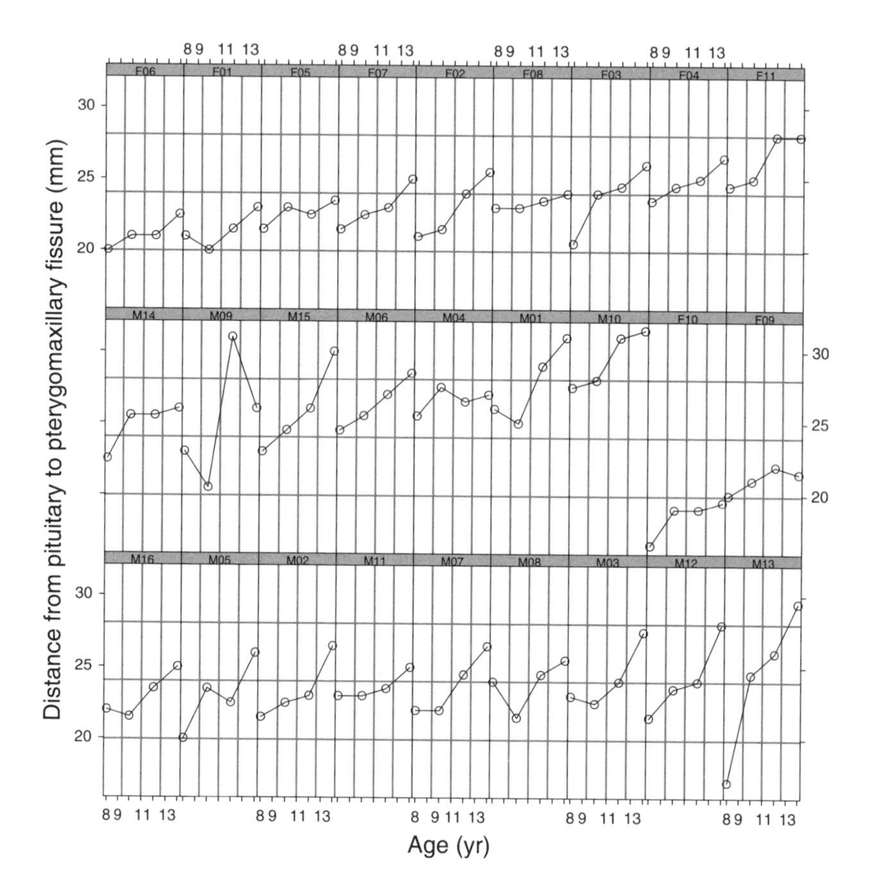

```
> plot(augPred(out6c))
```

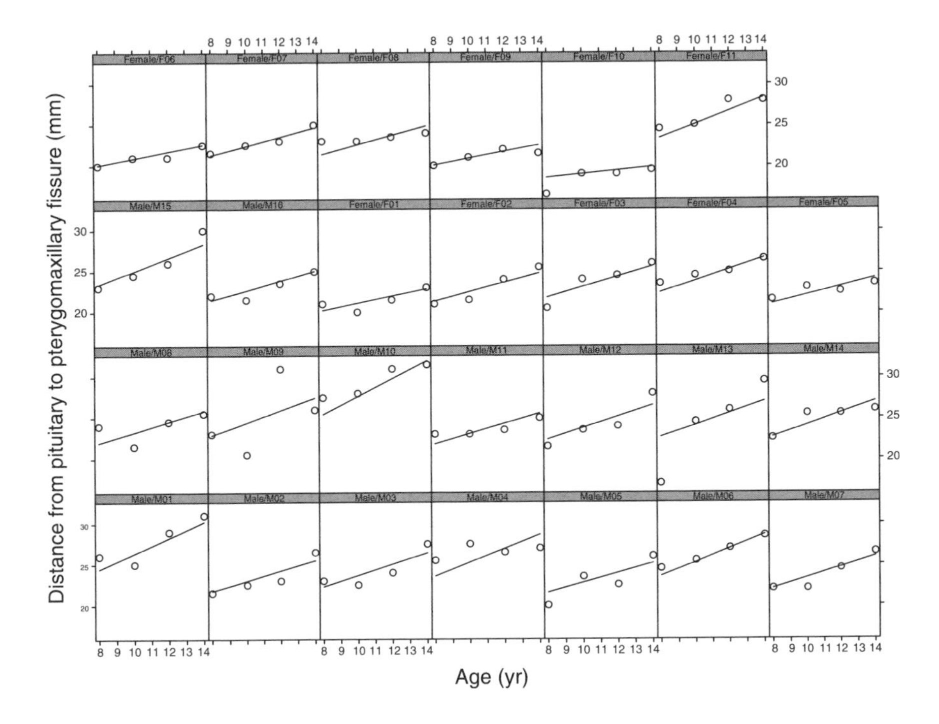

Yes, the graphs indicate random effects with respect to both Sex and Subject within Sex.

Problem 7.4

See Problems 7.1 and 7.2 (above) for model specifications. Answers to Problems 7.4 and 7.5 are summarized in Fig. B7.17.

Model 1:

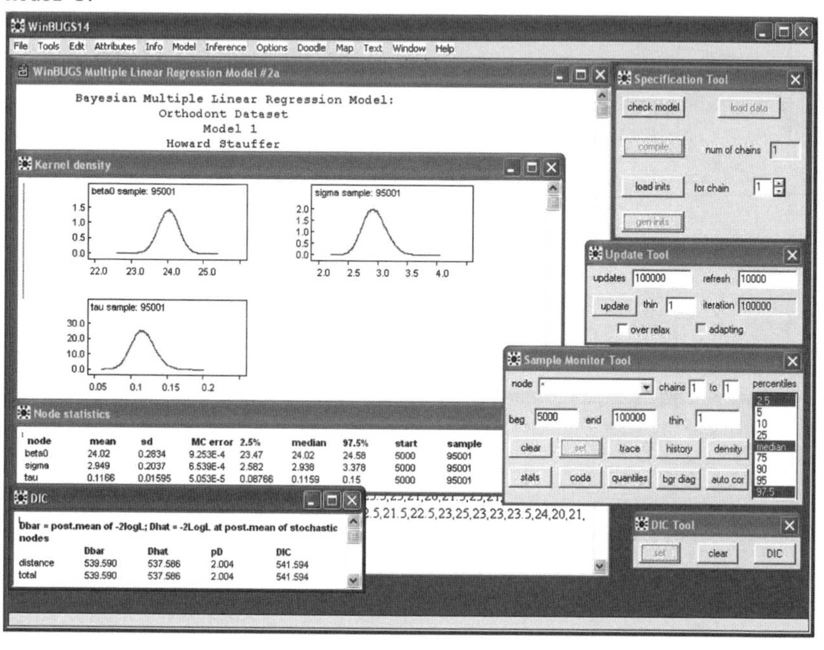

Model 2:

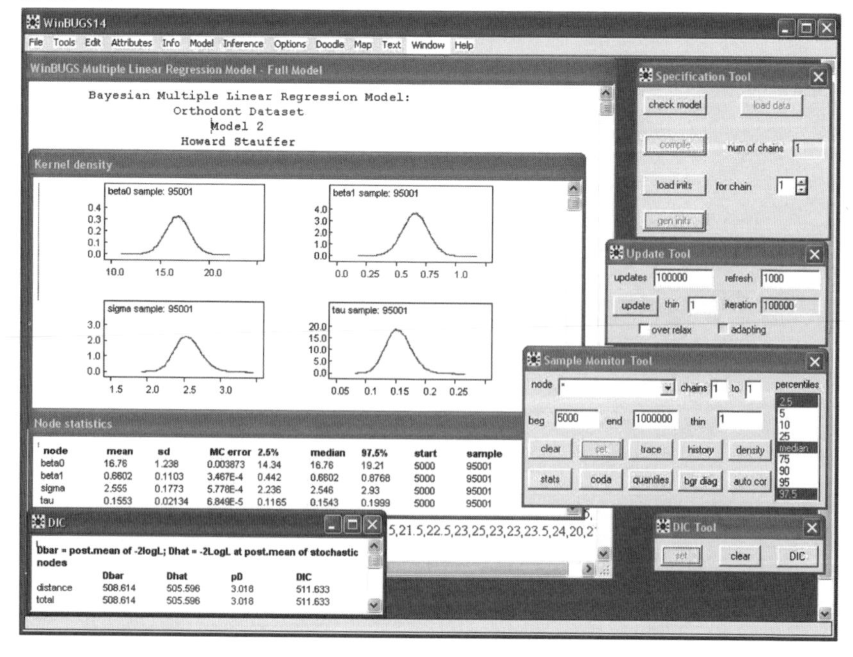

Model 2a:

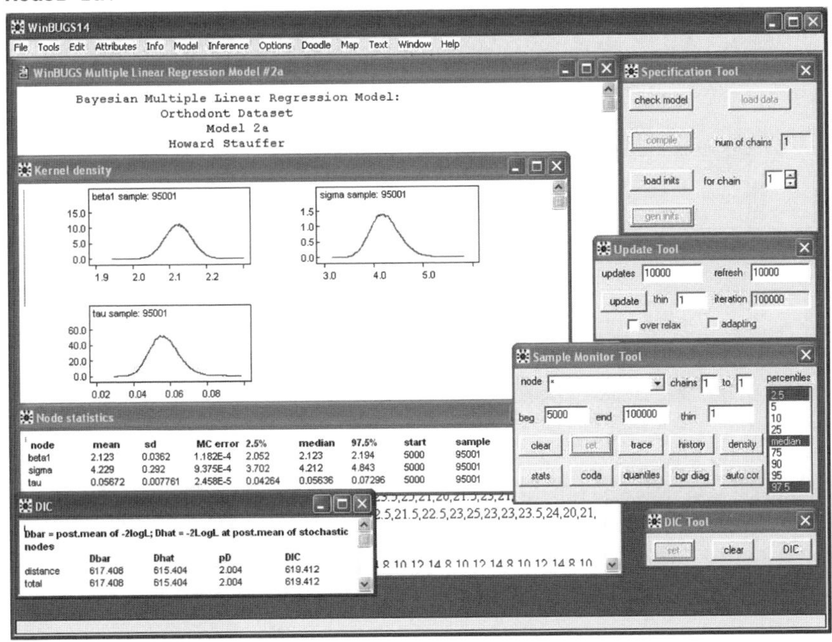

Model 3:

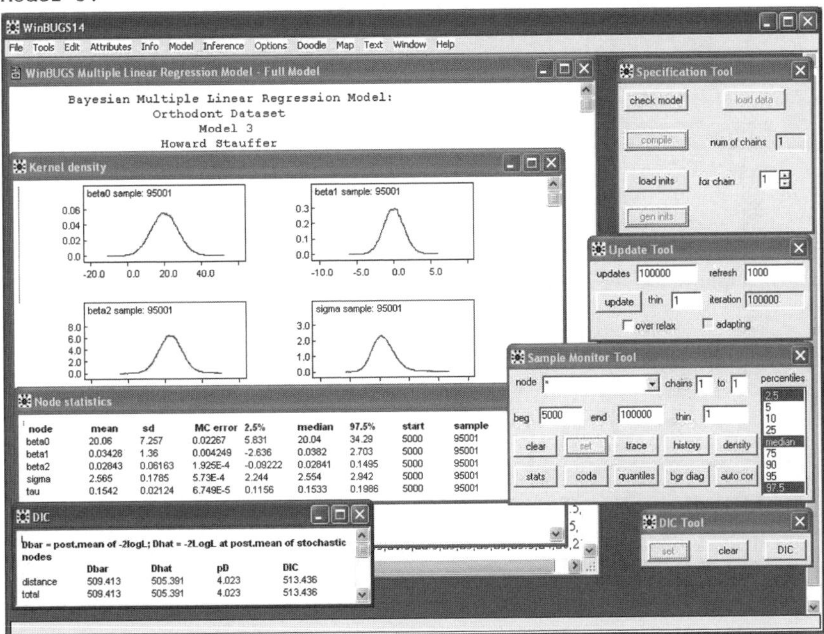

Model 3a:

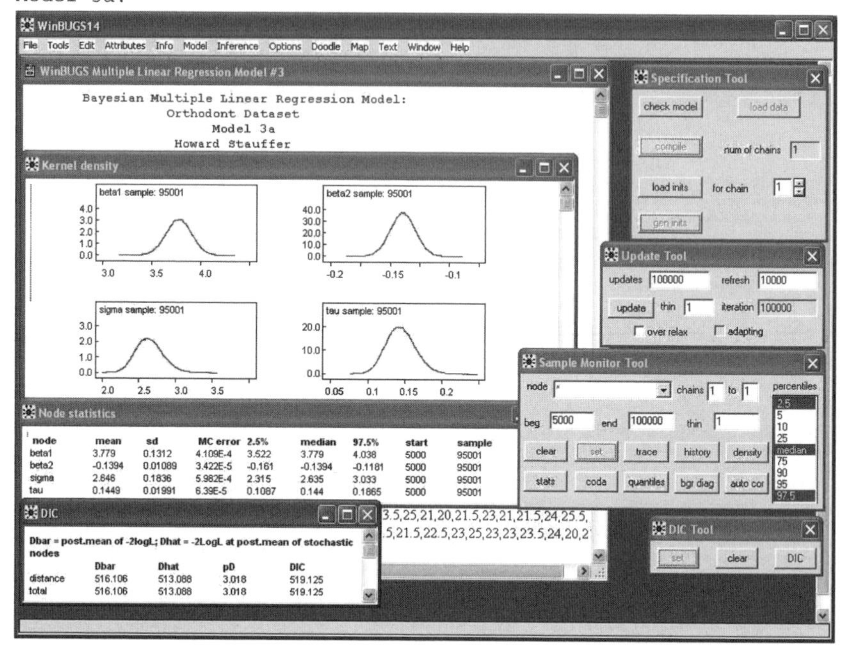

Model 3b:

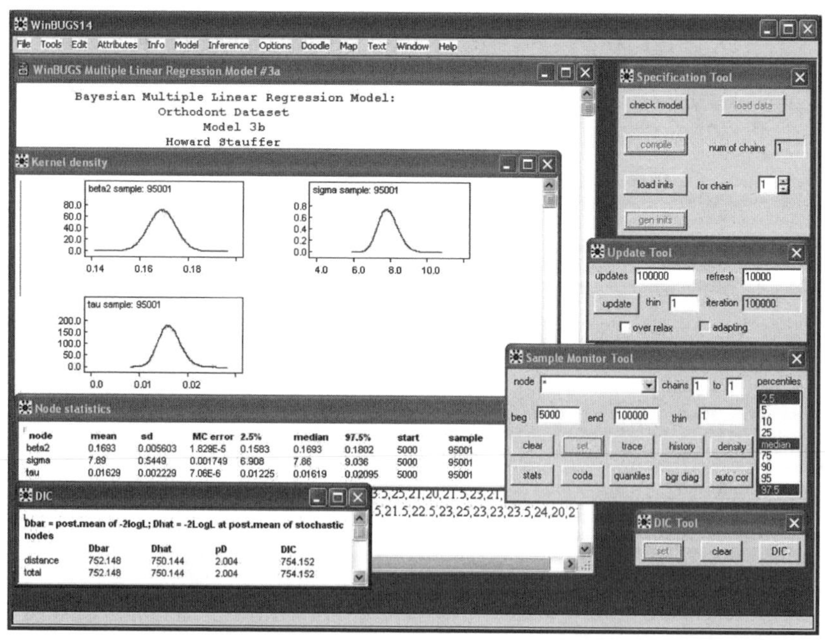

Problem 7.5

Model 4a:

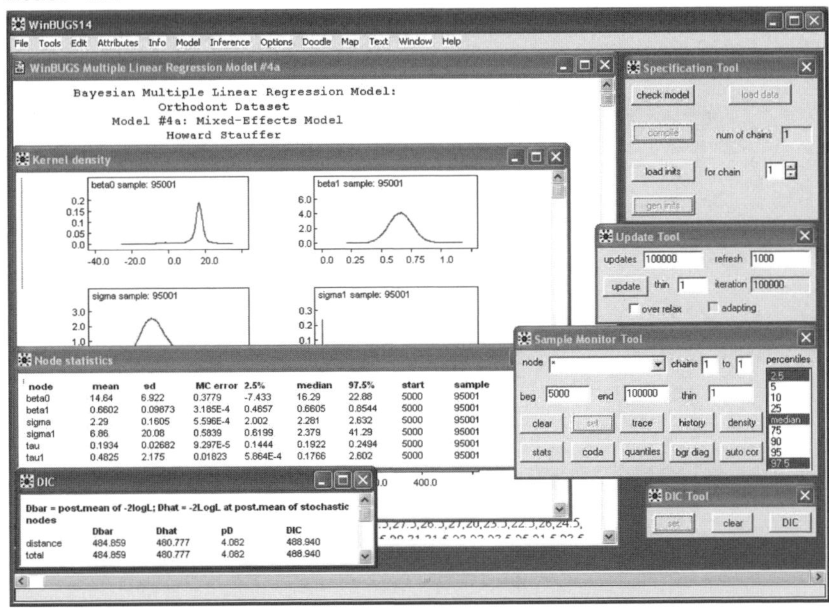

Model 4b:

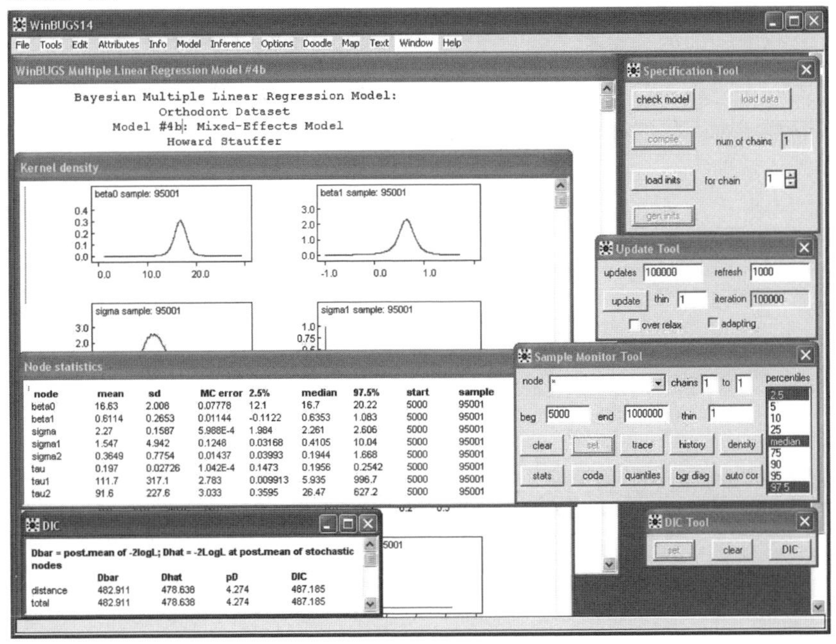

Model 4c:

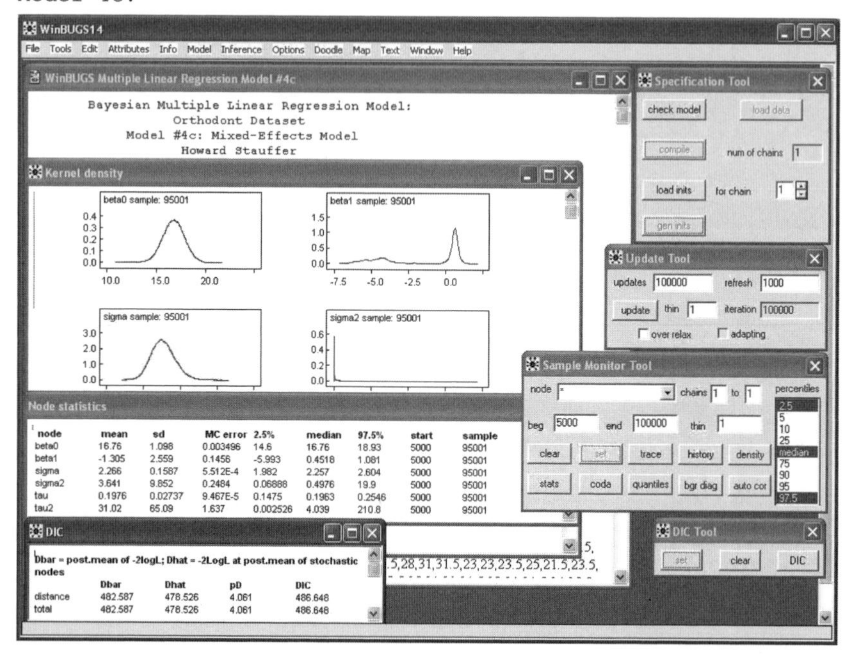

Model 5a:

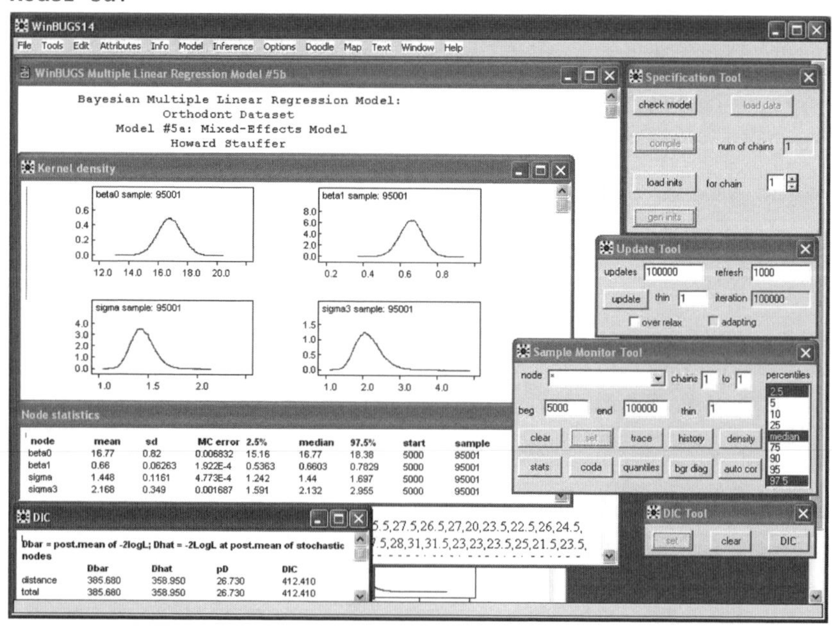

Model 5b:

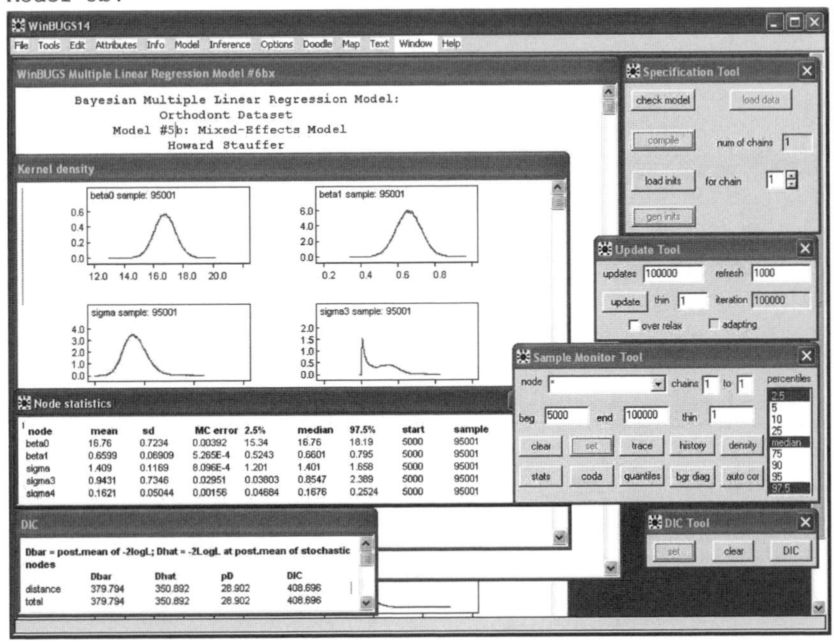

Model 5c:

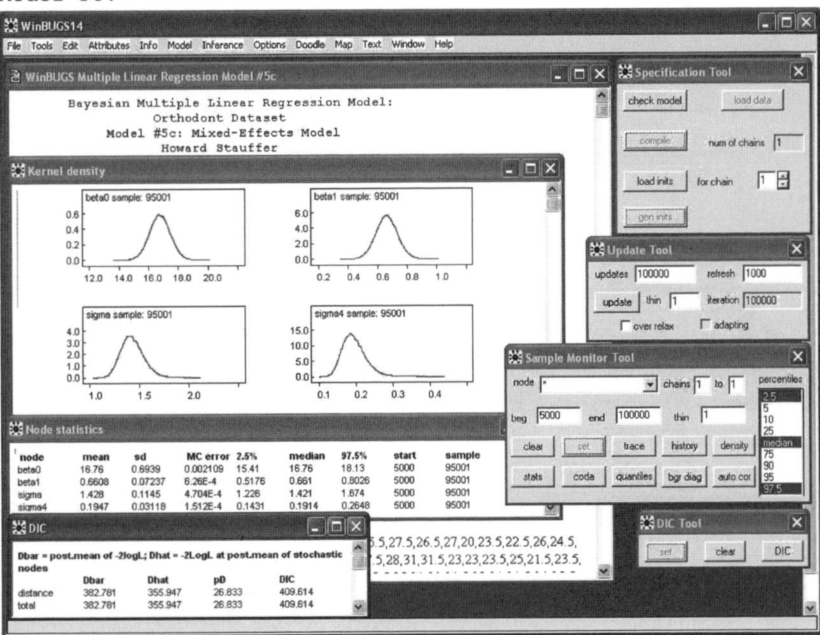

Model 6a:

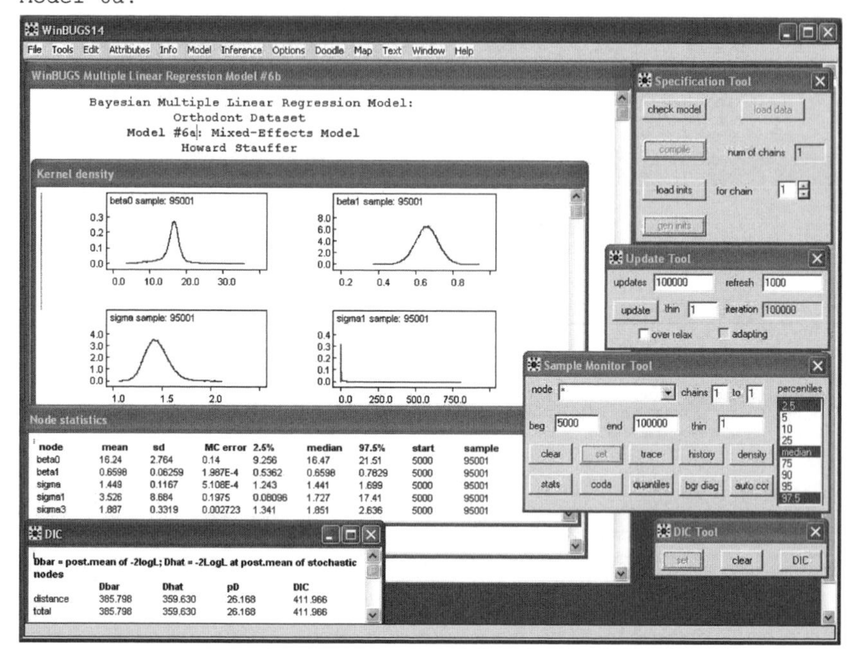

Model 6b:

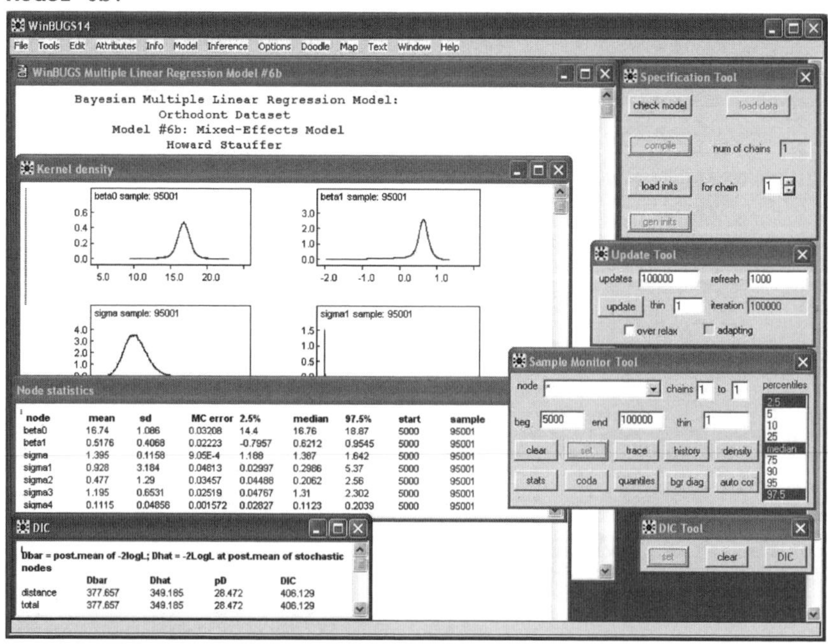

Model 6c:

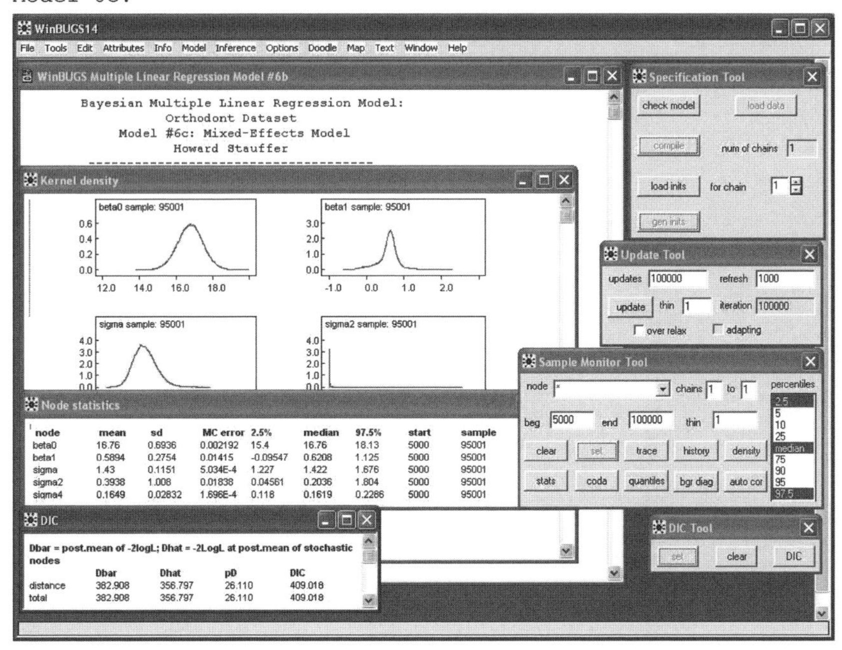

Conclusion: Model 6b is best-fitting, with distance ~ age, random = ~age|Sex/Subject. These results differ slightly from the frequentist results that favored model 6c. (See Figure B.7.17.)

Problem 7.6

Yes.

Figure B.7.17. Answers to Problems 7.4 and 7.5: comparative Bayesian statistics for linear regression and mixed-effects models.

<table>
<tr><th>Model</th><th>Covariates</th><th>k</th><th>AIC</th><th>AIC Weights</th><th>p_D</th><th>DIC</th><th>DIC Weights</th></tr>
<tr><td>1</td><td>Linear regression: 1</td><td>2</td><td>542.2815</td><td>< 0.00005</td><td>2.004</td><td>541.594</td><td>0.0000</td></tr>
<tr><td>2</td><td>Linear regression: 1, age</td><td>3</td><td>515.1695</td><td>< 0.00005</td><td>3.018</td><td>511.633</td><td>0.0000</td></tr>
<tr><td>2a</td><td>Linear regression: age</td><td>2</td><td>624.2148</td><td>< 0.00005</td><td>2.004</td><td>619.412</td><td>0.0000</td></tr>
<tr><td>3</td><td>Linear regression: 1, age^2, age</td><td>4</td><td>520.6967</td><td>< 0.00005</td><td>4.023</td><td>513.436</td><td>0.0000</td></tr>
<tr><td>3a</td><td>Linear regression: age^2, age</td><td>3</td><td>532.0595</td><td>< 0.00005</td><td>3.018</td><td>519.125</td><td>0.0000</td></tr>
<tr><td>3b</td><td>Linear regression^2: age</td><td>2</td><td>762.6865</td><td>< 0.00005</td><td>2.004</td><td>754.152</td><td>0.0000</td></tr>
<tr><td>4a</td><td>Mixed-effects on 1: ~1|Sex</td><td>4</td><td>497.0458</td><td>< 0.00005</td><td>4.082</td><td>488.940</td><td>0.0000</td></tr>
<tr><td>4b</td><td>Mixed effects on 1, age: ~age|Sex</td><td>5</td><td>498.7545</td><td>< 0.00005</td><td>4.274</td><td>487.185</td><td>0.0000</td></tr>
<tr><td>4c</td><td>Mixed effects on age: ~age-1|Sex</td><td>4</td><td>494.9654</td><td>< 0.00005</td><td>4.061</td><td>486.648</td><td>0.0000</td></tr>
<tr><td>5a</td><td>Mixed effects on 1: ~1|Subject</td><td>4</td><td>455.0025</td><td>0.0370</td><td>26.730</td><td>412.410</td><td>0.0242</td></tr>
<tr><td>5b</td><td>Mixed effects on 1, age: ~age|Subject</td><td>5</td><td>454.6367</td><td>0.0444</td><td>28.902</td><td>408.696</td><td>0.1552</td></tr>
<tr><td>5c</td><td>Mixed effects on age: ~age-1|Subject</td><td>4</td><td>453.0857</td><td>0.0964</td><td>26.833</td><td>409.614</td><td>0.0981</td></tr>
<tr><td>6a</td><td>Mixed effects on 1: ~1|Sex/Subject</td><td>5</td><td>452.0344</td><td>0.1631</td><td>26.168</td><td>411.966</td><td>0.0303</td></tr>
<tr><td>6b</td><td>Mixed effects on 1, age: ~age|Sex/Subject</td><td>6</td><td>453.2491</td><td>0.0888</td><td>28.472</td><td>406.129</td><td>0.5601</td></tr>
<tr><td>6c</td><td>Mixed effects on age: ~age-1|Sex/Subject</td><td>5</td><td>449.5305</td><td>0.5703</td><td>26.110</td><td>409.018</td><td>0.1321</td></tr>
<tr><td></td><td></td><td></td><td></td><td>1.0000</td><td></td><td></td><td>1.0000</td></tr>
</table>

REFERENCES

Akaike, H. 1973. Information theory as an extension of the maximum likelihood principle. In B. N. Petrov and F. Csaki, editors, *Proceedings of 2nd International Symposium on Information Theory*, Akademiai Kiado, Budapest, pp. 267–281.

Akaike, H. 1974. A new look at the statistical model identification. *IEEE Transactions on Automatic Control* AC 19:716–723.

Anderson, D. R., K. P. Burnham, and W. L. Thompson. 2002. Null hypothesis testing: Problems, prevalence, and an alternative. *Journal of Wildlife Management* 64:912–923.

Anderson, D. R., W. A. Link, D. H. Johnson, and K. P. Burnham. 2001. Suggestions for presenting the results of data analysis. *Journal of Wildlife Management* 65(3):373–378.

Bayes, T. 1763. An essay towards solving a problem in the doctrine of chances. *Philosophical Transactions of the Royal Society of London* 53:370–418.

Berger, J. O. 1985. *Statistical Decision Theory and Bayesian Analysis*, 2nd edition. Springer-Verlag, New York.

Berry, D. A., and D. K. Strangl, editors. 1996. *Bayesian Biostatistics*. Marcel Dekker, New York.

Bremaud, P. 1999. *Markov Chains: Gibbs Fields, Monte Carlo Simulation, and Queues*. Springer-Verlag, New York.

Burnham, K. P., and D. R. Anderson. 1998. *Model Selection and Inference*. Springer-Verlag, New York.

Burnham, K. P., and D. R. Anderson. 2002. *Model Selection and Multimodel Inference*, 2nd edition. Springer-Verlag, New York.

Carlin, B. P., and T. A. Louis. 2000. *Bayes and Empirical Bayes Methods for Data Analysis*, 2nd edition. Chapman & Hall, London.

Casella, G., and E. I. George. 1992. Explaining the Gibbs sampler. *The American Statistician* 46:167–174.

Cochran, W. G. 1977. *Sampling Techniques*, 3rd edition. Wiley, New York.

Cohen, J. 1988. *Statistical Power Analysis for the Behavioral Sciences*, 2nd edition. Lawrence Erlbaum Associates, Publishers, Hillsdale, NJ.

Congdon, P. 2001. *Bayesian Statistical Modeling*. Wiley, New York.

Conover, W. J. 1980. *Practical Nonparametric Statistics*, 2nd edition. Wiley, New York.

Cook, R. D. 1998. *Regression Graphics*. Wiley, New York.

Contemporary Bayesian and Frequentist Statistical Research Methods for Natural Resource Scientists. By Howard B. Stauffer

Cook, R. D., and S. Weisberg. 1999. *Applied Regression Including Computing and Graphics*. Wiley, New York.

Daniel, W. W. 1990. *Applied Nonparametric Statistics*, 2nd edition. PWS-Kent Publishing Company, Boston.

Dobson, A. J. 1990. *An Introduction to Generalized Linear Models*. Chapman & Hall, London.

Dowdy, S., and S. Wearden. 1991. *Statistics for Research*, 2nd edition. Wiley, New York.

Draper, D. 2000. *Bayesian Hierarchical Modeling*. Unpublished Notes, University of Bath, Bath, UK.

Draper, N., and H. Smith. 1981. *Applied Regression Analysis*, 2nd edition. Wiley, New York.

Efron, B. 1979. Bootstrap methods: Another look at the jackknife. *The American Statistician* 7:1–26.

Efron, B., and G. Gong. 1983. A leisurely look at the bootstrap, jackknife, and cross validation. *The American Statistician* 37:36–48.

Efron, B., and R. J. Tibshirani. 1993. *An Introduction to the Bootstrap*. Chapman & Hall, London.

Fisher, R. A. 1922. On the mathematical foundations of theoretical statistics. *Royal Society of London Philosophical Transactions* (*Series A*) 222:309–368.

Fisher, R. A. 1925a. *Statistical Methods for Research Workers*. 1st edition. Oliver and Boyd, Edinburgh, Scotland.

Fisher, R. A. 1925b. Theory of statistical estimation. *Proceedings of the Cambridge Philosophical Society* 22:700–725.

Fisher, R. A. 1934. *Statistical Methods for Research Workers*, 5th edition. Oliver and Boyd, Edinburgh, Scotland.

Fisher, R. A. 1958. *Statistical Methods for Research Workers*, 13th edition. Hafner, New York.

Gamerman, D. 1997. *Markov Chain Monte Carlo*. Chapman & Hall, London.

Gelfand, A. E., and A. F. M. Smith. 1990. Sampling-based approaches to calculating marginal densities. *Journal of the American Statistical Association* 85:398–409.

Gelman, A. 1992. Iterative and non-iterative simulation algorithms. *Computing Science and Statistics* 24:433–438.

Geman, S., and D. Geman. 1984. Stochastic relaxation, Gibbs distributions and the Bayesian restoration of images. *IEEE Transactions on Pattern Analysis and Machine Intelligence* 6:721–741.

Gill, J. 2002. Bayesian Methods. Chapman & Hall, London.

Gregoire, T. G. 1998. Design-based and model-based inference in survey sampling: Appreciating the difference. *Canadian Journal of Forest Research* 28:1429–1447.

GS+: *Geostatistics for the Environmental Sciences*. 1998. Gamma Design Software, Plainwell, MI.

Hansen M. H., and W. N. Hurwitz. 1943. On the theory of sampling from finite populations. *Annals of Mathematical Statistics* 14:333–362.

Hardin, J., and J. Hilbe. 2001. *Generalized Linear Models and Extensions*. Stata Press, College Station, TX.

Hastings, W. K. 1970. Monte Carlo sampling methods using Markov chains and their applications. *Biometrika* 57:97–109.

Hicks, C. R. 1993. *Fundamental Concepts in the Design of Experiments*, 4th edition. Oxford University Press, New York.

Hilborn, R., and M. Mangel. 1997. *The Ecological Detective*. Princeton University Press, Princeton, NJ.

Hocking, R. R. 1996. *Methods and Applications of Linear Models: Regression and the Analysis of Variance*. Wiley, New York.

Horvitz, D. G., and D. J. Thompson. 1952. A generalization of sampling without replacement from a finite universe. *Journal of the American Statistical Association* 47:663–685.

Hosmer, D. W., and S. Lemeshow. 2000. *Applied Logistic Regression*, 2nd edition. Wiley, New York.

Iversen, G. R. 1984. *Bayesian Statistical Inference*. Sage University Press, London.

Jeffreys, H. 1961. *Theory of Probability*, 3rd edition. Oxford University Press, Oxford, UK.

Johnson, D. H. 1999. The insignificance of statistical significance testing. *Journal of Wildlife Management* 63:763–772.

Johnson, D. H. 2002. The role of hypothesis testing in wildlife science. *Journal of Wildlife Management* 66(2):272–276.

Krause, A., and M. Olson. 2000. *The Basics of S and S-Plus*, 2nd edition. Springer, New York.

Kuehl, R. O. 1994. *Statistical Principles of Research Design and Analysis*. Duxbury Press, Belmont, CA.

Link, W. A., E. Cam, J. D. Nichols, and E. G. Cooch. 2002. Of BUGS and birds: Markov chain Monte Carlo for hierarchical modeling in wildlife research. *Journal of Wildlife Management* 66(2):277–291.

Manly, F. J., L. L. McDonald, and D. L. Thomas. 1995. *Resource Selection by Animals*. Chapman & Hall, London.

Manly, F. J., L. L. McDonald, D. L. Thomas, T. L. McDonald, and W. P. Erickson. 2004. *Resource Selection by Animals*, 2nd edition. Kluwer Academic Publishers, Norwell, MA.

Manly, B. F. J. 1994. *Multivariate Statistical Methods: A Primer*, 2nd edition. Chapman & Hall, London.

Manly, B. F. J. 1997. *Randomization, Bootstrap and Monte Carlo Methods in Biology*, 2nd edition. Chapman & Hall, London.

McCullagh, P., and J. A. Nelder. 1996. *Generalized Linear Models*, 2nd edition. Chapman & Hall, New York.

McCulloch, C. E., and S. R. Searle. 2001. *Generalized, Linear, and Mixed Models*. Wiley, New York.

Metropolis, N., A. W. Rosenbluth, M. N. Rosenbluth, A. H. Teller, and E. Teller. 1953. Equations of state calculations by fast computing machine. *Journal of Chemical Physics* 21:1087–1091.

Nagelkerke, N. J. D. 1991. A note on a general definition of the coefficient of determination. *Biometrika* 78:691–692.

Nelder, J. A. and R. W. M. Wedderburn. 1972. Generalized linear models. *Journal of the Royal Statistical Society of America* 135:370–384.

Neyman, J., and E. S. Pearson. 1928a. On the use and interpretation of certain test criteria for purposes of statistical inference. Part I. *Biometrika* 20A:175–240.

Neyman, J., and E. S. Pearson. 1928b. On the use and interpretation of certain test criteria for purposes of statistical inference. Part II. *Biometrika* 20A:263–294.

Neyman, J., and E. S. Pearson. 1933. The testing of statistical hypothesis in relationship to probabilities a priori. *Proceedings of the Royal Statistical Society* (*Series A*) 231:289–510.

Neyman, J., and E. S. Pearson. 1936. Contributions to the theory of testing statistical hypotheses. *Statistical Research Memorandum* 1:1–37.

nQuery Advisor version 5.0. 2002. *Statistical Solutions*. Saugus, MA.

PASS. 2002. *NCSS Statistical Software*. Kaysville, Utah.

Peskun, P. H. 1973. Optimal Monte-Carlo sampling using Markov chains. *Biometrika* 60(3):607–612.

Pielou, E. C. 1969. *An Introduction to Mathematical Ecology*. Wiley-Interscience, New York.

Pinheiro, J. C., and D. M. Bates. 2000. *Mixed-Effects Models in S and S-Plus*. Springer-Verlag, New York.

Potthoff, R. F., and S. N. Roy. 1964. A generalized multivariate analysis of variance model useful especially for growth curve problems. *Biometrika* 51:313–326.

Ramsey, F. L., and D. W. Schafer. 2002. *The Statistical Sleuth*, 2nd edition. Duxbury Press, Belmont, CA.

R Development Core Team. 2005. R 2.2.1: *A Language and Environment for Statistical Computing*. R Foundation for Statistical Computing, Vienna, Austria (ISBN 3-900051-07-0, URL http://www.R-project.org).

Rice, J. A. 1995. *Mathematical Statistics and Data Analysis*, 2nd edition. Duxbury Press, Belmont, CA.

Robinson, D. H., and H. Wainer. 2002. On the past and future of null hypothesis significance testing. *Journal of Wildlife Management* 66(2):263–271.

Ryan, T. P. 1997. *Modern Regression Methods*. Wiley, New York.

Sarndal, C. E., B. Swensson, and J. Wretman. 1992. *Model Assisted Survey Sampling*. Springer-Verlag, New York.

SAS Institute Inc. 1995. *Logistic Regression Examples Using the SAS System*, version 6. SAS Institute Incorporated, Cary, NC.

Scheaffer, R. L., W. Mendenhall III, and R. L. Ott. 1996. *Elementary Survey Sampling*, 5th edition. Duxbury Press, Belmont, CA.

Schwarz, G. 1978. Estimating the dimension of a model. *Annals of Statistics* 6:461–464.

Seber, G. A. F. 1977. Linear Regression Analysis. Wiley, New York.

Siegel, S., and N. J. Castellan, Jr. 1988. *Nonparametric Statistics for the Behavioral Sciences*, 2nd edition. McGraw-Hill, New York.

Sokal, R. R., and F. J. Rohlf. 1995. *Biometry*, 3rd edition. Freeman, New York.

Spiegelhalter, D., A. Thomas, and N. Best. 2001. WinBUGS, version 1.4. MRC Biostatistics Unit, Cambridge, UK.

S-Plus 2000, release 3. 2000. MathSoft Incorporated, Seattle, WA.

Stauffer, D. F. 1999. Linking populations and habitats: Where have we been? Where are we going? Invited plenary paper presented at symposium, Predicting Species Occurrences: Issues of Scale and Accuracy. Snowbird, UT, October 1999.

Stauffer, H. B. 1982a. Some sample size tables for forest sampling. Research Note 90. British Columbia Ministry of Forests, Victoria, BC, Canada.

Stauffer, H. B. 1982b. A sample size table for forest sampling. *Forest Science* 28(4):777–784.

Thompson, S. K. 1992. *Sampling*. Wiley, New York.

Thompson, S. K., and G. A. F. Seber. 1996. *Adaptive Sampling*. Wiley, New York.

Thompson, W. L., G. C. White, and C. Gowan. 1998. *Monitoring Vertebrate Populations*. Academic Press, San Diego, CA.

Wackerly, D. D., W. Mendenhall III, and R. L. Scheaffer. 2002. *Mathematical Statistics with Applications*, 6th edition. Duxbury Press, New York.

Winer, B. J., D. R. Brown, and K. M. Michels. 1991. *Statistical Principles in Experimental Design*, 3rd edition. McGraw-Hill, New York.

Wolfinger, R. Fitting nonlinear and generalized mixed models with PROC LNMIXED. 2000. 200 Joint Statistical Meetings Continuing Education Series, Indianapolis.

Zar, J. H. 1996. *Biostatistical Analysis*, 3rd edition. Prentice-Hall, Upper Saddle River, NJ.

INDEX

Contemporary Bayesian and Frequentist Statistical Research Methods for Natural Resource Scientists. By Howard B. Stauffer